LANDSCAPES, IDENTITIES AND D

Landscapes, Identities and Development

Edited by

ZORAN ROCA
Universidade Lusófona de Humanidades e Tecnologias, Portugal

PAUL CLAVAL
University of Paris I – Sorbonne, France

JOHN AGNEW
University of California at Los Angeles, USA

Routledge
Taylor & Francis Group

LONDON AND NEW YORK

First published 2011 by Ashgate Publishing

Published 2016 by Routledge
2 Park Square, Milton Park, Abingdon, Oxfordshire OX14 4RN
711 Third Avenue, New York, NY 10017, USA

First issued in paperback 2016

Routledge is an imprint of the Taylor & Francis Group, an informa business

British Library Cataloguing in Publication Data
Landscapes, identities and development.
   1. Human ecology. 2. Human beings--Effect of environment
   on. 3. Cultural landscapes. 4. Cultural property.
   5. Landscape archaeology. 6. Land use--Planning.
   7. Landscape changes.
   I. Roca, Zoran. II. Claval, Paul. III. Agnew, John A.
   304.2'3-dc22

Library of Congress Cataloging-in-Publication Data
Roca, Zoran.
   Landscapes, identities, and development / by Zoran Roca, Paul Claval and John Agnew.
      p. cm.
   Includes bibliographical references and index.
   ISBN 978-1-4094-0554-2 (hardback) 1. Landscapes--Social aspects. 2. Landscape
assessment. 3. Cultural landscapes. 4. Group identity. 5. Land use--Social aspects. I.
Claval, Paul. II. Agnew, John A. III. Title.
   GF90.R63 2011
   304.2'3--dc22

2010027831

ISBN 13: 978-1-138-26995-8 (pbk)
ISBN 13: 978-1-4094-0554-2 (hbk)

Contents

List of Figures

Unless otherwise stated, all illustrations are the authors' own material.

List of Tables

Notes on Contributors

Alessia de Nardi: Department of Geography, University of Padua.

Andrey Y. Reznikov: Faculty of Geography and Geoecology, University of St. Petersburg.

Anne-Sophie Devanne: Canada Research Chair on Regional and Territorial Development, Université du Québec à Rimouski.

Anu Printsmann: Centre for Landscape and Culture, Estonian Institute of Humanities, Tallinn University.

Athanasia Mavridou: Department of Geography, University of the Aegean, Mytilene.

Barbara Borkowska: Institute of Ethnology and Cultural Anthropology, University of Warsaw.

Barbara Drusi: Department of Economy and Engineering Applied to Agriculture, Forestry and the Environment, Faculty of Agriculture, University of Torino.

Belén Pérez Pérez: Asistencias Técnicas Clave, S.L., Seville.

Benedetta Castiglioni: Department of Geography, University of Padua.

Bodo Freund: Geography Institute, Humboldt-University, Berlin.

Carlos Zeballos: Research Institute for Humanity and Nature, Kyoto.

Caroline Borré: Jilin University, Changchun.

Cibele Queiroz: Department of Systems Ecology, Stockholm University.

Daniel Terrasson: National Research Programme 'Landscape and Sustainable Development', Cemagref, Cestas.

Edmunds Valdemārs Bunkše: University of Delaware, Newark; University of Latvia, Riga.

Enrico Borgogno Mondino: Department of Economy and Engineering Applied to Agriculture, Forestry and the Environment, Faculty of Agriculture, University of Torino.

Frederick A. Day: Department of Geography, Texas State University, San Marcos.

Gregory A. Isachenko: Faculty of Geography and Geoecology, University of St. Petersburg.

Hana Skokanová: Department of Landscape Ecology, The Silva Tarouca Research Institute for Landscape and Ornamental Gardening, Brno.

Hannes Palang: Centre for Landscape and Culture, Estonian Institute of Humanities, Tallinn University.

Ivančica Schrunk: Department of History, University of St. Thomas, St. Paul.

James W. Vaughan: Department of Geography, Texas State University, San Marcos.

Johannes Renes: Utrecht University, VU University Amsterdam.

John Agnew: Department of Geography, University of California, Los Angeles.

José António Oliveira: TERCUD – Territory, Culture and Development Research Center, Universidade Lusófona de Humanidades e Tecnologias, Lisbon.

Junzo Uchiyama: Research Institute for Humanity and Nature, Kyoto.

Kati Lindström: Department of Semiotics, University of Tartu.

Lionella Scazzosi: Department of Architectural Design, Polytechnic of Milan.

Marc Antrop: Department of Geography, Ghent University.

Maria de Nazaré Roca: e-GEO Research Centre for Geography and Regional Planning, Faculty of Social Sciences and Humanities, Universidade Nova de Lisboa, Lisbon.

Marie-José Fortin: Canada Research Chair on Regional and Territorial Development, Département Sociétés, Territoires et Développement, Université du Québec à Rimouski. Université du Québec à Rimouski.

Marina Frolova: Institute of Regional Development, University of Granada.

Marju Kõivupuu: Centre for Landscape and Culture, Estonian Institute of Humanities, Tallinn University.

Mark Jackson: School of Geographical Sciences, University of Bristol.

Mattias Qviström: Department of Landscape Architecture, Swedish University of Agricultural Sciences, Uppsala.

Mauricio de Almeida Abreu: Department of Geography, Federal University of Rio de Janeiro.

Minna Tanskanen: Department of Geographical and Historical Studies, University of Eastern Finland, Joensuu.

Paul Claval: University of Paris-Sorbonne.

Regina Lindborg: Department of Systems Ecology, Stockholm University.

Roberto Chiabrando: Department of Economy and Engineering Applied to Agriculture, Forestry and the Environment, Faculty of Agriculture, University of Torino.

Rogério Haesbaert: Geography Department, Universidade Federal Fluminense, Niteroi, Rio de Janeiro.

Ruth Beilin: Melbourne School of Land and Environment, University of Melbourne.

Sophie Le Floch: Cemagref, Groupement de Bordeaux.

Tania Rossetto: Department of Geography, University of Padua.

Tereza Stránská: Department of Landscape Ecology, Silva Tarouca Research Institute for Landscape and Ornamental Gardening, Brno.

Thanasis Kizos: Department of Geography, University of the Aegean, Mytilene.

Theano S. Terkenli: Department of Geography, University of the Aegean, Mytilene.

Tobias Plieninger: Ecosystem Services Research Group, Berlin-Brandenburg Academy of Sciences and Humanities, Berlin.

Tor Arnesen: Eastern Norway Research Institute, Lillehammer.

Veerle Van Eetvelde: Department of Geography, Ghent University.

Veronica della Dora: School of Geographical Sciences, University of Bristol.

Vlasta Begović: Institute of Archaeology, Zagreb.

William E. Doolittle: Department of Geography and the Environment, University of Texas at Austin.

Zoran Roca: TERCUD – Territory, Culture and Development Research Center, Universidade Lusófona de Humanidades e Tecnologias, Lisbon.

Acknowledgements

The editors and publisher would like to thank all entities (museums, municipalities, archives, government boards, university libraries, publishers, artists and private collectors) for permission to reprint the illustrations, figures and other material used in this book.

Every effort has been made to trace copyright holders of materials reproduced in this book. Any rights not acknowledged here will be acknowledged in subsequent printings if notice is given to the publisher.

The editors are indebted to many for their advice and assistance at various stages in the preparation of this book. Particular recognition goes to Isabel Canhoto of Universidade Lusófona for her competence and devotion throughout the process of producing the final typescript, as well as to Val Rose of Ashgate for her encouragement in conceptualizing this publication.

The editors also acknowledge the financial assistance from the Foundation for Science and Technology (FCT), Lisbon, for the preparation of and follow-up to the 23rd Session of the PECSRL and the Permanent European Conference for the Study of the Rural Landscape, entitled 'Landscapes, Identities and Development', Lisbon and Óbidos, Portugal, 1-5 September 2008, which was organized by Territory, Culture and Development Research Centre of Universidade Lusófona (TERCUD), Lisbon, and whose presentations and discussions inspired the production of this book.

Introduction

Zoran Roca and John Agnew

Landscapes are treasures of the past, frame contemporary everyday life, and affect future environmental, economic and cultural processes. As material custodians of both historical memory and the sense of place, landscapes encapsulate our attachments, emotions, perceptions and knowledge, as well as our interests, decisions and actions. By providing a context to the spatial fixes and flows of everyday life, landscapes are long lasting witnesses to the production and consumption of territorial identities. Changes in spatial fixes and flows, provoked by agents of varying geographical scope, are reflected in the constant regeneration of the natural, economic and cultural character of territories. Landscapes are pivotal in the recognition of these changes. As constitutive elements of and factors in the making of territorial identities, landscapes are the medium through which the existing and emerging identities of places and regions are generated, recorded, assumed and claimed.

Modern societies are marked by identity crises which, all too often, involve major landscape disruptions. In this context, landscapes are no longer just the 'scenery' of either local or globalized economies and cultures. They increasingly gain economic and cultural value in their own right. The art of sustainable development planning is to sideline some and favour other landscapes as instruments of economic and cultural transformation. In fact, the key dilemma today, as Paul Claval stresses in the concluding chapter of this book, is how to reconcile these changes with the preservation of valuable inherited features and with the (re)shaping of harmonious new forms. A collective academic response to this dilemma was attempted at the 23rd Session of the PECSRL – The Permanent European Conference for the Study of the Rural Landscape, entitled 'Landscapes, Identities and Development', held in Lisbon and Óbidos, Portugal, 1-5 September 2008, whose presentations and discussions provided the basis for this book.[1]

---

1 Established in 1957 at an inaugural conference in Nancy, France, organized by Xavier de Planhol, the Permanent European Conference for the Study of the Rural Landscape (PECSRL) has been one of the most stable European networks of landscape researchers. Initially it consisted mainly of historical geographers, but today its membership is diversified to include nearly 500 ecologists, social scientists, rural planners, landscape architects, human and physical geographers, historians, archaeologists, landscape managers, as well as other scholars and practitioners interested in European landscapes. More at http://www.

The objective of this book is to offer a state-of-the-art survey of conceptual and methodological research and planning issues in the area of landscape, identity, and development. Given the inter- and trans-disciplinary relevance and international scope of its focus on the theory and empirical examples of landscape, identity, and development, the book provides a compendium of the most recent findings and interpretations on:

- historical, current and prospective linkages between changing landscapes and natural, economic, cultural and other identity features of places and regions;
- landscape-related identities as local and regional development assets and resources in the era of globalized economy and culture;
- the role of landscape history and heritage as platforms for landscape research and management in European contexts, including the implementation of The European Landscape Convention; and
- the strengthening of the landscape perspective as a constitutive element of developing policies for sustainable development.

The affirmation of the natural, economic, cultural or other features of territorial identities has gained strategic importance in this era of globalized economy and culture. This applies equally to those places and regions that already benefit from favourable, attractive, 'globally competitive' identities based on sustainable growth and development, and to lagging, mostly peripheral, rural areas that suffer from environmental degradation due to land-use conflicts, and/or from weak economy and fading cultural authenticity due to overexposure to or indiscriminate adoption of globalized goods, services and ideas.

Symptomatically, regardless of the developmental experience, landscapes have been increasingly regarded and treated as important repositories of material and intangible resources. Landscape preservation and requalification have increasingly become synonymous, implicitly and explicitly, with the removal of undesirable identity features, and the strengthening of existing and/or the creation of new favourable ones, aimed at promoting economic and cultural emancipation and sustainable development. However, there is often a gap between a pro-identity/development rhetoric and the anti-identity/development reality, as evidenced in landscape negligence and degradation. Bridging this gap calls for grasping landscape change as a fundamental part of territorial diagnoses and strategic planning. Sustainable development policies, plans and projects call for assessments of landscape transformations, and this is why the scope and importance of theoretical and applied, both macroscopic and, especially, participatory landscape research needs to be reinforced and expanded.

---

pecsrl.org. On the 23rd Session of the PECSRL see http://tercud.ulusofona.pt/PECSRL/PECSRL2008.htm.

The book contains 30 thematic chapters ranging from theoretical research essays to detailed empirical studies by 48 authors and co-authors from 22 countries of Western, Eastern, Central, Scandinavian and Mediterranean Europe, the Americas, Australia and Japan, grouped in four sections, followed by the concluding commentary of Paul Claval. The four sections move progressively from the more general to the more specific and from the more theoretical to the most applied, as befits the overall conception of the book as a compendium of current landscape research, particularly by European authors or authors with a European focus, and the need to lay out the more abstract modes of thinking animating contemporary landscape research before examining examples of more detailed applied research. Each chapter is given a brief précis of its main ideas.

Part I: Landscape and Identity between Imagery and Reality

Lionella Scazzosi questions the theoretical framework and practical modalities of defining the 'limits' to landscape change in order to respect, preserve and transmit the specific character of places and peoples' identities.

Edmunds Valdemārs Bunkše proposes an enrichment of experiential possibilities in rural landscapes by enlarging landscape perceptions and narratives with a poetic sensibility represented by such words as 'ethereal', 'evanescent' and 'ineffable'.

By comparing England and Italy, John Agnew describes how the role of landscape in national identity formation should be explained by the specific trajectories of national-state formation and nationalism.

Anu Printsmann et al. demonstrate the dual character of landscape by contrasting the ideal, memorized and official culture with the idea of national landscape representations and imagery in Estonia.

Benedetta Castiglioni et al. show how landscape used as trans-cultural medium can promote the integration of young immigrants by making them more aware and less passive in the process of building a multiethnic society.

For Rogério Haesbaert, the new (trans)territorialities, with open and hybrid frontiers and enclosures, fences and walls, stimulate new multiterritorial social practices and symbols and, thus, construction of new hybrid and/or open identities.

In their discussion of 'landscaping' vs. 'terraforming', Mark Jackson and Veronica della Dora exemplify the emerging places for physical consumption and social segregation by reference to the megaprojects involved in constructing artificial islands in the Persian Gulf.

Part II: Landscape History, Heritage and Social Change

Johannes Renes opposes the vision of a distinction between modern, dynamic cultural landscapes, on the one hand, and 'traditional', stable landscapes on the other, and illustrates the complexity of continuities and change with case studies from across Europe.

Ivančica Schrunk and Vlasta Begović discuss the controversy of functional identity of a historical landscape of the Brioni islands in the northern Adriatic in terms of its competing significance for natural and cultural heritage, tourism and state affairs.

William E. Doolittle documents how the rural and agricultural landscape in Mexico was Europeanized in the Viceregal era (1521-1810) by the construction of monumental bridge aqueducts.

Recalling the importance of sugar cane production for colonial Brazil, Mauricio de Almeida Abreu presents the method and results of uncovering the presence of sugar mills in the landscape of Rio de Janeiro in the seventeenth century.

Barbara Borkowska shows how a new (post-Second World War) Polish community in a former German territory has gradually changed attitudes towards a town landscape that was shaped by a different culture, reflecting alien ideas and values.

Minna Tanskanen shows how turning points in Finnish socio-economic and political history provoked changes in mire or marsh landscapes and how today these landscapes increasingly mirror the effects of EU agricultural policy.

Based on a comparative study of changes in cropping patterns, land abandonment and built environment over forty years, Bodo Freund discusses land use and amenity quality in a mountainous area of Northern Portugal.

Part III: Landscape Assets, Resources and Services

Theano S. Terkenli discusses the lack of a developed landscape conscience among the Greek population and explores its causes by tracing the past and the emerging legal, historical, aesthetic and socio-cultural trajectory of the relationship of Greece with its landscapes.

Based on analyses of land abandonment on agricultural landscapes of Australia, Sweden and Portugal, Ruth Beilin and Regina Lindborg call for a radical rethinking of the way landscapes are managed in order to improve outcomes for both agriculture and biodiversity as nature conservation.

Gregory A. Isachenko and Andrey Y. Reznikov describe the vast natural landscapes inside Saint Petersburg, Russia, discuss the problems caused by applying the traditional approach to nature conservation to this metropolitan area and suggest new planning and monitoring solutions.

Frederick A. Day and James W. Vaughan explore geographic differences that underpin landscape modification in the evolution of the Austin-San Antonio Corridor as one of the most favoured locations for settlement in the US in the early twenty-first century.

Based on empirical analyses, Hana Skokanová and Tereza Stránská validate the assumption that NATURA 2000 sites in a Czech Republic region with a long agricultural history have been managed in the same way for centuries, which probably contributed to their high levels of biodiversity.

Arguing that the ecosystems 'services' are determined at the small (management) scale and the sustainability of an area at the large (landscape) scale, Tobias Plieninger and Thanasis Kizos propose sustainability indicators for two Mediterranean agroforestry systems at both scales.

Zoran Roca et al. bring forward a methodological framework and results of a field study of the topophilia-terraphilia interface in the contexts of claims for the (re)affirmation of landscape-related and other territorial identity features as local development resources.

Athanasia Mavridou and Thanasis Kizos point to the impacts of gardens in urban sprawl and propose an integrated method for evaluating their environmental, ecological and economic functions, tested in the suburban areas of the city of Mytilene, Island of Lesvos, Greece.

Part IV: Landscape Research and Development Planning

Daniel Terrasson explains how development of a research programme aimed at supporting landscape policies faces the usual obstacles in science-policy interfaces as well as the specificity of landscape as a political issue.

Tor Arnesen addresses landscape as a sign for an object (a piece of land) for an interpreter (community) in some respect or capacity and proposes four guiding methodological principles for the study of landscapes.

Based on studies of 'social acceptability' of wind farms projects in Québec, Canada, and in France, Marie-José Fortin et al. explain that participative planning processes advocated by the European Landscape Convention fail to integrate a fundamental aspect of landscape: its political dimension.

Marina Frolova and Belén Pérez Pérez address the evolution of the approaches to energy landscapes through comparison of different planning strategies (public consulting and forming public opinion) which dominated during different periods in Spain, first for hydro power, then for wind and solar power.

Carlos Zeballos et al. focus on the uses of GIS methodology and 3D modelling techniques and present primary results from research aimed at a holistic interpretation of the evolution of the historical landscape of the Lake Biwa area in central Japan.

Veerle Van Eetvelde and Marc Antrop present the experience of compiling a new typology of the landscapes of Belgium in a trans-regional and trans-border perspective by means of a new method using a combination of holistic and parametric approaches at two scale levels.

Based on a case study at the urban fringe of Lund, Sweden, Mattias Qviström discusses the possibilities for a combined analysis of the legacies of planning and the footprints of former landscape ideals in peri-urban landscapes.

Enrico Borgogno Mondino et al. explain a new GIS procedure for peri-urban landscape change detection and present the outputs of a case study in an area of Torino, Italy, aimed to quantify and visualize degradation of perceived landscape quality.

In the concluding chapter Paul Claval stresses how

> ... at a time when landscape experts are losing a part of their traditional authority […] it is important to stress that we need other forms of expertise: the knowledge of the economic and social forces which contribute to shape visible forms and often create a severe unbalance of power and many inequalities among those who inhabit or visit them.

Indeed, the trans-disciplinary character of landscape research and planning now provides comprehensive insights and policy advice on the design, implementation and assessment of developmental goals and interventions that imply the (re/de)generation of territorial identity features that once were derived from the singular authority of disparate 'experts'. Development policies and projects are not achievable without integration of the various dimensions of landscape study and advocacy examined in this volume. In this regard, the whole landscape–identity–development nexus can only be addressed through the participatory and collaborative activities of a host of scholars such as those who have contributed to this book. It is then both a statement of the current state-of-the-art and the beginnings of what must be a continuing enterprise.

PART I
Landscape and Identity between Imagery and Reality

Limits to Transformation in Places' Identity: Theoretical and Methodological Questions

Lionella Scazzosi

Introduction

Landscape

The present contribution is part of a process of theoretical and methodological layout that aims at defining some basic concepts regarding the quality of landscapes. A quality meant as the achievement of an *appropriate* relationship between innovation and conservation of the specific characters that places have inherited from the past (Scazzosi 1993, 2002c, 2004, 2008, 2009).

Recently, a vast literature has been published on the changes in the concept of landscape and on its polysemy. What we must point out here is the fact that today the concept includes both a physical – material understanding and a perceptive one, its broader and more complex sense (cultural and sensorial). It involves the whole territory – i.e. all spaces (rural, urban, peri-urban or natural) regardless of their quality (outstanding, ordinary, degraded) – as a physical object but also as the object of the 'cultural look' people cast on places, even places with no human presence.

The definition of landscape given by the European Landscape Convention (Florence 2000) synthesizes this long cultural evolution: 'Landscape means an area, as perceived by people, whose character is the result of the action and interaction of natural and/or human factors' (art. 1). And '[…] This Convention applies to the entire territory of the Parties and cover natural, rural, urban and peri-urban areas. It includes land, inland water and marine areas. It concerns landscapes that might be considered outstanding as well as everyday or degraded landscapes' (art. 2).

Landscape as physical object

In a landscape approach, places may be considered in their physical characters as a large and complex *artefact* (a word meant to underline the material and building aspects) and a large *architecture* (if we want to point out the formal organization of spaces). They result from the action man and natures have carried out over the centuries; they may be either directly and voluntarily built or be the outputs of indirect changes made by man on nature, even in less anthropized areas. Places are *systems* of spatial, functional, visual and symbolic relationships between its constitutive elements. They are the result of a project that has been desired, more or less consciously, by a collectivity that has been building, using and modifying the place over the centuries[1] and sometimes the result has been voluntarily carried out by a single person.

Landscape as meanings

Places are attributed very different meanings. Some have been emerging and taking force according to the readings that past and present intellectuals, painters, photographers, mass media have passed on to us and have finally entered and settled into the mechanisms of collective memory.[2] They partially result from the use and the meaning that local populations attributed to places and elements (meeting, rest, feast, religious practices). But some are also the projection of expectations, wishes, meanings that collectivities or socio-cultural contemporary groups attribute to them (naturalness, new arcadia, etc.).

Landscapes as 'open works' and palimpsest

Landscapes are dynamic. They are subject to continuous changes, either diffused and/or exceptional, by the action of man and nature over the centuries. That is why they may be called *open works*. Traces from the past interlace with those of the present and that modify them continuously and inevitably. Places bear dense and diffuse tangible traces of their history (even in the most contemporary urbanised areas) which are still visible if one knows how to read them. Places' characters and meanings contribute to the people's identity.

Limits to the 'openness' of the 'work'

To consider landscape 'open work' does not mean freedom/permission to transform, innovate or add without questioning the limits beyond which the inherited *work* is physically *destroyed* or *used* in an *instrumental* way. Therefore, if we want to respect, preserve and transmit the specific characters of places and the identity of people, there are some *limits* to these transformations. Such an awareness of inevitable changes means taking a position, consciously or not, in the face of specific characters place inherited from the past, including the recent one.

---

1   Landscapes are the expression of the physical asset of the productive, social, cultural organization of the various collectivities who made it up: see the tradition of human geography and of history (Bloch, Braudel, Sereni, Gambi).

2   Although many recent studies are very interesting, a milestone is the work by M. Halbwachs, *Memoires Collectives* (1950).

The concept of 'limit'

The concept of limit is a question that needs to be explored. Various disciplines are useful to understand its different meanings.

From an etymological viewpoint, and according to the linguistic dictionary, *limit* (*limite* in Italian, *limite* in French, *limit* or *boundary* in English, *límite* in Spanish, but *Grenze*, in German) comes from the Latin *limes-itis*. It means the *terminal lines of an area* (in the figurative sense too) or *divisory lines between* two areas (*Limes* was the defensive fortified boundaries of the Roman Empire from the first century AC). However, the concept also refers to the *value* that may determine the *entity* or the *extension* of some activities, actions, behaviours, performances, properties or characters (for example, speed limit).[3] At the same time, limit is used to suggest the *means* or the *reason* for a constraint or a restriction. The latter may be imposed or suffered but also self-chosen for general or occasional reasons, mainly dictated by moral or economic rules (*Dizionario Etimologico* 1999; *Dizionario Devoto Oli* 2004).

Limit is a fundamental concept of mathematical analyses, clarified in 1821 by A.L. Cauchy to give rigour and strictness to infinitesimal calculation.

From a philosophical perspective, the main and still essential reference was provided by Aristotle (*Metaphysics*, Book 5, 1022A), who clearly explained the different meanings of the word *limit*: (1) the *last point* of a thing; (2) the *form* of a spatial magnitude or of a thing that has a magnitude; (3) the *end* of each thing as point of arrival or, sometimes, as starting point; (4) the *substance* of each thing and the essence of each. In this sense, limit means condition (necessary substance or essence) that says what a thing 'cannot not be' and is the precise reason of the thing[4] (Abbagnano 1998; *Dizionario di Filosofia* 2002).

During the last decades, the fourth meaning of the word and of the concept of limit has been particularly developed and used to tackle specific problems of our time.

The concept of limit came into force during the 1960s at a time when ecological problems were becoming a burning issue. These were taking an important and diffused part in the mind of a growing number of scholars, politics, administrators and populations. In 1972 the important Report *The Limits to Growth* (Meadows et al. 1972) opened the international debate on the links between economic development, preservation of natural resources and populations' needs. The Brundtland Report to the United Nation Organisation (United Nation Organisation 1987) was a key moment to introduce the concept of 'Sustainable Development' as 'forms of progress that meet the needs of the present without compromising the ability of future generations to meet their needs': in particular, it develops the

---

3   See the entry *Limite* in *Dizionario Etimologico della Lingua Italiana* (1999) and the entry *Limite* in *Dizionario Devoto Oli della Lingua Italiana* (2004).

4   See entry: *Limite, essenza, sostanza* by Nicola Abbagnano, in *Dizionario di Filosofia* 1998, and in *Dizionario di Filosofia Garzanti* 2002.

relations between economic growth, protection of natural resources, populations' quality of life (with a special focus on physical well-being and survival). It stated the pillars of Sustainable Development: economy and environment. Now, the concept of *sustainability* implicitly contains the concept of limit.

The concept of landscape adds a new pillar to the concept of sustainability, cultural diversity, which various international and European organizations recognize as essential for people's well-being. At international level, in 2001, the *Universal Declaration on Cultural Diversity* (Unesco 2001) states cultural diversity is as necessary for humankind as biodiversity is for nature (art. 1).

At European Union level, the *European Spatial Development Perspective-*ESDP (Potsdam 1999) establishes some guidelines for territorial policies and also introduces the concept of preservation of cultural characters of the various European regions as a means to preserve the populations' identity and as an economic and social resource. It defines three pillars: Economy, Environment, Society.

At Council of Europe level, in 2001, the European Landscape Convention (Florence 2000) gathers and articulates the concepts and includes references to the physical and material value of places as well as a perceptive value in its broader meaning (i.e. sensitive and cultural) (art. 1). The European Conference of Ministers responsible for Regional Planning (CEMAT) adopted the *Ljubliana Declaration on the Territorial Dimension of Sustainable Development* (Lijubliana 2003), which confirms and further defines the concept of sustainability. It emphasizes the existence of a 'fourth dimension' in addition to the three objectives established by ESPD: 'cultural sustainability'.

Place quality is considered an essential condition to individual and social well-being, not only in its physical sense but also in a physiological, psychological and intellectual sense: it is the *expression* of the *specificity* of places and a factor of the *identity* of populations. A cultural dimension is attributed to the entire territory which includes the social perception that people have from their living places and the way they recognize their historic and cultural diversities and specificities. An innovative aspect of the approach is that principles are applied to any living place be it of outstanding value or ordinary and degraded, and that it proposes an organic integration of many cognitive and operational viewpoints in order to reach such quality. Such dimension is essential in order to respect and preserve their *identity*, and represents an individual and social enrichment.

The fourth pillar involves the cultural field but it has also direct economic implications: if the 'cultural assets' that give people their sense of identity are neglected – if no support is given to cultural, material and immaterial values – then people are faced with a loss of well-being and a loss in economic outputs (Throsby 2001). The fourth pillar is seen as a *resource* able to favour economic activities. Economists have conducted deep analyses on traditional areas of culture (visual arts, museums, monuments) and they are now working on some studies on landscape (an international congress on economy and landscape was organized for the first time in 2009).

The concept of limit contributes to develop – perhaps in this period more than in the past – the idea of self-regulation, self-limitation, self-organization of the world by man.

To better explore this point we can refer to semiologic disciplines which for the last decades have been promoting ample debate around the interpretation of texts. They have placed it within the philosophical problem of interpretation – to identify the conditions that determine any interaction between us and something that is *given*, i.e. the world that surrounds us (cognitive sciences, epistemology, semiology). The focus is on understanding how a *work* has been built and how its form might guide/limit the interpreting/transforming action carried out by contemporaries who are willing to *avoid any instrumental use and any destructive impact*. The focus is therefore placed on the characters of any interpretative activity, intended as the link between the contemporary user and the work, and on a wider concept of 'interpretation' related to action.

Umberto Eco, more than others scholars, has explored the problem of limits and the difficulty in finding some parameters to the unlimited interpretation of any *work* (be it a written text or the nature or the whole world understood as a document).[5] This does not mean that contemporary needs and activities should take second place but rather that attention should be focused on the relationship between the work and the interpreter (Eco 1990).

Consequences of the concept of limit in the transformation of landscapes

The idea of a necessary *respectful attitude* towards the specific physical and cultural characters of places during the inevitable process of their transformation recalls the concept of 'limit', with two consequences. First, knowledge is the basic step in any project of transformation; and second, the issue of limits contributes to define management policies.

Knowledge is the first step in any project of transformation

From a conceptual and operational viewpoint, we must invert the most diffused process used to design/plan and realize territorial transformation (roads, buildings, technological infrastructures/equipments, but also agricultural layout, or simple

---

5   On this theme, see U. Eco 1962, 1990. For an analysis on landscape as a text, see L. Scazzosi 2009. In synthesis, Eco claims that when in the presence of an existing 'work', men are free to neglect its specific characters and can make an instrumental, uncontrolled, misleading *use* of it, till destroying it. Any work possesses its own autonomy and specificity (*intentio operis*), whatever the intentions of whoever created it or transformed it over time (*intentio auctoris*). Any willingness to respect such specificities may entail restrictions and constraints to the unlimited interpretative possibilities – changes in the case of landscape – assumed by any contemporary interpreter (*intentio lectoris*) (Eco 1990).

small territorial equipments, advertising, etc.). At present, this process normally starts from major – and even exclusive – attention towards the functional and formal requirements of new works; they are studied and built whatever the territorial and landscape consequences on the specific characters of the context in which they will be inserted. We believe, on the contrary, that the first step should be to acquire a detailed, deep and articulated knowledge of the characters of places: the description of landscapes constitutes the preliminary phase of any landscape policy (preservation, innovation, requalification).

The word *compatible*, or *suitable*, or *match* isn't sufficient to express the idea of *respect* towards an inherited (constructed by previous generations) work (or place) and neither to answer the *needs* of the contemporaneous world: the word *appropriate* better expresses this concept. This principle has been stated in some recent international and national documents (Guidelines 2008, Part I, I.1.h; Di Bene 2005).

The issue of limits contributes to define management policies

From a methodological point of view, we have to clearly *separate* the concept of *description* from the concept of *assessment* of places and from *realization* of the policies for landscape quality (European Landscape Convention 2000: Art. 6c, Guidelines 2008; LOTO 2005). The process can be synthesized in the following steps:

a. *Places knowledge* – it is the starting point for any landscape policy. This action requires a *knowledge project* (not a gathering of data and maps), including *Characterization* (the present physical and cultural characters), *Dynamics* (natural and cultural, past and future, material and immaterial historical traces), *Social perception* (past and present);
b. *Assessment* (diagnosis of problems and potentialities, not a simple allocation of levels of value);
c. *Planning policies* (a mix of preservation, innovation, enhancement, rehabilitation) foresees the following goals definitions and actions: landscape quality *objectives*; *strategies* and *tools/actions* to realize these objectives; *management programme* through time; *monitoring* of landscape transformation and of the effects of the policies; *reformulating* new objectives.

Social participation is required at different moments during the process.

Such an approach is important for all landscapes (according to the European Landscape Convention) as well as for all protected sites (Unesco sites, for example, or any area protected by national or regional instruments).

Parameters to define the limits for an appropriate transformation of places: The cultural quality of places, in particular regarding the fourth pillar of sustainability

Some parameters allow one to define whether a place might be recognized as having outstanding quality: *integrity, authenticity, rarity, entirety/completeness*. They are used for the preservation and protection of few selected sites pertaining to historic and cultural heritage, to landscape, but also to nature, at all territorial administrative levels. In particular, the major texts on such concepts as integrity, authenticity, rarity, entirety/completeness come from Unesco and Icomos within the framework of the implementation of the World Heritage Convention (Paris 1972) and in the procedure documents for allowing sites entering the World Heritage List (Operational Guidelines for the Implementation of the World Heritage Convention regularly up-dated by the World Heritage Centre). The concepts of integrity and authenticity are now being discussed in Icomos, both generally for any type of heritage (Nara Document on Autenticity, 1994) and specifically for landscapes (Icomos-Ifla 2009) and, especially, for those where continuous changes are more evident (Organically evolved landscapes) and for their sub-category (Continuing landscapes), which largely corresponds to agricultural landscapes.[6]

Theoretical reflections drawn from the tradition of monument protection could be very useful here for landscapes. The documentary value that collective memory attaches to the artefacts of the past may also be used for landscapes in all their material and non-material components: any evidence of the human and natural history, even the most recent one, has to be seen just as a *document*. All sites and artefacts belonging to the past or that we consider part of a culture that is no longer contemporary, even if still recent (for example, the works of the Modern Movement Architecture) should be considered *historical documents*. The concept of historical document is different from the concept of *antiquity* character (or historical *substance* or historical *readability*): that is the immediate evidence of the historical character of an element or area. The surviving signs of the past are still visible in current landscapes and they can be read in different ways and classified in various degrees and kinds of permanencies, either material or immaterial.[7] As

---

6   The three Categories of UNESCO Cultural landscapes are: (1) Clearly defined landscapes intentionally designed and created by men (for example, gardens and parklands); (2) Organically evolved landscapes, with the sub-category of Continuing landscapes and Relict or fossil landscapes; (3) Associatives cultural landscapes (Fowler 2003; Unesco 2003).

7   For example: soil design of places (morphological feature, centuriation, land parcelling, settlement localization, alignments, road tracks, water and channel network); physical features where old materials and building techniques prevail (terraces, a row of century – old trees, an ancient wooded area); way of use (productive, recreational); visual, functional and spatial links (as between castles network, villa with gardens and its farmland property); symbolic links (as between churches towers); cultivation techniques or traditional maintenance systems (as for a trained vineyard, an olive or fruit); giving meaning to elements and places (as places of local memory, linked to festivals, historic

for landscape, this means that we should consider not only each single artefact but also the relations among them all: location, links with other systems, elements, landscapes role.

The character of antiquity may be seen as a value from some cultural perspectives and not from others. The long history of theory and practice of monument restoration clearly shows the different attitudes that were taken in evaluating such characters. Differences or common attitudes can be noted among various socio-cultural groups in a single country or among various countries.

A peculiar and quite recent type of quality assessment consists in attaching value to highly *symbolic* sites even though no particular artefacts are present therein (battlefields, sites which are described in literature and represented in paintings, sites linked with religious meanings and ceremonies, use habits/behaviours). These sites belong to the collective memory of regional, national and international communities or to the collective memory of local populations and are important for their identity. The Unesco Convention for the Safeguarding of the Intangible Cultural Heritage (2003) and the theoretic debate and operational policies on this issue are useful references.

At the same time many studies try to analyse historic, contemporary and forthcoming aesthetic criteria to evaluate place quality. References can be found in *History of Beauty* (Eco 2004) and in *History of Ugliness* (Eco 2007), or in the recent reflections on the cultural perception of nature (Künster 2003), or in the many studies on the changing perception of places through the centuries in artists (Shama 1995; D'Angelo 2001; Milani 2005).

Parameters to assess and guide transformations of places

Landscape policies are articulated into different types of actions: protection, innovation, re-qualification, management. In practice, action on landscape mainly includes more than one of these actions on a single territory or on some parts or elements of it. Concepts and parameters are theoretically built in order to assess the impact of actions on the landscape context, their possible negative effects and their potentialities.

A recent Italian operational document, the *Landscape Report* (Di Bene 2006), provides, in its 'Technical Annexes', a conceptual reference list aimed at giving a first methodological classification that could favour more homogeneous and motivated activities to define and assess the impacts of landscape changes. The list may be useful for protected sites as well as for the whole territory. It records some parameters to read criticism of landscape transformations. It is culturally based on studies and methodologies on landscape but also on ecology and environment and

---

events, local cultural traditions, and places celebrated by high culture through past and recent iconography, photos, texts written by intellectuals and travellers).

even on other operational disciplines, which were developed at international and European level.[8] This list is open to additions, improvements and specifications.

The 'Annexes'[9] list the main types of 'changes' – a world that does not imply a negative connotation – which may impact the physical characters of existing landscapes.[10] They refer to the main categories used in reading the formal and material characters of places (for example, forms, features, colours, materials, building techniques). These include changes in the morphology; in the natural or anthropic skyline; in the typological, material, colours, building characters of the historic settlements (urban, scattered, agricultural/rural), settlements boundaries; in the vegetable components (tree felling, hedgerow elimination); in the perceptive, scenic or panoramic setting; in the settlements-historic setting; in the ecological functions, highlighting their impact on the landscape; in the agrarian, agricultural and cultivation pattern; in the basic features of the agrarian territory (distinctive characters, distribution of settlements, functional networks, small vegetable furniture, land plot fabric).

The Annexes also list the major types of 'alterations' – here the world implies a negative connotation of the concept – which includes a loss of quality in landscape systems.[11] Alterations may be:

- *Intrusion* (inclusion of elements irrelevant to the landscape system's distinctive, perceptive or symbolic characters; for instance, an industrial plant built in an agricultural area or in a historic settlement);
- *Division* (for instance a new road running through an agricultural system or an urban or scattered settlement, splitting it);
- *Fragmentation* (for instance, gradual inclusion of irrelevant elements into an agricultural area, dividing it into parts no longer linked);

---

8 For more information on the parameters used in methods to study and evaluate the whole territory as a landscape in the different European countries, see Scazzosi 2002a and Scazzosi 2002c. For other parameters to read landscape, anthropic and environmental risk, see the documents on instruments for preventing damages drawn up in the sector of the protection of historical heritage, in particular the *Carta del rischio del patrimonio culturale*, a methodology drawn up and experimented by the Italian Istituto Centrale del Restauro (ICR) since 1990.

9 In the Annexes, the parameters used to give a site the quality of outstanding value (diversity, integrity, visual quality, rarity, degradation) are only a few and refer to the most common and consolidated ones used internationally. Indeed, they refer to those aimed at identifying World Heritage sites and sites protected by many European countries' national regulations, which are actually founded on a basis of culture and laws in favour of monuments and 'natural beauties' since the early 1900s.

10 See DPCM 12/12/2005 *Relazione Paesaggistica – Allegato tecnico*, note 8 (di Bene 2006).

11 See DPCM 12/12/2005 *Relazione Paesaggistica – Allegato tecnico*, note 8 (di Bene 2006).

- *Reduction* (gradual reduction, elimination, change, replacement of parts or of basic elements of a system, such as a network of canals for agricultural irrigation, historic building amidst rural buildings, etc.);
- *Gradual elimination* of visual, historical-cultural, symbolic relations linking both the landscape context and the area as well as other elements in the system;
- *Concentration* (too many high impact developments within a rather limited area);
- *Disruption* in the ecological and environmental processes at a large or a local scale;
- *De-structuring* (when a landscape system is modified through fragmentation, reduction of the basic elements, elimination of structural, perceptive or symbolic relations);
- *Loss of characterization* (when basic characters of a landscape system are modified).

The list refers to terms and concepts used in ecology that are adequate for the definitions and cultural and symbolic aspects of landscape.

The alteration evaluation also has to consider other parameters, connected to time and to the quantity of the effects: alterations may be *totally* or *partially* damaging, may have *reversible* or *irreversible* effects, can *increase* or *diminish* over time.

Some other useful parameters used in instruments for planning and managing places refer totally or partially to the basic concepts of the SWOT analysis (Strengths, Weaknesses, Opportunities and Threats), largely used in developing economic, general or business strategies, but also in landscape planning and management. They may concern landscapes' *potentialities*, *threats* and *restricted uses*: *fragility/vulnerability*, *sensitivity* (for example, to economic pressures), *vitality*, specific *potentialities* (such as quietness, wilderness) or the different capacity in supporting changes: capacity of absorption, stability, instability, etc.[12] And finally they are also parameters linked to the knowledge of *running, foreseen, foreseeable dynamics*. Other important parameters concern the analysis of the values that become apparent when planning changes in landscape and in the conflicts that implies. An important reference is the systematization that occurred within geography disciplines: economic, amenity and security values (Jones 2009).

Architecture disciplines also bring some important contributions through an ongoing reflection on the quality of new architecture – meant as new building, adding or modifying existing buildings, public spaces, small and widespread arrangements. The Italian and the international debate are trying to determine the exact meaning of the quality of architecture, how to reach it and guarantee the

---

12   See DPCM 12/12/2005 *Relazione Paesaggistica Allegato tecnico*, note 2 (di Bene 2006).

architectural interventions on the territory are appropriate to the specific characters of places.[13] Quality of architectural and landscape innovation takes on several forms. In 2004, the Italian Ministry for Cultural Heritage and Activities opened a debate that focused on some parameters that are valuable for places although they were originally thought for buildings only. The first and simplest is the parameter for assessing the *performance function* (the capacity to guarantee the performance of each single component, such as the roof, the openings). The second, more complex, consists in assessing the *functionality* (the capacity to complete a function such as being a residence, a business, a recreation place). The third parameter, more difficult and widely discussed, is the *aesthetic* assessment, the capacity to express contemporary values (DARC 2003a, b, c). Another parameter should be added that would assess the kind of relationships that exist with the pre-existences, (or context), and which would be called *relational* assessment.

Relation with the context has already been often explored and discussed during the history of architecture.[14] Today the issue under discussion is how to go beyond an excessive self-referentiality of architecture so that new interventions could build a conscious and explicit link between innovative aspects (meant as any voluntary action to change places) and the material and immaterial characters of inherited places. Some conceptual formulations have tried to articulate and clarify the different kinds of links between the new and the old. For instance, they speak of replacements, pre-valence (domination, evidence), integration (not interference, uniformity, mimesis, coherence), hidings (masking, concealment) (Barocchi 2002). Other concepts and couples of opposite concepts are also useful such as *continuity/discontinuity, harmony/contrast, coherence/de-contextualization* or *estrangement, in-scale/out-of-scale.*

Who can or who must define assessment and limits

The relativity, plurality and variability of values, which has been largely proved in history (just think of the changes in the artistic values attached to monuments at

---

13 See the Resolution of the Council of Europe on the architectural quality of rural and urban environment (Brussels, 12/01/2001); the Law on the quality of architecture, France (1997); Bill on the quality of architecture, Italy; Law by Regione Emilia Romagna, Italy (Regional Law of 15 July 2002, n. 16, *Rules to requalify historic-artistic monuments and to promote the architectural and landscape quality of the territory*). The Italian Ministry for cultural heritage and activities, through its General Direction for architecture and contemporary art (DARC), organized, after several preparatory meetings, an international conference on *The Quality of contemporary architecture in European towns and territories* (21-23 November 2004), with the participation of institutions of various European countries (France, Spain, Finland, Belgium, Holland, Germany) that have been tackling such issues in the last decades (DARC 2003a, b, c).

14 For example, after the Second World War in Italy the journal *Casabella*, edited by Ernesto Natan Rogers, focused particularly on this issue and had international echoes.

different historical periods), have been pointed out in recent documents, such as the *Krakow Charter for cultural heritage* (Krakow 2000) and the *Nara Document of Authenticity* (Nara 1994). According to them, cultural heritage, of which landscape is a sensory expression, cannot be defined in univocal and absolute terms over time and in the various geographical-cultural areas.

The European Landscape Convention focuses on people's *perception* as an indispensable and founding step in each single choice (art. 1). It focuses on the need/opportunity to share any transformation choice with people. Landscape quality means, above all, conservation of the specificities of places as they determine the identity of people who built them, who live in them, who bring changes day after day and are necessarily responsible for the changes they bring. As a consequence, the minimum paradigm for accepting any interpretation-change in places must depend on people's agreement. As, indeed, people gather all former interpretations that have been passed on by tradition as well as contemporary interpretations, either expressed or concealed.

This does not mean that we are moving towards an absolute relativism or towards a lack of general and shared criteria at a global level or at the level of wide cultural areas.

The process through which people give sense to places and elements and attribute values or disvalues to them, in order to make choices for the transformation of places and implement and manage them, is extremely complex. It has not been much studied or experimented yet. It concerns the different kinds of relationships between the new and the old, in the light of the different cultural meanings and values that socio-cultural groups attribute to past (material and immaterial) heritage. But also in the light of conditions and requisites that contemporariness has put forward. The word 'people' is to be understood in its wider sense. Indeed it includes all types of categories (social, economic, cultural and other) and all types of levels (from local to supranational). So it must not be seen as an issue mainly limited to local population. Moreover, it also involves technicians, politicians and administrators with cultural and operational responsibilities either in reading and building value systems or in the democratic decision-making process.

Our present historical period is characterized by diffused, deep, epochal territorial changes. But there is a growing diffused awareness-raising around the idea that these changes are not creating quality places, that a negative process of homologation of the characters of places is underway and that there is a feeling of indifference towards the material characters and meanings of places.

As a matter of fact, there is a growing cultural acceptance of the principle according to which any innovation should coherently integrate with what already exists, and respect both contemporary needs and the specific constitutive and identitarian characters of places – since such identitarian role is one of the most significant needs of contemporariness.

Scientific, technical, cultural groups and communities are also able to point out some examples of bad changes in landscape. This process involves a still larger number of individuals and takes different forms. At an international level, some important debates are underway after Unesco's position against the project of building high constructions near historic sites (ex. in Vienna's historic town centre, which is inscribed in the World Heritage List). In Italy, for example, national and regional laws and regulations have been created and some public administrations have carried out significant actions; some new or existing associations focus on landscape quality and support political and cultural battles.

On the other hand, it seems more difficult to reach a unanimous and shared definition of good solutions for managing inevitable transformation of landscapes. Some public initiatives resort to the instrument of identifying, awarding and disseminating *good practices*. There is, for example, the first edition of the European Prize for Landscape, launched by the Council of Europe. Or some activities focusing on research and awareness-raising among specialists and the large public, which are organized through EU international cooperation programmes (INTERREG, Culture 2007, etc.).

From a theoretical point of view, the discussion on the limits of interpreting a text has pointed out several key-points (Eco 1990, 1992). If it seems impossible to clearly define some criteria for the quality of places, it is possible to indicate methods to achieve it. In this perspective, awareness-raising, education, good practice awards, exchange of ideas and solutions are crucial activities to be carried out. And above all, people participation may be the key to the whole process.

Some conclusions

The themes of *limits* to changes and of *appropriate* transformations should be dealt with on the basis of multi-disciplinary contributions (sociology, preservation of historic and cultural heritage, ecology, history, geography, juridical studies, architecture and planning, agronomy, economics) in an interdisciplinary perspective and in an effort to link theory to practice. Still many questions are open, like the difficult problem of defining and measuring assessment parameters, the issue of the instruments for reading and representing the characters of places, the challenging objective of getting different disciplines' approaches – in fact today larger efforts are being made to find convergences between cultural and ecological-environmental approaches (Farina 2006).

Interdisciplinarity – both theoretical and practically applied in case studies that would confirm and enrich theoretical hypothesis – would be a very useful and significant method of research. It would bring significant and shared progresses in the difficult debate around *appropriate* transformations in contemporary places. A debate that tries to answer the question of how to connect inherited landscapes to future landscapes, which is a crucial issue for the contemporary world.

References

Abbagnano, N. 1960. Limite, essenza, sostanza, in *Dizionario di Filosofia*. Torino: Utet. 3rd ed., updated and enlarged by Fornero, G. 1998.

Barocchi, R. 2002. *Manualetto di progettazione rispettosa del paesaggio, ad uso dei professionisti e delle commissioni edilizie integrate – ISPAR Istituto per lo studio del paesaggio e dell'architettura rurale*. Mariano del Friuli: Ed. della Laguna.

D'Angelo, P. 2001. *Estetica della natura*. Bari: Laterza.

DARC. 2003a. *Quality of contemporary architecture in European towns and territories* (Bologna, 21-23 November 2003), published by the Italian Ministry for Cultural Heritage and Activities, General Direction for Architecture and Contemporary Art DARC (proceedings forthcoming).

DARC. 2003b. *Quality of contemporary architecture in European towns and territories. Buyers, contexts and diffused quality* (Rome, 15 May 2003, Proceedings of the 1st preparatory workshop) published by the Italian Ministry for Cultural Heritage and Activities, General Direction for Architecture and Contemporary Art (DARC).

DARC. 2003c. *Quality of contemporary architecture in European towns and territories. Design of the infrastructure and project quality* (Milan, 22 September 2003, Proceedings of the 2nd preparatory workshop) published by the Italian Ministry for Cultural Heritage and Activities, General Direction for Architecture and Contemporary Art (DARC).

di Bene, A. and Scazzosi, L. (eds) 2006. *La Relazione Paesaggistica. DPCM 12-12-2005. Finalità e contenuti/The Landscape Report. 12-12-2005. Purposes and contents/Le Rapport Paysager. 12-12-2005. Objectifs et contenus*. Rome: Gangemi (Italian, English, French version).

Dizionario Etimologico della Lingua Italiana (1999). Bologna: Zanichelli.

Dizionario Devoto Oli della Lingua Italiana (2004). Florence: Le Monnier.

Dizionario di Filosofia (2002). Milano: Garzanti.

Donadieu, P., Küster, H. and Milani R. 2008. *La cultura del paesaggio in Europa. Manuale di teoria e pratica*. Florence: Olschki.

Eco, U. 1962. *Opera Aperta, Forma e indeterminazione nelle poetiche contemporanee*. Milan: Bompiani.

Eco, U. 1979. *Lector in fabula*. Milan: Bompiani.

Eco, U. 1990. *I limiti dell'interpretazione*. Milan: Bompiani.

Eco, U. 1992. *Interpretation and overinterpretation*. Cambridge: Cambridge University Press. Italian version: *Interpretazione e sovrainterpretazione*. Milan: Bompiani, 1995.

Eco, U. (ed.) 2004. *Storia della bellezza*. Milan: Bompiani.

Eco, U. (ed.) 2007. *Storia della bruttezza*. Milan: Bompiani.

Farina, A. 2006. *Il paesaggio cognitivo. Una nuova entità ecologica*. Milan: Franco Angeli.

Fowler, P.J. 2003. *World Heritage Cultural Landscapes 1992-2002*, Unesco World Heritage Papers 6. Paris: Unesco.

Guidelines for the implementation of the European Landscape Convention – Recommendation CM/Rec(2008)3 of the Committee of Ministers to member states of the Council of Europe (Adopted by the Committee of Ministers on 6 February 2008 at the 1017th meeting of the Ministers' Deputies).

Halbwachs, M. 1950. *Memoires Collectives*. Paris: Presses Universitaires de France.

Icomos-Ifla International Scientific Committee on Cultural Landscapes. 2009. *Operational Guidelines of World Heritage Centre*, update 2009.

Jones, M. 2009. Analysing landscape values expressed in planning conflicts over change in the landscape, in *Re-Marc-able Landscapes – Marc-ante Landschappen. Liber Amicorum marc Antrop*, edited by Veerle Van Eetvelde, Marjanne Sevenant and Lisa Van Velde. Ghent: Academia Press.

Künster, H. 2003. *Geschicte des Waldes. Von der Urzeit bis Gegenwart*. München: Beck.

LOTO Landscape Opportunities for Territorial Organisation – INTERREG Project (2005). *The landscape management of territorial transformations: Guidelines and pilot actions/La gestione paesistica delle trasformazioni territoriali: Linee guida e casi pilota*. Milan: Lombardy Region Tipografia Ignizio Sondrio (English and Italian version).

Meadows, D., Meadows, D., Randers, J. and Behrens, W.W. 1972. *I limiti dello sviluppo, Club di Roma*. Milano: Mondadori.

Milani, R. 2005. *Il paesaggio e' un'avventura*. Milan: Feltrinelli.

Pelissetti, L.S. and Scazzosi, L. (eds) 2009. *Giardini storici. A 25 anni dalle Carte di Firenze: Esperienze e prospettive*. Florence: Olschki .

Scazzosi, L. 1993. *Il giardino opera aperta*. Florence: Alinea.

Scazzosi, L. (ed.) 2002a. *Leggere il paesaggio. Confronti internazionali/Reading the Landscape. International comparisons*. Rome: Gangemi (Italian and English version).

Scazzosi, L. 2002b. Leggere e valutare il paesaggio. Confronti/Reading and assessing the landscape. Comparisons, in *Leggere il paesaggio. Confronti internazionali/Reading the landscape. International comparisons*, edited by L. Scazzosi. Rome: Gangemi, 19-59 (Italian and English).

Scazzosi, L. 2002c. Valutare i paesaggi, in Clementi, A. (ed.) *Interpretazioni di paesaggio*. Rome: Meltemi, 217-241

Scazzosi, L. 2003. *Landscape and Cultural Landscape: European Landscape Convention and Unesco Policy*, in UNESCO, *Cultural Landscapes: The Challenge of Conservation*, World Heritage Papers 7. Paris: Unesco, 55-59.

Scazzosi, L. 2004. Reading and Assessing the Landscape as Cultural and Historical Heritage. *Landscape Research, Journal of the Landscape Research Group*, 29(4), October, 335-355.

Scazzosi, L. 2008. Il paesaggio opera aperta: Conservare/trasformare, in *La cultura del paesaggio in Europa. Manuale di teoria e pratica*, edited by P. Donadieu, H. Küster and R. Milani. Florence: Olschki.

Scazzosi, L. 2009. Giardini e paesaggi opera aperta. I limiti delle trasformazioni, in *Giardini storici. A 25 anni dalle Carte di Firenze: Esperienze e prospettive*, edited by L. Pelissetti and L. Scazzosi. Florence: Olschki.

Shama, S. 1995. *Landscape and Memory*. London: HarperCollins.

Throsby, D. 2001. *Economics and Culture*. Cambridge: Cambridge University Press. Italian transl. *Economia e cultura*. Bologna: Il Mulino (2005).

Unesco. 2001. *Universal Declaration on Cultural Diversity.* Paris: Unesco.

Unesco. 2003. *Cultural Landscapes: The Challenges of Conservation*, Unesco World Heritage Papers 7. Paris: Unesco.

United Nation Organisation – Brundtland Commission. 1987. *Report 'Our Common Future'*, in *World Commission on Environment and Development* (WCED) *Report*, published by the United Nation Organisation.

Chapter 2

The Ineffable, Ethereal, and Evanescent as Values of Local, National, and European Identities

Edmunds Valdemārs Bunkše

How do you know but ev'ry Bird that cuts the airy way,
Is an immense world of delight, clos'd by your senses five?
(William Blake)

... ineffable splendor of the light divine in which the soul has
reveled before its descent to earth.
(Vincent Cronin)

All of you, talk, talk, talk.
But here we live, below,
And do not see,
How spruces bloom
Atop
And then wonder,
As cones fall
On our heads.
(Imants Ziedonis)

The aim of this chapter is to enrich experiential possibilities in ordinary rural landscapes by broadening landscape perceptions and narratives. To think self-consciously within the domain usually inhabited by poetic souls, as suggested by the words 'ineffable', 'ethereal', and 'evanescent'. There is a certain magic, an indefiniteness, to these words and the experiences that they denote. They exist between the world of myth and fact.

Today we possess facts reaching to infinity. Myth has long been banished from normal European discourse, save for those few individuals and groups in some remote, hidden corners of the Continent. It is no longer possible to believe in myths; however, knowledge of myths may enrich the experience of landscape when something momentary or not immediately explicable is encountered.

Experiences and conversations, in particular rural landscapes, have prompted my thinking: In Jutland (Denmark), in Latvia, and in Greece (during the PECSRL meeting there). In the first two instances, it was concern about what remarkable landscapes or objects could be shown to visiting geographers who came from countries with famous landscape features. In the case of Greece, it was the discovery of fascinating ordinary rural landscapes that are rarely included in tourist publications, which favour the stereotypical, spectacular, obvious, and emphasize 'the sights'. The thinking and the perceptions of tourism are materialistic, based on the grand narratives of the beautiful, sublime and picturesque, set forth during the second half of the eighteenth and early nineteenth centuries.

In a world where famous landscapes and landscape features are consumed just like Toyotas and Audis, ordinary, unspectacular rural landscapes hold little interest. The goal is not do away with consumption of 'the sights'. Attractions are, after all, an economic mainstay of tourism. The purpose is to interest people in ordinary rural landscapes, which open up to those who seek an existential bond with the landscape by letting the landscape, not a tour guide, speak to them; by letting the landscape come, as it were, to them, to envelop them, if only for a moment, thus to move beyond the grand narratives of landscape aesthetics, 'landscape feature branding', materialism, and consumerism. To make finding or creating little local narratives important, which is the charm of discovery in ordinary rural landscapes (and which made for rich rural lives and discourses before the mass media). In my own life, I have encountered that richness by reading a long, thorough, often humorous account of a childhood on a farm by the Latvian writer Jaunsudrabiņš (1914); and by spending two childhood summers on a Latvian farm in total freedom (Bunkše 2004a). These and similar experiences in nature have led me to not seek local curiosities, such as the 'Bockstensmannen' (roughly: Swamp man) near Varberg in Sweden (on the Kattegat). But rather to seek landscapes that offer possibilities to experience the subtle, momentary, mysterious, possibly the mythical (even though it must be admitted that the 'Swamp man' does represent a medieval murder mystery).

The little narratives that an individual discovers in ordinary rural landscapes stand in sharp contrast to the meta-narratives that the poet Ginsburg calls mindless, mechanical instructions that come from 'The Industrial Moloch' of which the mass media is a part (Codrescu 1990: 62). Indeed tourism and 'acts of modernization' may threaten the authenticity of local narratives (Chambers 2010: 101). In rural childhood, little narratives once were common, as it was for the poet Czeslaw Milosz, who found himself in a mythical dream by a riverbank in Lithuania. It was a magical experience of myth, fear, awe, nostalgia 'to which the world of facts will add nothing' (Codrescu 1990: 62-63). But adults too can still experience little narratives, if they give up the Moloch as a source of information. Barry Lopez, a gifted writer of landscapes and nature, has created little mythologies in a small book of magical landscape essays entitled *Winter Count* (Lopez 1981). More on that later.

Intersubjectivity with landscapes and with nature

It is the broader, underlying purpose of this chapter to have the rural landscape connected to the inner human being – to diminish the distance of a visitor, urban settler, or migrant from it. In other words, to enable such people to begin regarding the wider rural landscape as home. To find possibilities by which 'a person can accept a landcape as part of the self and to nurture it' (Kursīte 2008: 39). Inter-subjectivity is a term that describes cultural change among people. It involves sharing opinions, insights, knowledge and the like, through conversations, stories, even snippets of ideas. It means sharing subjectivities. Why not share subjectivities with a rural landscape?

In an ever more migratory world, an important issue is not only inter-subjectivity of visitor and native, outsider–insider, but also the inter-subjectivity of the person and landscape, so that a little local narrative comes into being, based on shared 'values' of person and landscape. This means that the landscape is not passive; it shares in the experience in some active way, for such is inter-subjectivity.

A case in point comes from the uneasily settled Kaliningrad, where in the Dunes of Courland she experienced 'an ineffable moment united with space ...' (Kursīte 2008: 39). In sharp contrast, during an ethnographic expedition in 2003, Kursīte recounts the following dialogue with a woman born in Russia, in the Ivanova region in 1913, who migrated after WWII;

> Q: 'How do you like the climate and nature here?'
> A: 'I like nothing here. Nothing. What's good here? Nothing's good here.'
> Q: 'But the sea, the bay ...'
> A: 'No, I don't like the sea, nor any wider waters. In my youth I had not seen the sea, nor any wider waters. A little river and that's all.'

The illustration is of an extreme case in an unusual region of sand dunes, sea, bay and sky – a landscape that would seem to be ideal to ethereal experiences, great and small. There is no hint of inter-subjectivity, no little narrative of the landscape as something positive. The dialogue represents an archetypal story of migration, in which the past and the homeland hold the only value. Does it always have to be thus? Does this story apply to an ex-urbanite seeking rural nature somewhere? Not perhaps to everyone, for in the latter case migration is voluntary. Does it apply to an urban cottage owner? Probably not, but one has to suppose that the visual and sports landscape is of paramount importance. What can be changed? How can such people be drawn closer to the land? Or to the seascape?

From mute senses to ethereal perceptions

In this chapter I build on a paper entitled 'Feeling is believing, or landscape as a way of being in the world' (Bunkše 2007b). It represents work that I have been doing with all the human senses as counterpoint to the all-pervasive visual dominance in the contemporary world (with aural experiences of lesser, secondary importance). The word 'feeling' means both emotion and the sense of touch as a more trustworthy sense than seeing. The emphasis is on the human being in the landscape, experiencing it as a unity through all the senses; and the harmony, disharmony, or neutrality of the relationship between inner and outer landscapes.

The most startling discovery recounted in that essay was the scent of sage in a mundane landscape in the NW corner of the State of Nevada, with few hints of the picturesque. Like the woman in the Dunes of Courland, I felt no connection with the landscape, felt it to be the penultimate landscape of exile. Yet, every time I passed through it, spent time in it, the scent of sage, emanating from scruffy sage bushes, beguiled me. Eventually it became a landscape of great fondness for me.

It was not a visual or verbal message that pulled me into it, it was the scent of all-enveloping sage. An inter-subjective exchange seemed to occur between the landscape and myself. A gladness of being there followed. Diane Ackerman (1991) calls the sense of smell, taste, and touch the earthbound senses [Tuan (1975) uses the word 'proximate'], in contrast to sight and hearing. Oliver Sacks (2003) adds yet a sixth earthbound sense, that of *proprioception,* i.e., how the entire human body senses the world (the older term is 'kinesthetics').

Earthbound senses are landscape and place specific. They are highly subjective. What is sensed is difficult to describe or broadcast to a wider world, to an audience that has no specific experience of the particular scent or feel of the landscape. Yet what is touched, tasted, and especially smelled leaves lasting memories of those landscapes. Moreover, these sensory characteristics of landscapes are not passive presences – they are active, they touch and engage the senses, even if the human being may not always be consciously aware of them (a slick log on a mountain river crossing can change that in a millisecond!). And since it is not possible to broadcast them electronically, or even describe them sufficiently (Marcel Proust's evocation of a madeleine is a masterly exception), as experiences they are part of the little narratives of landscapes and places. The experiences are personal, subjective, deep, they are private micro-narratives. And as such, inherent in them is the possibility to develop affective relationships with the landscape. If the fall from a slick log is survived without dire consequences, it may become part of the narrative of a memorable journey in a memorable landscape.

In 2008, at a conference in Rīga, I tried to broaden landscape aesthetics to include earthbound senses, but soon realized that this was a false trail. That it was an existential, not aesthetic issue. In its original sense aesthetics has been a search for truth and beauty. In this the visual sense has always been dominant and communicable, particularly in architecture. Indeed, there was so much agreement about the form that aesthetic beauty should take that in architecture 'an aesthetic language dating back to the Temple of Apollo at Delphi ended up gracing the familial homes of Edinburgh accountants and Philadelphia lawyers' (de Botton 2006: 32). Clearly experiences of earthbound senses have aesthetic dimensions, but since it is difficult to communicate them, aesthetic categories are difficult to create.

But they do represent experiences of landscape. Some of these experiences can be replicated, especially in the culinary arts. And to a degree they can be communicated by good storytellers, poets, composers and writers able to share their existential experiences. Particularly if a story is told in the actual landscape of the experience. Communicating earthbound senses is a path into the experience of landscape that is open to exploration. Even more so is the synthesis of fact and myth.

Barry Lopez and the synthesis of fact and myth

> *"[...] in the end we become part of the landscape", taking on the*
> *characteristics of a sacred place.*
> (Vine Deloria, Jr.)

> *This they tell, and whether it happened so or not, I do not know;*
> *but if you think about it, you can see that it is true!*
> (Black Elk)

It was Barry Lopez who inspired me to go in this direction a few years ago. But then I did not know how to proceed in my own way. Lopez has written many thoughtful non-fiction books about landscapes and geography; about human relationships with the land and waters and the creatures of the living world. *Arctic Dreams. Imagination and Desire in a Northern Landscape* (1986) is his magnificent major work. But it was a small book, of fictional stories, *Winter Count* (1981), which I used in my landscape and literature course, that convinced me it is possible to have myth-like perceptions of landscapes, in which there is a tenuous zone between fact and myth – a zone that makes it possible for the imagination to grasp the charm of myth, if not accept it all as fact.

It is, of course, impossible to recount Lopez's stories here, except for some impressionistic highlights. In one, on a snowy night, into the wide, lit-up canyon of New York's Park Avenue, a group of 15 or 20 'Great blue herons were descending slowly against the braking of their wings, their ebony legs extended to test the depth of the snow which lay in a garden that divided the avenue, [...] They had landed as if on a prairie [...]'. They did not stay long. The story is entirely plausible. But that is not what appeals to the imagination. 'They had landed as if on a prairie' is part of the magic of the story. The contrast with Park Avenue's self-conscious artificiality is extreme. It is pushed back deep into prehistory, especially by the elegant motions of the birds. They and the snow had altered the avenue into a primeval state.

In another story, Lopez uses Indian and White men's tales and dates of observations to create a synthesis of fact and myth. The story is about the existence and disappearance of a herd of large white buffalo, perhaps in the winter of 1845. With a wind-like moaning song, they are said by the Indians to have climbed up the Medicine Bow Mountains, never to be seen again. In 1925, a geology student finds skeletons of two large buffalo at a height they never frequented. At the end of the story the teller wakes up in his tent, both legs broken. I did not attempt to explain that to the students. I had them draw their own interpretations. To me it was symbolic of retribution for the destruction of the buffalo and the lands of the Indians.

Ineffable, evanescent, ethereal

The subtle and difficult attempt to communicate the experiences of the earthbound senses and the mythological stories of Lopez led me further afield, to investigate the tenuous, fleeting, but beguiling experiences of landscapes represented by the words: ineffable, evanescent, and ethereal. I was inspired by Lopez's little mysteries, but his tales belong to another continent, another people. For me the mysteries come from Europe, mainly from Latvia, from my own cultural past, situated on the Baltic Sea.

Let us focus first on the words 'ineffable', 'evanescent', and 'ethereal'. *Ineffable* means that something is too great or extreme to be expressed into words; beyond description, or begging description, indefinable. *Evanescent* means passing away, vanishing, gradually, imperceptibly; transient. *Ethereal* denotes something light, delicate, not of this world (Britannica 1959). I chose these words because in their meanings they are linked to subtle, mysterious aspects of landscape experiences, to the *inter-subjectivity of landscapes and the human being.* What follows are accounts of personal moments of being that illustrate these notions. They are separated in time and space because arriving at this story took time.

Two forest epiphanies

"Those are not the footsteps of a giant," again and again the boy tried to assure himself. "There are no giants. The sounds are made by the trees bending and knocking against each other in the storm. It only sounds like the footsteps of a giant." However, the moment that he had thus convinced himself that there was nothing supernatural about it, there they were again – slow, rhythmic, giant footsteps, which came ever closer, ever closer, from the direction of the sea ...
The twelve year old boy was gaunt. He had just been made a boy scout and it was his turn to stand watch at the camp, which was situated on a bluff above the Baltic Sea, in a pine and deciduous forest. He stood up against the trunk of a large pine and trembled in the wind, which was driving large, sparse raindrops before it. He had to be on guard from three until four in the morning – the darkest hours of the night. The hour seemed endless, but it did pass and he went to wake his mate. He told him nothing about the giant. (Bunkše 2004b)

The 12-year-old boy experiencing that ineffable moment was myself. The second epiphany was recorded many years later, in 1990.

> In the natural and cultural milieu of Latvia, the eastern and northern Baltic, if one should so wish, it is still possible to find connections with the medieval and even paganistic, [...] Such a mystical connection I found on the night of the summer solstice, when for a moment I was alone in a small forest clearing. The sky was unusually clear, translucent, bluish green. Nothing stirred, not a breath of air. It was quiet – so quiet that the sudden, impulsive flight of a fly was like the 'plop' of a frog in a Japanese haiku, with which the poet evokes the essence of silence. At that moment I felt a direct connection with the people who had lived here for centuries – with forest clearers, hunters, robbers, seafarers. And with our witches, who were burnt at the stake. For during that summer I had seen eyes, which made me believe that not all the witches had been burnt. (Bunkše 2007a)

It was a fleeting, ethereal moment that a native to the place, culture and history was able to experience. A mystical moment too, for I felt completely immersed in the continuum of nature and culture. A sharing of fact and myth not unlike that evoked by Lopez. What both stories have in common is the unexpected, inter-subjective connectedness with vastness: in the former, with the landscape and nature; in the latter, with the long reach of time into the past, connected to the present. In both cases there is knowledge of historical narratives associated with the ethereal experience. The storm and the solstice were facts, as was the burning of witches; the giant and countless generations celebrating the sun were fairy tales and mythology. Both were experiences that contained elements of ineffable, evanescent and ethereal. They came unbidden (although John's night, as the summer solstice celebration is called in Latvia, suggests unexpected experiences), they were fleeting, and they were mysterious. I am a secular individual, but at the times of those experiences I felt in the presence of something vast and even sacred. It is true that we Latvians, because of late Christianization, are somewhat closer to our pagan past than some other Europeans. Perhaps closeness to a pagan past gives freer rein to our imagination?

Evanescent, ethereal surprises, or learning the subjectivities of landscape

I have recently designed a cottage, located on a desolate coast of NW Kurzeme (Courland). The beaches have few people on them, there are few cars on the gravel highway leading along the coast, and inland is a vast stretch of country consisting of large tracts of empty forests and bogs. There are the typical trees and brush: birches, spruces, pines, some junipers, the ever-present alders and swamp willow bushes in the wet spots. An abandoned fisherman's place (and there are many) may have a linden tree or a forgotten apple tree. Slowly I am learning this landscape, peeling back its mysteries inland from the coast. There are straight pines and then there are pines that seem to have emerged from nightmares. One stands outside my westerly window. It has none of the elegance of many of the windblown pines in the dunes, which could pass for full-scale Japanese bonsai. This one is a dark tree which turns black when it rains. Many of its branches grow from the main trunk very near the ground, curve upward to form a U, then have branches going off in all directions. Most of the ends of the branches have been broken off by gales. They may be the size of a human wrist, with sharp tips of wood splinters. It is an ominous tree, an evil tree, if one believed in pathetic fallacies. It stands in sharp contrast to the two gentle *vigas* – Latvian for meadows, which are located seaward from the cottage.

Meadows, that during the course of a summer produce hair-grass that responds in waves to the gentlest zephyrs. I never cut this grass, for I find it beautiful. But the black pine (some locals call it gypsy pine) reminds me that all is not as it seems to be here. It is a mysterious presence. The surprise came during my first summer there. I had placed a tent where I wanted the cottage to be. Taking into consideration views, sunsets, trees, birch groves, it was situated on the edge of the *viga*, which I knew to be a former fishermen's grazing meadow. A crude feng shui idea. Old foundations indicated that most, but not all, fishing folk had built on the landward dunes – *kangars* in Latvian. One late afternoon in early July I was returning from the sea. It was about 10pm, the sun was still over the sea, but the *vigas* were already in the shadow of the trees that grow on the seaward dunes. To my great surprise, a thin film of fog had spread over the *vigas*, waist high in places and gathered in a slight knot around the tent. My first thought was: 'Why didn't it make any noise getting here? How could it suddenly be here?' And then: 'How can there be fog when the trees on the *kangars* are still in the sun?' The next thought was: 'The souls of the departed are said to appear in low, rolling fogs of November'. It wasn't November!

Of course I do not believe in dead souls coming around as fog. Just as I do not believe that whistling will bring wind for a stranded sail boat. Still, I never whistle for wind. But as I watched that fog late into the night, moving, rolling, gathering in one side of a *viga*, then another, then slowly rolling down the middle, I thought of dead souls and could see why an American Indian might say, as they do on such occasions, 'It might be true'. I built my cottage with considerable difficulty on the *kangars*, with its big trees to be cleared, and now watch with ever growing fondness the silent coming, stretching, and rolling about of the souls of the departed. I named the place *Smilgumiglas* (Hair-grassfogs) and firmly believe that the fog rises from the hair-grass. Even though I suspect that the physics of it is quite different. But I do not ask about it. Why disturb a dialogue with the landscape?

Art and landscape

What I have presented are contemporary experiences, together with myths, fairy tales, and history in an ordinary landscape of abandoned fishermen's cottages and grazing lands, with the sea on one side, a vast, mostly empty forest, streams and bogs on the other. By spending time there, by letting the landscape 'speak' (Spirn 1998), it was possible to open up an area of being between material landscapes and mystical ones. Tourists who journey into rural landscapes cannot help but notice surface textures and smells. But most go in search of visible, known landscapes that can be recorded and taken home as souvenirs. Most will not receive invisible messages from landscapes, as does Ranah P.B. Singh (1994), for whom 'The visible and invisible messages have been a real source of meaning in my life'. It takes knowledge to open up to the invisible landscape and its hidden stories and meanings. As Anne Whiston Spirn writes, after a short stay in a landscape, 'I could read only the most obvious stories and not the subtleties [...] *Culture can prevent eyes from seeing and ears from hearing*' (Spirn 1998: 36, my italics).

And there are subtleties, especially in a seacoast landscape, that mean endless discovery. There are the different scents and feel of air in the hollows between two shoreline dunes. Walking barefoot on sand, meadow-grass, pine needle strewn surfaces, or even on freshly turned earth by the tusks of wild boar, offers radical changes in sensations, in kinesthetics. Winds shift and change endlessly in small bits of landscape. One learns of this by the Baltic Sea when searching for some warmth behind a dune or in a wind blown hollow in the sand. It is probably quixotic to ask for meaning in invisible landscapes in a world dominated by economic, political, and materialistic processes that also define relationships with landscapes. Such a quest should not be regarded as a return to the Romantic era, with its Gothic tales and narratives of genius loci and mutability of prospects. Nor need it be a renewed effort to collect folktales and folklore, as Elias Lönnrot did in Finland and Krišjānis Barons in Latvia.

But a new era seems to be forming, for the material world is saturated with visual images and sensational technological feats in creating visual fantasies, as Andy Warhol so well recognized, as did the philosopher Richar Kearney (1988). There are still undiscovered mysteries in rural landscapes. Art and science blossomed during the romantic period in an exuberant and effective way (Holmes 2008). Today, too, art and artists are needed to evoke, what the young Latvian writer Inga Ābele (2008) calls '*esmes* mirklīši' – moments of being of an ineffable, evanescent, or ethereal nature.

Such a moment came one morning at Smilgumiglas, when a visiting fairy-tale writer remarked that some of the birches in my birch grove in a distant section of a *viga* had moved closer to the cottage during the night. I saw no change, but good-naturedly accepted it as a fairy-tale writer's natural vision. I began to watch the grove though, and by and by, over several months, some of the birches did indeed appear to move closer, then again be further away. In my rational mind I finally realized that during the course of a day (let alone in the moonlight), the sun cast light and shadow on the bodies of the birches, which seemed to change their relationship to one another. That they did, indeed, 'move'. (A child would probably have accepted the observation immediately.) The reality was experiential, not material. It nourished my spirit as an original way of changing one's perceptions and feelings about the landscape. The strange gypsy pine and the lonely feeling of the place also helped. And the fact that not long thereafter I discovered a werewolf pine deep in the adjacent forest, the said powers of which I do not intend to test. But it is there, on a barkan-like dune by a dark, deep river, its many, twisted roots completely exposed to light in a mysterious way. Thus there is beauty and mystery in that landscape. There is nourishment for the spirit; there are original ways of focusing feelings by 'seeing' the invisible. As I said above, the earthbound senses are difficult to express, and so too are the experiences of the ethereal, ineffable, and evanescent. Here is where poetry, music, theatre, sculpture, literature, can find original ways of sensing and expressing the experienced landscape. As Kearny (1988: 368) writes of poetry, '[...] it is a creative letting go of the drive for possession, of the calculus of means and ends [...] Poetics is the carnival of possibilities, where everything is permitted, nothing censored'. It is the opposite of the kind of deadness in culture that the landscape architect Spirn (1998) complains of as masking the true discovery of landscape.

Possibilities within the framework of the European Landscape Convention

The Convention, if I understand it correctly, offers a new era for locals and artists, together with official experts to facilitate connections with ordinary rural landscapes by outsiders. It means engaging locals who know local landscapes and their 'languages'. (An example might be a life-long Latvian hunter, who showed me a segment of a forest landscape with the signs of wild boars and wolves living out their life dramas.) The questions for the locals would no longer be so much about lore (or folklore), but about their personal landscape sensibilities and knowledge of mysteries of felt and sensed landscapes. About harmonizing inner and outer worlds through mutual inter-subjectivities. If I could get to know the subtle mysteries of an out-of-the-way, ordinary rural landscape of Greece – which I do not – I would definitely choose to spend time there.

But mysteries are disappearing:
How will the legend of the age of trees
Feel, when the last tree falls in England?
When the concrete spreads and the town conquers
The country's heart; when contraceptive
Tarmac's laid where farm has faded,
Tramline flows where slept a hamlet,
And shop-fronts, blazing without a stop
from Dover to Wrath, have glazed us over?

Simplest tales will then bewilder
The questioning from children, What was a chestnut?
What was autumn? They never taught us.

(From *The Future of Forestry* by C.S. Lewis)

Note: All translations from Latvian are by the author.

References

Ābele, I. 2008. *Paisums* [Rising tide]. Rīga: Dienas Grāmata.

Ackerman, D. 1991. *A Natural History of the* Senses. New York: Vintage Books.

Britannica World Language Dictionary 1959. Chicago: Encyclopaedia Britannica, Inc.

Bunkše, E.V. 2004a. *Geography and the Art of Life*. Baltimore and London: Johns Hopkins University Press.

Bunkše, E.V. 2004b. Softly heaves the glassy sea: Nature's rhythms in an era of displacement, in *Reanimanting Places: A Geography of Rhythms*, edited by Mels, T. Farnham: Ashgate.

Bunkše, E.V. 2007a. *Intīmā bezgalība* [Intimate immensity]. Rīga: Norden.

Bunkše, E.V. 2007b. Feeling is believing, or landscape as a way of being in the world. *Geografiska Annaler*, 89B(3), 219-231.

Chambers, E. 2010. *Native Tours. The Anthropology of Travel and Tourism*. Long Grove, Ill: Waveland Press Inc.

Codrescu, A. 1990. *The Disappearance of the Outside. A Manifesto for Escape*. Reading, MA: Addison-Wesley Publishing Company, Inc.

De Botton, A. 2006. *The Architecture of Happiness*. New York: Pantheon Books.

Holmes, R. 2008. *The Age of Wonder: How the Romantic Generation Discovered the Beauty and Terror of Science*. New York: Pantheon Press.

Jaunsudrabiņš, J. 1914. *Baltā grāmata* [White Book]. Rīga: Valters un Rapa.

Kearney, R. 1988. *The Wake of Imagination*. Minneapolis: The University of Minnesota Press.

Kursīte, J. 2008. *Sfumato nesfumato*. Rīga: Madris.

Lopez, B. 1981. Winter Herons, in *Winter Count*. New York: Vintage Books, 15-25.

Lopez, B. 1981. Buffalo, in *Winter Count*. New York: Vintage Books, 27-35.

Lopez, B. 1986. *Arctic Dreams. Imagination and Desire in a Northern Landscape*. New York: Charles Scribner and Sons.

Sacks, O. 2003. A Neurologist's Notebook. The Mind's Eye. *The New Yorker*, 28 July, 48-59.

Singh, R.P.B. 1994. *The Spirit and Power of Place; Human Environment and Sacrality. Essays dedicated to Yi-Fu Tuan*. Varanasi: National Geographical Society of India.

Spirn, A.W. 1998. *The Language of Landscape*. New Haven and London: Yale University Press.

Tuan, Yi-Fu. 1975. Place, an Experiential Perspective. *Geographical Review*, 65(2), 151-165.

Chapter 3

Landscape and National Identity in Europe: England versus Italy in the Role of Landscape in Identity Formation

John Agnew

In the literature in English on landscape and identity, England is often taken as a paradigm case of the significance of a certain idealized landscape as symbolic of national identity (see, for example, Lowenthal 1994). It is invariably a rural landscape. Recent research in Scandinavia, Finland, and Switzerland seems to back this up (Kaufmann and Zimmer 1998; Jones and Olwig 2008). In this broad overview of the role of landscape in political identities I wish to challenge the English model by showing how much of a singular social construction it has been, representing in fact a narrow regional ideal within England if that, and suggesting through use of the Italian case a more complex association between landscape and national identity that sometimes privileges identities other than the national (or, more specifically, ones that fail to become national) or that have contradictory connotations because of the power of past associations that point away from the national. My message, more evocative than definitive as befits a broad overview, is that the role of landscape in national identity should be related to the specifics of national-state formation rather than presumed to be invariant across all cases. In other words, the *politics* of landscape in particular cases is what should concern us, not identifying and celebrating landscape elements that presumably represent the natural flowering of all particular national identities.

Landscape and national identity

National identities are based on the creation of 'imagined communities' among people who do not know most of their co-nationals or much of the national territory other than that which they encounter in the course of their lives. Although some national identities have old roots in places within present-day national territories, national identities as they are known in Europe today are usually traced to the period in the late eighteenth and early nineteenth centuries when political elites 'invented' traditions of group occupancy of a given national territory and began to associate this with popular rather than purely monarchical sovereignty (Anderson 1983; Hobsbawm and Ranger 1983).

Based on the creation of national communication networks and vernacular literatures in national idioms, the circulation of common stories about national origins and tribulations, the casting of national definitions of taste and opinion, and commemoration of the heroic and tragic sides of a common past, national identities became basic components of self-identities for the burgeoning middle-classes and segments of the working classes across the whole of Europe by the close of the nineteenth century. Everything from the orientation of railway networks around capital cities to military conscription and mass elementary education conspired to produce political identities in which the national was increasingly dominant in relation to both other geographical scales and to social identities such as class or religion.

None of this happened, however, without intensive political struggle and rhetorical dispute. In his classic book *The Country and the City*, Raymond Williams (1973) showed that literary and cultural production, in the form of anthems, political pamphlets, and novels, derived much of its aesthetic appeal from conflicts arising from the changing geography of the times. New classes, interests, and ideas about national identity contended with existing ones in a struggle for control over territory. There was nothing inevitable about the outcome of this process. It took quite different paths in different countries. We err when we insist on making all cases conform to the history of a particular ideal-type, whether that is England, France or even the United States. The role of landscape imagery in national identity is a good example.

Since the nineteenth century dominant images of landscapes in Europe for outsiders and nationalizing intellectuals alike would seem to have been 'national' ones. Often these are quite specific vistas turned into typifications of a 'national landscape' as a whole. Quaint thatched cottages in pastoral settings (England), cypress trees topping a hill that has been grazed and ploughed for an eternity (Italy), dense village settlements surrounded by equally dense forests (Germany), high-hedged fields with occasional stone villages (France) constitute some of the stock images of European rural landscapes conveyed in landscape painting, tourist brochures, school textbooks and orchestral music. Ideas of distinctive national pasts are conjured up for both 'natives' and 'foreigners' by these landscape images. They are 'representative landscapes', visual encapsulations of a group's occupation of a particular territory and the memory of a shared past that this conveys (Graham 1994). They can also be thought of as one way in which the social history and distinctiveness of a group of people is objectified through reference (however idealized) to the physical settings of the everyday lives of a people to whom we 'belong', but most of whom we never meet. Yet, these landscape images are both partial and recent. Not only do they come from particular localities within the boundaries of their respective nation-states (respectively, southern England, Tuscany, Brandenburg and Normandy) their visualization as somehow representative of a national heritage is a modern invention, dating at most to the nineteenth century. The history of these landscape images, therefore, parallels the history of the imprinting of certain national identities onto the states of modern Europe.

The English case

The agents of every modern national state aspire to have their state represented *materially* in the everyday lives of their subjects and citizens. The persisting power of the state depends on it. Everywhere anyone might look would then reinforce the identity between state and citizen by associating the iconic inheritance of a national past with the present state and its objectives. Yet this association is harder to achieve than might first appear. In cases where the past can be readily portrayed as monolithic and uniform, as with the English, consensus about a national past with unbroken continuity to 'time immemorial' suggests a comfortable even casual association is easily accomplished. But nowhere else in Europe is landscape 'so freighted as legacy. Nowhere else does the very term suggest not simply scenery and *genres de vie*, but quintessential national virtues' (Lowenthal 1991: 213). Even in England, however, not all is as it seems. The visual cliché of sheep grazing in a meadow, with hedgerows separating the fields and neat villages nestling in tidy valleys dates from the time in the nineteenth century when the landscape paintings of Constable and others gained popularity among the taste-making elite (Rose 1994). Nevertheless, though 'invented', the ideal of a created and ordered landscape with deep roots in a past in which everyone also knew their place within the landscape (and the ordered society it represents) has become an important element in English national identity, irrespective of its fabulous roots in the 1800s (see, for example, Bishop 1995).

Figure 3.1 John Constable: The Hay-Wain
Source: National Gallery, London. Reprinted with permission.

Elsewhere in Europe, capturing popular landscape images to associate with national identities or inventing new ones has been much more difficult. The apparently straightforward English case is potentially misleading, therefore. This is one good reason for turning to cases other than the English. It suggests a simple historical correlation between the rise of a national state, on the one hand, and a singular landscape imagery, on the other, however insecure this may now be in the face of economic decline, North-South political differences, the revival of Celtic nationalisms challenging the English hubris to represent something they alone now call 'British', the immigration of culturally distinctive groups unwilling to abandon their own separate identities, and the devastation wrought on the rural landscape by agribusiness. National-state formation elsewhere in Europe took a very different direction from that of England; not to say that everywhere else it was the same. Two aspects of the difference are vital.

One was the complex history of local and urban loyalties in many parts of Europe, particularly those later unified as Italy and Germany (see, for example, Thom 1999). In these contexts there was often a long history of city independence and local patriotism with little or none of the early commercialization of agriculture and industrialization that swept English rural dwellers into national labour markets and national-social class identities at the very same time a state-building elite was strengthening and extending national institutions. The image of a bucolic past tapped the nostalgia of those experiencing the disruptions of industrialization, reminding them that all had not changed. Such landscapes could still be found, even if no longer experienced on a day-to-day basis. Later industrialization often also involved less disruption of ties to place. In particular, as electricity replaced steam-power, industries moved to areas of existing population concentration rather than, as in the case of the English coalfields, requiring that people move to where the industry was.

Another aspect was the external orientation of the English state and economy. Not only were English merchants, industrialists and travellers increasingly dominant within the evolving world economy of the nineteenth century, they were often nostalgic for what they had left behind when they travelled abroad and needed to compare what they saw with a datum or steady point of view. This led many of them to idealize an England in their mind's eye that was largely the product of a merging of their own experience and the renderings of England in paintings and other visual representations. This produced a unified vision that was much harder to achieve in those contexts where influential people had less empire, travelled little and thus had less need of a single, stable vision (Baucom 1999).

Italy as a counterpoint

Italy provides a good case for examining the connection between landscape and identity. Not only was Italy at the centre of the 'visual revolution' of the Renaissance in which visual representation became a vital part of the modern means of communicating the meanings and significance of religious and political messages. It has also been a state in which the process of state formation was long delayed by the existence of alternative foci of material life (in particular, city-based economies), the home-base of the papacy and the Roman Catholic Church, and local cultural identities alternative to that of the 'Italian'. It may be at the opposite pole to the English case, insofar as creating a match between a representative landscape and an Italian national identity was a difficult and obviously 'artificial' process from the start. It thus draws attention to the *process* of linkage between identity and landscape in more complex ways than would an examination of the English case. Also, much has already been written about the English case (see, for example, Daniels 1993 and Lowenthal 1991), so looking at the other extreme of European experience as a whole (a late-unifying state with much internal cultural heterogeneity) has much to recommend it.

An Italian state only formed in the second half of the nineteenth century. Although this tardiness in establishing a single state for the peninsula and islands had numerous causes, the strong municipal, city-state and regional-state governments (particularly in the North) held off the forces pushing the country towards unification. Most importantly, in the late Middle Ages and during the Renaissance, at the same time that the great western monarchies were consolidating territorial states in England, Spain and France, the politics of northern and central Italy was characterized by a fragmented mosaic of city-states and localized jurisdictions of a variety of types, from principalities to republics. It was the

> extraordinary energy and growing capacity of urban centers [that] led paradoxically to the early elimination from central and northern Italy's political firmament of any *superior* – king, emperor, or prince. The cities transformed themselves precociously into city-states with corresponding territorial dimensions and political functions. (Chittolini 1994: 28)

It was from northern Italy and, initially at least, by northern Italians that Italy was made. It was the traditions of the city-states and the Europeanness of the Savoyard regime which gave unified Italy its monarchy that provided the 'new' Italy with its mythic resources. The South, and the zones the Austrians had controlled in the North, had been 'won' from foreign domination but foreign domination, particularly in the South, was seen as having created a society which was now doubly disadvantaged: geographically marginal to Europe and politically marginal to the 'high' Italy of Renaissance city-states from which the new territorial state could be seen as having descended.

Foreign political-constitutional models, particularly those provided by England, France and the new Germany, were also important to the nationalizing intellectuals who set up shop in Rome after the final annexation of that city to the new state in 1870. Acceptance by other Europeans as a rising Great Power became a particularly important element in national policy that was to last until 1945. This meant taking very seriously what foreigners found exceptional in Italy. The new state could then build on foundations that would lead to respect from the others. It was to ancient Rome, both Republican and Imperial, and to certain Renaissance landscape ideals articulated by foreign visitors to Italy as well as by local savants that the visionaries of the new state turned. Both of these represented powerful images that would serve double duty: to mobilize the disparate populations of the new state behind it and impress outsiders with the revival of a glorious past, only now in an Italian rather than a Roman or a Renaissance form.

The Macchiaioli

Turning first to the Renaissance inspiration, the Italian *Risorgimento* (revival – through – unification) of the mid-nineteenth century was largely concerned with reestablishing Italy as a centre of European civilization, as 'it' had been during the Renaissance. Florence was, of course, the preeminent centre of the Renaissance; long since consigned to the role of *città d'arte* or storehouse for all that Italy had been. It was in Florence in the 1850s that a group of landscape painters set about putting their talents into service for the new state. The so-called Macchiaioli painters (from the various meanings of *macchia*: spot, sketch, dense underbrush) set about defining a representative landscape for Italy. Not surprisingly, they saw Tuscany as the prototypical Italian setting. This was what they were most familiar with. Of course, it had great Renaissance connotations. It also fit the foreign (particularly English) Romantic attachment to Tuscany and other locales in northern and central Italy, as expressed by the early nineteenth-century generation of poets and writers (Churchill 1987). The city of Dante, Michelangelo and Machiavelli was the appropriate center for a national revival. The Macchiaioli used their Renaissance forebears and European contemporaries (particularly English painters) as their guides. They expressed their nationalism through a search for images that could be used to tie the noble past to the developing present.

> They searched the riverflats along the Arno, the orchards and farms of the suburbs of Florence, the hill pastures around Pistoia, and the wild Maremma region (with its thick *macchie* of scrub pine and underbrush) for motifs appropriate to their fresh viewpoint. Their topographical specificity and personal response were totally integrated in what might be called a *macchia-scape* – the landscape that retained the sincerity of vision they admired in the Tuscan artists of the *Quattrocento*, but that also conveyed the modernity and nationalism of contemporary Italian life. (Boime 1986: 38)

This culture of the Risorgimento associated closely with Florence and Tuscany was not to outlast it. Even as Florence became (temporarily) the capital of the new Italy in 1865 and was beginning to assert its position as a national cultural centre, the Macchiaioli started to lose their social cohesiveness and their common political commitments (Boime 1994). Tuscany was not a smaller version of the whole of Italy. Tuscan history was not national history. Florence was not to be the permanent capital of the country. Their images did not stick, much like the promise of the Risorgimento itself. Quickly, the Macchiaioli were redefined as precursors of Impressionism or simply another school of provincial Italian painters. Other schools of painting enamored more of modernity than of traditional rural landscapes eclipsed the Macchiaioli. Only during Fascism were they once again raised as proponents of an idealized Italy, this time, of course, as precursors of the chauvinistic and ultra-nationalist vision of an older rural Italy beloved of the most reactionary Fascists.

Figure 3.2 Giovanni Fattori: The Hay-Rick
Source: Museo Civico Giovanni Fattori, Livorno. Reprinted with permission.

Rome and its ruins

A better known attempt at creating a representative landscape for Italian national identity than that of the Macchiaioli came to fruition after unification was achieved. This involved looking to the ancient past of Rome as the seat of empire to find inspiration for a new Rome around which the new Italy could be built. The selection of Rome as the capital certainly suggests that the Roman past was in the minds of Italy's unifiers even before unification was finally achieved. 'For me Rome is Italy', wrote the great hero of unification, Giuseppe Garibaldi, in his memoirs (cited in Treves 1962: 78). As early as 1861, although not yet part of the new state, Rome was declared as capital. The annexation of Rome and its surrounding region not only provided the last chunk of the national territory claimed by Italian patriots but also a 'neutral' city not associated, as were Turin, Milan, and Florence, with the local elites who had taken hold of the process of Italian unification (Caraccioli 1956: 36). In other words, as Birindelli (1978: 23) puts it: Rome 'became the capital not for the qualities that it had but for the ones it was missing'. This political advantage plus the obvious associations with a 'glorious' past gave Rome crucial points over its competitors. Rome's international visibility also counted. Italian unification was more the result of international diplomacy than of nationalist revolt. Consequently, attracting outside support was critical. By way of contrast, German unification during the same period (1850-70) was much more internally oriented. The choice of Berlin reflected both the Prussian dominance of the new state and the Prussian state's prior commitment to economic and military growth as manifested in the growth of Berlin itself. Rome was so different. Rather than a center of national prestige or strength, Rome was widely viewed in the new state as a 'parasitic' city that consumed but did not produce (Scattareggia 1988: 43). It was an ecclesiastical city without either manufacturing industry or modern bureaucracy.

Figure 3.3 'Excavating the Roman forum', 1890
Source: Public domain.

After unification, the reorientation of the axis of the city and the placing of monuments created a new secular image for the city at odds with the ecclesiastical one that had hitherto predominated. In particular, the placement of the Vittoriano (the monument to Vittorio Emanuele II) and its construction in white Brescian marble at odds with the brown tones of surrounding buildings provided a new visual anchor for the city. Via Nazionale and its western extension, Corso Vittorio, ploughed a new east-west axis through the historic centre, making Piazza Venezia, in front of the Vittoriano, the central hub for traffic as well as the new symbolic centre of the city. Other changes, such as the embankment of the River Tiber, the straightening of streets and the 'regularization' of piazzas into Euclidean shapes, and the clearing of archaeological sites to set them off monumentally, also represented successful attempts at both remaking the city and associating the changes with the glories of the angular and rational city built by the ancient Romans before the 'decadence' of later times.

Fascism continued what had begun under the liberal regime. Two new anchors to the city as a whole emerged over time: the Foro Mussolini to the northwest of the historic core (where the Olympic Stadium now stands) and the EUR complex to the southeast (built beginning in 1937 for an exposition that was never held, finished in the 1950s). Possibly Mussolini's most important act in terms of the manipulation of urban space for political purposes was the transfer of his office from Palazzo Chigi to the Palazzo Venezia in Piazza Venezia in 1929. Thereafter, Piazza Venezia became the key space in Rome for performing the ceremonies and the ritual speech-making that were the hallmark of Italian Fascism. Broadcast to central piazzas in towns and cities throughout Italy, Mussolini's speeches from the balcony of Palazzo Venezia created a sense of national 'togetherness' that Italy had never had previously and, apart from when the national football team takes the field, has not enjoyed since (Atkinson and Cosgrove 1998).

However successful individual architectural projects might have been, the overall impact of both Liberal and Fascist attempts at making over Rome as a unifying landscape image or the new Italy was severely limited (Agnew 1988). For one thing, Rome was naturally policentric. The city in 1870 had a complex structure from its variegated past of eras of expansion and contraction. One consequence was that it lacked a single pre-existing monumental centre that could be captured for the new national identity. The city was still the seat of the Pope, who, until 1929, refused to recognize the new state. As the headquarters of the Roman Catholic Church, Rome was still symbolically connected to the world 'in between' ancient Rome and modern Italy that the architects of the new Italy had wanted to erase from popular memory in order to celebrate the arrival of the new state. Another problem with Rome as the setting for a representative landscape for Italy as whole was that there was too much past present in the city to offer singular interpretations of what was there. As a city of layers of ruins built up over the centuries, Rome lends itself to the image of Eternal City. But this image is at odds with that of a new national identity. The eclectic mixture of epochs and influences in the physical fabric of the city leads towards universalistic more than nationalistic interpretations. Rome is a city for the ages, and for all (at least, Christian) peoples (Agnew 1995: chapters 2, 3 and 4).

Italy's differences

Italy's very Europeanness worked against achieving a long-lasting association between a particular landscape ideal and an Italian national identity. It remained forever associated with the glories of ancient Rome and the Renaissance, phenomena that the whole of Europe (or, even more expansively, the whole of Western Civilization) claimed as parts of their heritage. Italianizing these also suffered from a number of features of Italian geography and society that point up the difficulties of realizing singular landscape ideals.

The first is the obvious one that Italy does not have an 'integrative' physical geography. Its geographical identity as a singular unit is undermined by strong separations between the Po basin in the north and the mountainous spine/coastal plain pattern and islands to the south. As a result the physical landscapes 'available' for expropriation are remarkably varied, reflecting both the terrain, the climate and vegetation of a peninsula stretching from the heart of continental Europe almost to the shores of Africa. This range, working against the effective integration of the modern state, also produced a widely accepted continental/Mediterranean dichotomizing of Italian population and society that made a singular landscape image a difficult proposition to accept.

Certain features of Italy's historical geography also worked against the successful creation of a national landscape ideal. One of these is the absence of a demographically dominant city, such as London in England or Paris in France, to subordinate the country to a singular vision. Rome was only the fifth city of the new state in 1871, exceeded in population by Naples, Milan, Genoa, and Palermo. As the capital city it grew vigorously, but it still is politically and culturally predominant only in its immediate hinterland and in parts of the South. Indeed, it suffers from a very negative reputation in other parts of the country, particularly in the North, where it is associated with corrupt politics and inefficient bureaucracy. Related to this is the continuing importance of local and regional identities in Italian culture and society. Dialect differences, local economic interests, and attachment to local customs and traditions remain very strong in Italy. Unlike in England, Scotland, France, and Germany, and more like Spain, class and status distinctions in Italy are expressed in local as much as in national terms of reference. 'Folk' religious beliefs with strong localist connotations have remained strong – and resilient – in some regions even in the face of massive urbanization and social change (see, for example, Filipucci 1996; Pratt 1996 and Cartocci 1994). A powerful *campanilismo* or localism has persisted, therefore, rather than faded away in the face of pressures for nationalization (see, for example, Levi 1996). It has also proved impervious to ready co-optation, as in the German case, into reinforcing a larger national identity. At the moment in northern Italy a political movement (the Northern League) is attempting to use local identities as the basis for either a program of radical federalism or secession from Italy.

Moreover, Italian unification was never able to successfully capture the religious beliefs and practices of the Italian population. From 1870 until 1929 the state remained alienated from the Church, denied access to its spiritual authority. Fascism tried to 'sacralize' the state by building an alternative civic religion. But this had to coexist with existing religious affiliations and the physical presence of the Pope (Penco 1977; McCarthy 2000; Kertzer 2000). As a result the ritualistic power of the Italian state remained compromised, never able to obtain that symbolic investment in its attempts at designating certain sites as sacred to the nation and landscapes as representative of the spirit of the people that seem to arise so effortlessly in, say, the English or German cases.

Perhaps most significant in accounting for the absence of a singular landscape ideal, however, Italy has lacked the single dominant heroic event or experience upon which many singular landscape ideals are based. In England the shock of industrialization produced a romantic attachment to a rural/pastoral ideal that has outlasted the original historical context. In the United States the myth of the frontier and the subjugation of the 'wilderness' has likewise served to focus national identity around themes of survival, cornucopia and escape from the confines of city life. In Italy only the recycling of the idea of the *Risorgimento* serves a similar purpose. The problem is that the *Risorgimento* has multiple messages that have varied from the start, depending on which side of the unification process one chooses to emphasize. Its landscape legacy is likewise divided: Florence versus Rome. When allied to disputes over later mythic episodes such as the impact of Fascism, the Resistance to Fascism (1943-45), and the political unrest of the period 1968-85, the net effect is to produce multiple interpretations of Italianness and its essential landscape that persist but mutate over time following dramatic events rather than a stable, singular interpretation that serves to knit all Italians together (Vacche 1992).

References

Agnew, J. 1988. The Impossible Capital: Monumental Rome under Liberal and Fascist Regimes, 1870-1943. *Geografiska Annaler B*, 80, 229-40.

Agnew, J. 1995. *Rome*. New York: Wiley.

Anderson, B. 1983. *Imagined Communities: Reflections on the Origins and Spread of Nationalism*. London: Verso

Atkinson, D. and Cosgrove, D. 1998. Embodied Identities: City, Nation and Empire at the Vittorio-Emmanuele II Monument in Rome. *Annals of the Association of American Geographers*, 88, 28-49.

Baucom, I. 1999. *Out of Place: Englishness, Empire, and the Locations of Identity*. Princeton, NJ: Princeton University Press.

Birindelli M. 1978. *Roma italiana, come fare una capitale e disfare una città*. Rome: Savelli.

Bishop, P. 1995. *An Archetypal Constable: National Identity and the Geography of Nostalgia*. Cranbury, NJ: Fairleigh Dickinson University Press.

Boime, A. 1986. The Macchiaioli and the Risorgimento, in *The Macchiaioli: Painters of Italian Life, 1850-1900*, edited by E. Tonelli and K. Hart. Los Angeles: Wight Art Gallery, UCLA.

Boime, A. 1994. *The Art of the Macchia and the Risorgimento: Representiung Culture and Nationalism in Nineteenth Century Italy*. Chicago: University of Chicago Press.

Caraccioli, A. 1956. *Roma capitale: Dal Risorgimento alla crisi dello stato liberale*. Rome: Rinascita

Cartocci, R. 1994. *Fra Lega e Chiesa. L'Italia in cerca di integrazione*. Bologna: Il Mulino.

Chittolini, G. 1994. Cities, 'City-States', and Regional States in North-Central Italy, in *Cities and the Rise of New States in Europe, A.D. 1000-1800*, edited by Charles Tilly and Wim P. Blockmans. Boulder, CO: Westview Press.

Churchill, K. 1987. *Italy and English Literature, 1764-1930*. Totowa, NJ: Barnes and Noble.

Daniels, S. 1993. *Fields of Vision: Landscape Imagery and National Identity in England and the United States*. Princeton, NJ: Princeton University Press.

Filipucci, P. 1996. Anthropological Perspectives on Culture in Italy, in *Italian Cultural Studies*, edited by David Forgacs and Robert Lumley. Oxford: Oxford University Press.

Graham, B.J. 1994. No Place of Mind: Contested Protestant Representations of Ulster. *Ecumene*, 1, 258.

Hobsbawm, E. and Ranger, T. (eds) 1983. *The Invention of Tradition*. Cambridge: Cambridge University Press.

Jones, M. and Olwig, K.R. (eds) 2008. *Nordic Landscapes: Region and Belonging on the Northern Edge of Europe*. Minneapolis: University of Minnesota Press.

Kaufmann, E. and Zimmer, O. 1998. In Search of the Authentic Nation: Landscape and National Identity in Canada and Switzerland. *Nations and Nationalism*, 4, 483-510.

Kertzer, D.I. 2000. Religion and Society, 1789-1892 in *Italy in the Ninetenth Century*, edited by John A. Davis. Oxford: Oxford University Press.

Levi, C. (ed.) 1996. *Italian Regionalism: History, Identity and Politics*. Oxford: Berg.

Lowenthal, D. 1991. British National Identity and the English Landscape. *Rural History*, 2, 213.

Lowenthal, D. 1994. European and English Landscapes as National Symbols, in *Geography and National Identity*, edited by D. Hoosen. Oxford: Blackwell.

McCarthy, P. 2000. The Church in Post-War Italy, in *Italy Since 1945*, edited by P. McCarthy. Oxford: Oxford University Press.

Penco, G. 1977. *Storia della Chiesa in Italia*. Milan: Jaca.

Pratt, J. 1996. Catholic Culture, in *Italian Cultural Studies*, edited by D. Forgacs and R. Lumley. Oxford: Oxford University Press.

Rose, G. 1994. Place and Identity: A Sense of Place, in *A Place in the World? Place, Cultures and Globalization*, edited by D. Massey and P. Jess. Oxford: Open University/Oxford University Press.

Scattareggia, M. 1988. Roma capitale: Arretratezza e modernizzazione (1870-1914). *Storia Urbana*, 42.

Thom, M. 1999. City, Region and Nation: Carlo Cattaneo and the Making of Italy. *Citizenship Studies*, 3, 187-201.

Treves, P. 1962. *L'idea di Roma e la cultura italiana del secolo XIX*. Milan-Naples: Ricciardi.

Vacche, A.D. 1992. *The Body in the Mirror: Shapes of History in Italian Cinema*. Princeton, NJ: Princeton University Press.

Williams, R. 1973. *The Country and the City*. New York: Oxford University Press.

Chapter 4

The Dual Character of Landscape in Lahemaa National Park, Estonia

Anu Printsmann, Marju Kõivupuu and Hannes Palang

Introduction

The aim of this chapter is to demonstrate the dual character of landscape in Lahemaa National Park (LNP), one of the icons of the Estonian identity, heritage, nature conservation and tourism business. We explore the correlation between the ideational Estonian national landscape, on the one hand, and the physical landscape of Lahemaa, on the other.

Ideational landscapes are both imaginative (in the sense of being a mental image of something) and emotional (in the sense of cultivating or eliciting some spiritual value or ideal) (Knapp and Ashmore 1999). Ideational landscapes can be expressed by representations: in spoken or written language and by pictorial or graphic means. Representations can convey diverse meanings of the same physical landscape since landscape values are not hidden in the landscape or representation itself, but in the minds of the viewer. Ideational landscape representations serve as showcases for tourists but are often perceived by locals as outward, awkward, lagging behind and lacking dynamism and holistic perception (see Antrop 2000). The silenced everyday landscapes of past and present may turn into embodied physical idyll as time goes by: some of the historic poverty or niche landscapes can gain the aura of nostalgia for when times were simpler. The chosen set of valuable elements of reality should be spruced up to meet expectations, at least externally to attract visitors. Ideational landscape as a representation has a dialectically constitutive interaction with the physical landscape, termed circular reference (Olwig 2004).

This relationship between ideational and physical landscape is influenced by processes of political, economic, cultural and environmental alterations affecting both differently, diverging them sometimes more, sometimes less. This enables analytical discernment between remembered and memorized landscapes. The former denotes how physical landscapes are (locally) collectively recollected, while the latter stands for (nationally) selected valuable characteristic representative traits (Jones 1991) of reminiscence to be (re-)presented, enacted and performed as authentic, aspiring thus the reassurance of national identity by ingratiating the heritage tourism industry. Although remembrance is done by individuals, it is still society which determines what is worth keeping in mind. Undoubtedly, some inertia is involved in the processes of remembrance and oblivion.

Data that calls forth this kind of contemplations comes from an extensive fieldwork project, Basic Research for Management Plan of Lahemaa National Park and Inventory of Coastal Villages (2007-2009). In order to fully appreciate our conclusions, how socially determined time makes the dual character of landscapes of LNP, degrading its integrity, we need to delve more specifically into the history of Lahemaa and into Estonian conceptions of landscape protection and heritage.

Inventing Lahemaa

Despite the fact that LNP has enjoyed a very specific role for Estonians for decades, it is a rather newly constructed concept.

The name Lahemaa ('Land of Bays', but which can also be translated as 'Cool Land') was invented by the Finnish geographer J.G. Granö in 1922, soon after Estonia became independent, to boost national identity through a new regionalization of Estonian landscapes. At that time people were used to the regionalization based on the system of church-parishes, while Granö used four features – landforms, vegetation, waters, and human artefacts – to define his landscape regions. Since this was done mostly without consulting lay persons, it took time for the people to get used to the new landscape names (Figure 4.1).

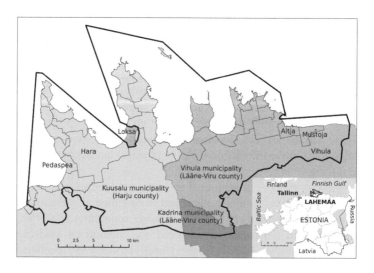

Figure 4.1 The location of Lahemaa National Park and its coastal villages
Note: Former church-parishes are proxies for contemporary municipalities.
Source: Estonian Land Board, 01.01.2009.

The LNP was established in 1971 as the first national park in the Soviet Union and at the time it embodied the idea of Estonianness (Smurr 2008), disguised by a combination of united nature and heritage protection. The main arguments brought forth while creating national parks were 'untouched' nature (see Mels 2002) or presence of landscapes of outstanding cultural value. The LNP clearly belongs to the latter category, which has been its success and misery. The LNP was created to protect the characteristic North-Estonian landscapes and the national heritage of the area, and to preserve the harmonious relationships between man and nature (Encyclopedia about Estonia). That also provided the pretext to use and promote the old name proposed by Granö, as the delimited area was placed in different administrative units. On the one hand, Lahemaa could be seen as an elitist project of its founders, as the expression of their nationalist feelings (see Smurr 2008), but since this construct was quickly brought about by the public, it started to re-create the national identity. On the other hand, delimitation of the national park gave rise to the formation of the Lahemaa identity, which since then has also been promoted in scientific and popular literature, still and moving pictures, escalating tourism.

The LNP covers over 725 km², of which 251 km² is sea (Encyclopedia about Estonia). It comprises mainly four large peninsulas, which are separated from each other by four bays scattered with erratic boulders. The national park is not a uniform entity as it features forests and settlements, amongst the oldest in the country; it is rather represented as a model of Estonia where everybody can find a piece for themselves (see Mels 2002). The foundation of a national park enabled limitation of the Soviet ruler-led planning and large-scale amelioration, thereby preserving the local peculiarities. This has also created the impression that nothing much has changed since the 1970s.

The area of Lahemaa has served as summer resort from the end of the nineteenth century. Houses in coastal villages developed 'summer rooms' to accommodate people who wanted to spend their vacations in the peaceful seaside. This semi-legal extra income became especially handy during the Soviet era, when the traditional fishing livelihood was extinguished (Figure 4.2). The national park contained a number of centralized summer holiday homes for different enterprises and pioneer camps. In the mid-1970s barbed wire was removed from the seashore – something that had existed there ever since the whole area was claimed a border zone soon after the Second World War to avoid people fleeing over the sea to Finland.

The 'land of bays' became a 'cool land' – a bit more 'open' in Soviet terms, perceived as traditionally Estonian and soaked with symbolic capital. Due to its proximity to the capital Tallinn, the area was favoured by intellectuals, art and literary people making it a sort of an elitist place. Nowadays it is still prestigious to own a summer home there. And lately, a lot of summer houses have been built.

With Estonian re-independence, the LNP's vision as resistance collapsed, the hard-earned identity was lost and the characteristic landscapes became threatened by lifestyle changes, e.g. suburbanization in so-called cowboy-capitalism conditions. The LNP administrators and authorities have no clear guidelines from above on how to maintain 'traditionality' (the result of many reforms, e.g. the

Figure 4.2 The Altja village net-sheds

Note: Top = the original net-sheds, ca 1930s; Bottom = the net-sheds restored in the 1970s to create the authentic milieu illusion of nostalgic poverty. Their role was just decorative – there were simply no fishing nets to be dried and stored in the sheds since access to the sea was still limited by Soviet ideology.

Source: Top = Jaan Vali's collection; Bottom = Marju Kõivupuu, 29.07.2008.

responsibility for planning was shifted in 2003 from the Ministry of Environment to the Ministry of Internal Affairs) and are unable to hand them out to the locals themselves. The Planning Law (2002) speaks of (spatial) planning, not landscape planning. The number of these unresolved historical-social-political-economic issues has a direct or indirect influence on the locals and their landscapes.

Firstly, the changing social and political circumstances have brought along a building boom and a demand for a 'dream summer home' in a secluded coastal area, with a sea-view, though some offers raise suspicion. New colours and not-local materials and technologies spread, which often do not harmonize with the 'traditional' way (Figure 4.3). As former subsistence gardens are replaced by front lawns, the formerly large plots are divided among friends, resulting in very tiny land parcels which do not follow the general village settlement logic. The re-emerged institution of private property favours gated communities with limited admission. Also, access to the sea becomes increasingly limited, as the requirement for free passage along the coast is not respected.

Figure 4.3 A newly built house with new materials and design on a former common ground in Pedaspea village

Source: Anu Printsmann, 05.03.2008.

Secondly, life in a protected area has always caused problems for the locals. Although the upkeep of the villages is above average, the biggest problem in Lahemaa is the lack of information about heritage preservation: how one is supposed to maintain, restore, rebuild it, what is allowed and where to get financial support for it. Some heritage acts contradict common sense; sometimes people feel that the heritage is created only in theory and that in everyday practice this has no reasoning whatsoever.

Thirdly, during Soviet times the rules of double protection – border zone and national park – were harsh, but could be outmanoeuvred and there was some regulation that worked. These days the situation is fuzzier, and customary law and other regulations do not work. Besides, the attitudes towards the Soviet military heritage are not clear – they are the painful reminder of history but could also serve as tourist attraction (Figure 4.4).

Figure 4.4 Contested Soviet military heritage in Hara harbour
Source: Marju Kõivupuu, 13.05.2009.

**Figure 4.5 An 'inappropriate' summer home bearing hints of
farm-house style in Pedaspea village**
Source: Anu Printsmann, 05.03.2008.

Fourthly, the number of people living in Lahemaa all year round is too small to take care of the land. However, the conflict between the locals and the summer residents is almost non-existent. Only sometimes is the abundance of tourists seen as a hindrance, but this shuttling is seasonal and follows certain routes, which brings along better roads for the locals. Summer residents appear to be even tetchier towards tourists. The building of summer houses has never stopped, except that by now several of them, built in the 1970s and 1980s, seem 'unsuitable' by national park officials as they deviate from the 'traditionalist' style, although they are sort of derivate of ethnic style (Figure 4.5).

The current Protection Rules of the LNP state that the national park has been created for conservation, research and promotion of the natural and cultural heritage characteristic of North-Estonia, including ecosystems, biodiversity, landscapes, folk culture, and sustainable use of nature. In addition to the natural values, also the rich cultural heritage should be researched and preserved; in Lahemaa this consists mostly of villages, the most peculiar of which are the coastal villages with their architecture and settlement systems.

The character of the coastal villages

The 'settlement' of Lahemaa goes back to 3-2 millennia BC. People then pulled in on coastal areas and especially on the mouths of rivers, rich in fish. The archaeological findings until the thirteenth century tell us about the gradual development of agriculture on the alvars where the thin but fertile soils were easier to cultivate. Presumably, the coastal areas were used as fishing and hunting grounds in summer (Kahk and Tarvel 1992; Lang 2000; Tarvel 1983).

Most of the coastal villages originate from the turn of the fifteenth and sixteenth centuries, when the area devastated by war, plague and hunger witnessed the formation of the estate system, whereby many people from liquidated inland villages had to move coast-ward and join existing villages or establish new ones. Many old farmsteads originate from the seventeenth century; they later changed their shape, but have played an important role as a core element in furthering development of the structure of a village (Tarvel 1983). Although no new villages emerged, the population of Lahemaa's coastal areas had reached 6000 by mid-eighteenth century and almost doubled during the nineteenth (Linnus 1985). A number of separate free farms were created (Kahk and Tarvel 1992).

Troska (1974) describes Lahemaa as a sparsely populated area where villages and estates, both of average size, gathered in groups around arable areas, and small groups of separate farmsteads are also located nearby. Coastal villages and sparse farms are located at the end of the bays. Tarvel (1983) distinguishes, on economic grounds, between fishermen villages and coastal agricultural villages. He also describes coastal life in the nineteenth century as flourishing due to fishing, when taxes were paid in fish; trade was done with inland peasants and later with towns and across the Gulf with Finland. Shipbuilding, smuggling trade and seafaring brought innovations to the area, and provided wealth to the coastal villages (Figure 4.6). After a railway line from St. Petersburg to Tallinn was built in 1870, estate and townspeople started spending their summers there.

After Estonia became independent in 1918, a land reform was carried out (see Mander and Palang 1994) which gave the land from estates to peasants; population growth continued, state policies supported agriculture and rural life, the economic

Figure 4.6 So-called sea-captain's house in Altja village, a valuable *milieu* area also assigned as a hallmark of Lahemaa
Source: Marju Kõivupuu, 28.07.2008.

growth of the 1930s increased traffic and also accessibility and brought in tourists. This was followed by urbanization and the development of summer resorts.

The Soviet period can be characterized by the significant decrease in population, change in previous lifestyle, collectivization and the following urbanization (see Mander and Palang 1994). An urban mentality and cultural life still flourished due to famous art and literary people, especially during summertime.

The period after regaining independence in 1991 brought along land reform, market economy and the collapse of the resort economy, impoverishment and emptying of villages, continuing urbanization and urban sprawl. Restituted farmhouses are used only for seasonal habitation.

Thus, by adapting to larger socio-political transformations the pride of the LNP, the coastal villages, changed their appearance and character while still some selected 'traditional' elements can be distinguished, e.g. smoke ovens and fences (Figure 4.7). People's care for their surroundings is expressed in adhering to current garden-fashion trends, i.e. installation of carts and their wheels, stone pyramids, flowers in carts/boots/stumps, etc. A surge for roots is articulated, for example, in the display of distinctive farm labels (Figure 4.8) and an interest in local lore (e.g. Elstrok 1995; Kindel 2004; Piikmann 2008; Sandström 1996), which could not be published during the Soviet period.

The ideational understanding of Lahemaa's landscapes guides outsiders to consider some things as appropriate and others not so, but the local perspective is much more versatile. Ideational landscapes are based on representations of a couple of selected elements, while at the same time they should be liveable to those few locals who have chosen to live there with modern commodities (Figure 4.9). Outsiders' craving for authenticity originates two main options: First, outwardly constructed ideational memorized landscapes are re-presented and re-enacted as well as possible whereas 'around the corner', everyday landscapes make no special claim to be part of the LNP, producing polarized landscapes; second, since remembered landscapes are impossible to bring back to life, as the systems that created them have disappeared, the solution is to create something completely new and unique, to add a new layer onto the landscape.

The decisions regarding what to consider heritage, what deserves protection and how to plan new plots and buildings in a cultural landscape should be negotiated by all parties and levels in order not to ruin what is deemed distinctive to Lahemaa but would be liveable in contemporary society (see Kõivupuu et al. 2010).

Landscape protection and heritage

While the problematic of authenticity is ubiquitous in all heritage-related discourses, whether details or the overall impression is more important, whether to protect objects or an idea (Gustavsson and Peterson 2003), whether only outside built heritage or inside as well, whether to create museum by 'freezing' any development followed by cleansing 'improper' elements or living landscapes, whether to protect for or from people (see Palang and Fry 2003; Schama 1995), Estonia makes no special case here. Still, Estonia possesses some specific features worth mentioning to better explain the hesitations and hardships of negotiating identity, heritage and the dual character of landscape.

Figure 4.7 Fence in Mustoja village
Note: It is merely decorative now, but it still features a distinctly local building style and it has been put into use carrying the sign 'private property'.
Source: Anu Printsmann, 29.01.2008.

**Figure 4.8 Setting out a farm name label 'Allika' expresses the devotion
to one's home, region and roots, Altja village**
Source: Anu Printsmann, 29.01.2008.

First, the term *landscape* in Estonian originates from the artists of the early twentieth century and has been referring mostly to nature or rurality (see Palang et al. 2000; Peil et al. 2004; Printsmann and Palang 2008, for longer discussion).

The first protected area was set up in 1910 for protecting birds and their habitats. The first law of nature protection was passed in 1957 (Sepp et al. 1999). Between the 1960s and the 1980s nature conservation also partly covered cultural heritage protection as a form of nationalistic counter-movement against the totalitarian regime. It peaked with the formation of the first and only heritage reserve outside an urban area in 1987, under the pretext of protecting one of the largest grave fields threatened by phoshorite mining and immigrant workers (Sooväli et al. 2008). This also marked the division of conservation activities into nature and heritage protection undertaken by different ministries. With the incompatible encumbered concepts of 'landscape as nature' and 'heritage as culture' problems occur everyday on the local, governmental and international arenas. Landscape-related management problems are shared within and between the Ministry of the Environment, the Ministry of Internal Affairs and the Ministry of Agriculture. The National Heritage Board belongs to the Ministry of Culture. This spread of responsibility for landscapes may partly explain the reluctance of any one part of government to take the lead in signing the European Landscape Convention (Printsmann and Palang 2008).

The 1980s' focus on heritage protection as anti-governmental activity was substituted in the 1990s with the concern for the environment, as without censorship the statistics of environment degradation were disastrous. Since the 1990s, the 'biologization' of nature conservation and especially the regulations of Natura 2000 have caused nature conservation legislation to focus mostly on species and habitats and much less on landscapes (Sepp et al. 1999). Even national parks' first priority is to protect species and habitats. This often means maintaining

Figure 4.9 A shed adapted to contemporary needs in Mustoja village
Source: Anu Printsmann, 29.01.2008.

semi-natural areas through extensive grazing or mowing in wooded or coastal meadows benefitting landscapes as well. Heritage protection concentrates more on monuments than landscapes. A monument, as defined by the Heritage Conservation Act (2002), may be classified as archaeological, architectural, artistic, industrial or historical – meaning cultural, not natural (Printsmann and Palang 2008).

As seen above, landscape and heritage can only be understood in their historic-cultural context (see Jones 1991; Palang et al. 2006), and in the Estonian case they are very much ideology-laden, politicised and administratively separated. Still, recent advances were made to protect built heritage, also from the Soviet period.

The main issue in Estonia regarding heritage is what period should be taken as a yardstick against which traditionality can be measured (see Palang et al. 2006). Due to socio-economic reasons, the 1930s are often considered the 'golden age' of private property in pre-Soviet independent state (see Alumäe et al. 2003); this also constitutes an era remembered by senior members of the society – a 'sunny childhood'. The other period often mentioned is peasant-Estonia, a meaning which was lost in the nineteenth century with the national awakening movement. In the public discourse, everything is traced back to the ancient period of freedom before the thirteenth century. Since then the Estonian area has belonged to the Danes, Germans, Swedes, Poles and Russians, clearly distinguishing what is 'Estonian' and what is not. Heritage thus becomes deeply contested – not only time-wise but also as reconciling confronting legacies (Figure 4.10).

Apparently, the acceptance of 'the Other' needs some time but once it becomes publicly accepted, it is inseparable from the mainstream heritage discourse or sometimes even underpins it, as something that Estonians have 'endured'. It is hard to predict when the ideational landscape of Lahemaa can adjust to the Soviet military and summer house layer (which started before the Soviet era) in addition to the lagging shabby fishermen villages – an expected image that is not met in everyday reality. Heritage must come from the past serving contemporary ends, but the overall Estonian national consensus should modernize its nostalgic views on landscape character incorporating the rebelliousness and multi-layerness of the past.

Figure 4.10 Vihula Manor in Vihula village in the Vihula municipality

Note: One of the examples of contested heritage whose appreciation has changed over time is landed estates, with their magnificent manor houses built by the Baltic Germans who remained landlords through seven centuries. These manors represented 'the Other'; attempts were made to erase by burning these landmarks of foreign power, most famously in the socialist workers' 1905 riots. In 1919 they were nationalized and given as reward to soldiers, or used as hospitals or schools; today they glorify the cover of every tourism brochure.

Source: Marju Kõivupuu, 06.08.2008.

Concluding remarks

The dual character of landscapes in the LNP is embedded in the discrepancy between the ideational Estonian national landscape and Lahemaa's ever-changing everyday physical landscape. As the expected memorized landscape does not coincide with local landscape it is degrading the integrity of the national park (circular reference) as well as indicating the national incapability of reconciling with the recent past as the changes brought about by the twentieth century are not integrated. The Estonian culture tends to live in the past, the socially determined time is preventing the yardstick against which to measure traditionality for memorized landscapes, heritage and identity to be updated. It is the pre-previous layer (meaning when enough time has passed) in the landscape (Palang et al. 2006) that becomes romanticized or mythologized (Alumäe et al. 2003). Estonian landscapes can be characterized as contested, but with independence there is no need for counter-landscapes: 'Ours' has become mainstream heritage. This lack of subtext in landscapes is unprecedented and uneasy to accept.

That the elitist project of the LNP was implemented is a miracle of its own and as such it gave a strong push towards national identity. In a way Lahemaa has served as a small model for Estonia, reflecting all heritage- and identity-related problems (especially as the first national park in the USSR has now lost its feeling of resistance, it is much more difficult to stand *for* something than against), changes and turns, mental and material, administrative and political; on the other hand, it does not embrace Estonia as a whole – South-Estonians do not connect with it completely.

The constructed entity of Lahemaa lacks regional identity. The ideational LNP should be clearly distinguishable from the rest of the country but this is not always the case. On a first approach Lahemaa represents a coastal idyll; on a second approach, it embodies traditional architecture and fences, rustic village life, peaceful sea-captains' houses, simple romantic net-sheds, famous restored manors, erratic boulders, sea-coast and forests amplified by artists and writers, renowned as a summer holiday place – an abundance of symbols that explains the difficulties to determine a fixed local identity. What Lahemaa embodies is really hard to grasp – to put a finger on a most typical of houses is impossible. Conversely, what Lahemaa does *not* comprise is very clear: Loksa – an industrial town – does not fit into that perceived cultural image of Lahemaa, and indeed Loksa is cut out of the national park.

Moreover, Lahemaa covers three former rather distinct church-parishes. The made-up geographical form was not fulfilled with a uniform cultural content, as seen in the photos above. Territorially, Lahemaa still consists of two parts. The western part – closer to Tallinn – is more multi-cultural and richer, there are plenty newer buildings, the residents seem more pretentious and less interested in the 'national park stuff'. It looks less traditional and therefore deserves less attention from the administration. The eastern part looks poorer, more traditional and less pretentious and is more oriented towards the preservation of the estate and also folk culture. Yet, it is an outward look, as the locals remember and live the landscape in continuities according to customary borders whether these are appreciated by political administration or not.

Generally, the focus of preservation is much more on material – both architecture and nature – than on mental (folklore). Tradition means something that has been passed on from one generation to another. It involves some sort of continuity. A tradition (e.g. lifestyle) could be broken, spoiled, kept, maintained, but protected? Traditionality is always constructed retrospectively at the present moment. Traditionality, then, is not so much the unchangeability of a feature, but rather its persistence. And persistent things are things that change flexibly. What is not functional is lost. In this way tradition is a process that balances changes, sustains persistent ways of thinking, expressing, and acting. So traditionality should be understood not only as heritage of the past to the present but as continuity constructed at present leaning on the past. Authentic illusion (Gustavsson and Peterson 2003) has been more appreciated than postmodern play with a collage of elements from different periods. The pursuit for authenticity in what is perceived as heritage by locals (no uniform group), newcomers, real estate developers, administrators, officials, domestic and foreign tourists and scientists creates discrepancies within their understandings of how to appreciate, manage, protect and plan development.

This generates a very creative, albeit mechanical and technical, approach to authenticity: visitors are offered the selectively memorized 'traditionalist' elements that they expect, and the rest of life continues as usual, degrading the integrity of the national park. Cultural heritage is changing from an intrinsic value to an illusion of authenticity, a tourism destination. The way out of this schizophrenic situation should come from the administration of the national park; they should develop a clear vision and goals – a new face for the LNP, really – and then communicate them to the general population – to make an elitist project also work nowadays without 'freezing' it too much.

Due to historical reasons we find ourselves in a situation where the identity of Estonia is confusing, the identity of the LNP is contested, landscape and heritage are understood differently, too rapid changes have unbalanced the protection of nature and culture; we should fight the dead hand of the past and contemporize ideational landscapes. The dual character of landscape, the discrepancy between (Estonian national) ideational and (local) physical, memorized and remembered landscape will always be there but it is time to incorporate at least twentieth century developments.

Acknowledgements

For their help in preparing the chapter our special thanks go to Juta Holst, Rasmus Kask, Kaarel Kõivupuu, Tõnu Laasi, Elo Lutsepp, Joosep Metslang, Jana Paju, Jaan Vali, Heiki Pärdi, the Estonian Land Board, and all the Lahemaa people we met during the fieldwork.

Financial support came from the European Union through the European Regional Development Fund (Centre of Excellence in Cultural Theory) and the Estonian Ministry of Education target-financed project SF0130033s07 Landscape Practice and Heritage.

References

Alumäe, H., Printsmann, A. and Palang, H. 2003. Cultural and historical values in landscape planning: Locals' perception, in *Landscape Interfaces. Cultural Heritage in Changing Landscapes*, edited by H. Palang and G. Fry. Series: H.

Décamps, B. Tress and G. Tress (eds) Landscape Series I. Dordrecht: Kluwer Academic Publishers, 125-145.

Antrop, M. 2000. Background concepts for integrated landscape analysis. *Agriculture, Ecosystems & Environment*, 77(1-2), 17-28.

Elstrok, H. (ed.) 1995. *Kõrvemaast põhjar annani: Kuusalu kihelkonna kirjanduslik-kodulooline antoloogia*. Tallinn: PrinTall.

Encyclopedia about Estonia [Online] Available at http://www.estonica.org/eng/lugu.html?menyy_id=529&kateg=10&nimi=&alam=71&tekst_id=530 [accessed: 3 August 2009].

Granö, J.G. 1922. Eesti maastikulised üksused. *Loodus*, 5, 277-278.

Gustavsson, R. and Peterson, A. 2003. Authenticity in landscape conservation and management – the importance of the local context, in *Landscape Interfaces: Cultural Heritage in Changing Landscapes*, edited by H. Palang and G. Fry. Series: H. Décamps, B. Tress and G. Tress (eds) Landscape Series I. Dordrecht: Kluwer Academic Publishers, 319-356.

Jones, M. 1991. The elusive reality of landscape: Concepts and approaches in landscape research. *Norsk Geografisk Tidsskrift – Norwegian Journal of Geography*, 45(4), 229-244.

Kahk, J. and Tarvel, E. (eds) 1992. *Eesti talurahva ajalugu 1*. Tallinn: Olion.

Kindel, M. 2004. *Mööda maad ja piki randa. Kohajutte Kuusalu kihelkonnast*. Veljo Tormise Kultuuriselts.

Knapp, A.B. and Ashmore, W. 1999. Archaeological landscapes: Constructed, conceptualised, ideational, in *Archaeologies of Landscape: Contemporary Perspectives*, edited by W. Ashmore and A.B. Knapp. Social Archaeology. Oxford: Blackwell, 1-30.

Kõivupuu, M., Printsmann, A. and Palang, H. 2010. From inventory to identity? Constructing the Lahemaa National Park's (Estonia) Regional Cultural Heritage, in *The Cultural Landscape & Heritage Paradox: Protection and Development of the Dutch Archaeological-historical Landscape and its European Dimension*, edited by J.H.F. Bloemers, H. Kars, A. van der Valk. Amsterdam: Amsterdam University Press, 115-131.

Lang, V. 2000. *Keskusest ääremaaks. Viljelusmajandusliku asustuse kujunemine ja areng Vihasoo-Palmse piirkonnas Virumaal*. Tallinn: Ajaloo Instituut.

Linnus, J. 1985. Lahemaa rahvastik 18. sajandil, in *Lahemaa uurimused 2. Rahvuspargi asustusajalugu ja etnograafia*, edited by I. Etverk. Tallinn: Valgus, 29-50.

Mander, Ü. and Palang, H. 1994. Changes of landscape structure in Estonia during the Soviet period. *GeoJournal*, 33(1), 45-54.

Mels, T. 2002. Nature, home, and scenery: The official spatialities of Swedish national parks. *Environment and Planning D: Society and Space*, 20, 135-154.

Olwig, K.R. 2004. 'This is not a landscape': Circulating reference and land shaping, in *European Rural Landscapes: Persistence and Change in a Globalising Environment*, edited by H. Palang, H. Sooväli, M. Antrop and G. Setten. Dordrecht: Kluwer Academic Publishers, 41-65.

Palang, H. and Fry, G. 2003. Landscape interfaces, in *Landscape Interfaces: Cultural Heritage in Changing Landscapes*, edited by H. Palang and G. Fry. Series: H. Décamps, B. Tress and G. Tress (eds) Landscape Series I. Dordrecht: Kluwer Academic Publishers, 1-14.

Palang, H., Mander, Ü., Kurs, O. and Sepp, K. 2000. The concept of landscape in Estonian geography, in *Estonia. Geographical Studies 8*, edited by J.-M. Punning. Estonian Geographical Society. Tallinn: Estonian Academy Publishers, 154-169.

Palang, H., Printsmann, A., Konkoly Gyuró, É., Urbanc, M., Skowronek, E. and Woloszyn, W. 2006. The forgotten rural landscapes of Central and Eastern Europe. *Landscape Ecology*, 21(3), 347-357.

Peil, T., Sooväli, H., Palang, H., Oja, T. and Mander, Ü. 2004. Estonian landscape study: Contextual history. *BelGeo*, 2-3, 231-244.

Piikmann, V.J. 2008. *Lahemaa rannakülade lugusid*. MTÜ Eru Lahe Rannarahva Selts. OÜ Infotrükk.

Printsmann, A. and Palang, H. 2008. Estonia, in *Landscape as Heritage. The Management and Protection of Landscape in Europe, a Summary by the COST A27 Project LANDMARKS*, edited by G. Fairclough and P.G. Møller. Bern: Institute of Geography, University of Bern, 61-76.

Sandström, H. 1996. *Lahemaa randlased*. Tallinn: Eesti Entsüklopeediakirjastus.

Schama, S. 1995. *Landscape and Memory*. New York: Alfred A. Knopf Publishers.

Sepp, K., Palang, H., Mander, Ü. and Kaasik, A. 1999. Prospects for nature and landscape protection in Estonia. *Landscape and Urban Planning*, 46, 161-167.

Smurr, R.W. 2008. Lahemaa: The paradox of the USSR's first national park. *Nationalities Papers*, 36(3), 399-423.

Sooväli, H., Palang, H. and Tint, S. 2008. Rebala Heritage Reserve, North-Estonia: Historical and political challenges in maintaining the landscapes, in *LANDMARKS – Profiling Europe's Historic Landscapes*, edited by C. Bartels, M. Ruiz del Árbol, H. van Londen and A. Orejas. Bochum: German Mining Museum, 43-49.

Tarvel, E. 1983. *Lahemaa ajalugu*. Tallinn: Eesti Raamat.

Troska, G. 1974. *Põhja-Eesti külad XIX sajandil*. Väitekiri ajalookandidaadi kraadi taotlemiseks. Tallinn: ENSV Teaduste Akadeemia Ajaloo Instituut.

Chapter 5

Young Immigrants and Landscape: Cultural Mediation and Territorial Creativity

Benedetta Castiglioni, Tania Rossetto and Alessia de Nardi

Introduction

Within a wide range of meanings generally associated with the word landscape, both in the scientific context – in a field in itself transdisciplinary or interdisciplinary – and in common language (Zerbi 1993; Zanetto et al. 1996; Pedroli et al. 2006; Minca 2007; Papotti 2008), an influential shared meaning is today that of 'an area, as perceived by people, whose character is the result of the action and interaction of natural and/or human factors'. This definition opens the European Landscape Convention, a key document to all those who are interested in landscape, both in terms of study and research, management and administration, sensitization and valorization. Such a conception of landscape emphasizes the necessary involvement of populations in what concerns landscape, no longer considered as an objective datum but as a result of perception. Values, meanings and the whole realm of the immaterial are therefore essential parts of the relationship which binds the population itself to the territory.

As is known, another key idea suggested by the Convention concerns landscape extension: if every 'portion of territory' is perceived, then landscape is everywhere, not only in exceptional or natural places, as the relevant binding regulations often led us to consider. In any place, even in ordinary and everyday life places – or perhaps even more there than elsewhere – the population is related to the landscape which it contributes to build and modify. In this sense, landscape is ubiquitously considered by the European Convention as an 'important part of people's life quality' and a 'key element in the individual and social welfare' (Luginbühl 2006).

In such a context, enquiry on landscape – and on its social representations – provides an interesting perspective for a wider enquiry on population, cultural diversities and place belonging. Modern societies, characterized by mobility, migration flows and diasporic conditions, demand a new perspective through which to look at the increasing role landscape plays in scientific production, policies and common practices.

Individuals and groups, instead of finding themselves encapsulated within the well-defined and confinable tesserae of a cultural mosaic, find themselves immersed in a 'cultural flow' where diversity undergoes a disarticulation and a subsequent re-articulation, shaping novel 'habitats of meaning' (Hannerz 1991; Semprini 2003). Diversity resulting from processes of migration into community life contexts may produce unbalance and instability, but the cultural process, mostly observable on a micro-social scale, tends to form areas of negotiation, spaces of comparison among plural perspectives, new assemblies of original cultural materials, promoting the formation of a new 'common sense' necessary to the ordinary social existence. Within such processes, the sharing of physical spaces, which constitute the stage for individual and collective life, gains special relevance. Although, as many suggest, the space of *flows* has today in fact replaced the space of *places*, place is still the cornerstone of the same mobility experiences typical of the migrant's condition. Moreover, the existence and the diffusion of practices which bind attachment to place and openness to the new, the need for rooting and the experience of diversity is nowadays acknowledged (Papastergiadis 2005). Accordingly, the diversity imported through migratory processes constitutes a tank of opportunities, a reserve of new interpretations which could allow our societies to better tackle contemporary complexity. Immigrants, therefore, can be thought of also as a cultural resource for eliciting innovating strategies. In this perspective, it seems increasingly urgent to investigate the ways in which migrants relate to places, by employing such an extraordinary conceptual device as landscape, both as a material and symbolic horizon for spatial practices and as a physical substratum where meanings, values and emotional contents are cast.

The key questions leading the research presented in this chapter raise the following issues: Does landscape affect the development of a sense of belonging to place in the foreign population? Does landscape play any role in formulating cohabitation strategies in societies characterized by migrant presence? And may the immigrant component, in its interaction with the hosting society, be a source of suggestions and directions for setting the agenda of landscape policies?

Recent research, carried out in marginal territorial contexts (Rossetto 2008), showed the richness of such questions and the still unexplored potential of such an approach. The relationship established by foreign groups and individuals with particular local contexts highlights aspects of unexpected creativity: just consider some habitability formulas (Tosi 2002) – as for instance the re-use of abandoned heritage estates – or spare time practices (Giusti 2008), which tend to re-activate often neglected public spaces and environmental resources. Yet, beyond the concrete and material sphere, we need to call attention – and the specific reflection on landscape urges us to do it – to the whole reservoir of sensations, mental images and expectations that foreigners associate with their own places of life in the different stages of their migration experience. Therefore, even just this immaterial sphere of *mindscapes* realizes that kind of 'territory re-invention' (Brandalise 2007) which we need to pay more attention to.

Based on such assumptions, research on second generation migrants seems significant for the greater degree of complexity and ambivalence that it bears on the relationship between mobility and rootedness. Moreover, doing research with second generations means introducing the education factor in the above mentioned themes, in close connection with recent contributions to 'children's geographies' (Holloway and Valentine 2000). The European Landscape Convention itself (art. 6) proposes education on landscape as a strategic measure to be implemented (Pedroli and Van Mansvelt 2006; Castiglioni 2009).

The Italian geographer Giacomo Corna Pellegrini has underlined the beneficial tie between immigration, education and landscape: by drawing connections between the estrangement experienced by immigrants and the *dépaysement* now prevailing in indigenous communities, he calls for a common educational path through landscape, which is 'what you see and what you must try to grasp beyond what is visible' (Corna Pellegrini 1999: 24). If landscape is not to be confused with panorama, the geographer continues, it is still not wrong to start with panorama in order to understand landscape: 'on the contrary, this may generate a useful interaction between the different subjects in the formative moment' (ibid.). Landscape provides the cultural exchange with a visual mediation and offers immigrants the reference for the establishment of a significant relationship with the place of arrival.

Referring more specifically to the minors' educational context, Alberto Agosti (1999) underlines the inclination of geographic knowledge to become instrument for cultural interaction: a geography meant not so much as a preset description of places but as an open process, socially constructed – just like geographic reality itself is – helps develop a flexible attitude, open to diverse viewpoints and to the creative interchange of experiences. Hence the need to inquire of young immigrants about their own subjective geographies, and search for apposite situations of interaction with 'the natives' in order to make phenomena of negotiation of the landscape idea – from the most conflictive to the most dialogic ones – explicit. Mental representations of places, once compared, can create crucial opportunities for knowledge and revision of self and others: personal experiences, those connected to the local dimension and those connected to mobility phenomena, can be made explicit more easily. The comparison between perspectives, in its constant engaging with the near and the far, also helps prevent some paradoxes of the education to multiculturalism that sometimes tends to emphasize diversity when it assigns individuals to cultural – and territorial – distinct and absolute realities (Aime 2004).

Promoting landscape as a 'mediator' seems to be the proper response to the need to strengthen didactic instruments for the management of the transcultural dynamics that characterize school contexts (Gabb 2006). Landscape can become an essential educational resource: its pedagogic vocation is reflected in its proposing itself as an elective context for the activation of the dialectics between identity and alterity, 'in its giving room to the needs for rootedness as well as to the wish of discovering [...], of possessing or belonging to places; of getting disorientated and finding new meanings' (Zanato Orlandini 2007: 44). The dialogic and hermeneutic nature of the landscape experience grants an open semiosis, where the voices of the 'others' can intervene: in the frequent numbing of the ordinary landscapes of our daily routine, letting other people observe 'our' landscape allows a very useful cognitive decentring where visions and innovative projectualities can spread out from. Hence, landscape configures itself as a third, transitional space, placed as a mediator between 'us' and 'others' (Castiglioni and Ferrario 2007: 421), open to a continuous negotiation of meanings, within which diversities may be re-articulated and multiple belongings practised.

Thanks to its accessibility and its capability of immediately referring to the emotional, affective, existential aspects of the relationship with places, landscape has been placed at the centre of some research activities on foreign minors, illustrated here. Employing landscape in didactic and research contexts within educational institutions implies a projection towards the extra-scholarly world. In this way we face the often debated issue of the different degree of hospitality shown inside and outside schools towards foreign students, summarized by the expression 'students in classroom, foreigners in town'.

However, the research studies briefly presented below do not constitute only a survey, they also raise questions as to the responsibilities of the institutional actors who should be concerned with orienting territorial policies towards the promotion of a more inclusive society. Which messages and which territorial images are transmitted to the new inhabitants? How do you respond to the demand for territorial knowledge coming from the immigrant component? What chances are provided to allow it to participate in landscape planning, designing and making?

Differences between Italian and foreign youngsters regarding their perceptions about the place of life: First results of a quantitative survey

Inside a wider European project (SIOI – Social Integration of Immigrants, by UE Interreg IIIA program), the study of cultural processes involving migrants in North-eastern Italy has represented a very important occasion to test this approach to landscape.

The research, conducted in close collaboration between demographers and geographers, focused on a population aged 14 to 16, comparing the life place perceptions of Italian teenagers to those of boys and girls coming from abroad. Included in the second wave of the ITAGEN2, the first quantitative national survey on second generation immigrants in Italy (Dalla Zuanna et al. 2009), the research aimed to shed light on the ways in which teenagers relate to their referential landscape, and to detect the existing connections between such modes and other characteristics of their daily life.

Divided into two complementary parts, the research on the one hand adopted quantitative methods using data deriving from the telephone survey proposed for the second wave; on the other, it developed a qualitative methodology, whose results will be discussed in the next subsections.

Being completely experimental, the design of the telephone questionnaire for the part concerning landscape required particular care: quantitative means of analysis of landscape perception normally refer, in fact, to an extensive use of the photographic image; forcing respondents to use just verbal language – and in the most essential and concise way – promoted attentive reflection on the objectives which were meant to be achieved as well as on the terminology to be adopted. However, the results achieved by the interviews addressed to 1,350 Italian girls and boys and to 500 foreigners suggested some important remarks, thanks in particular to the possibility of crosschecking these data with the rest of the information elicited by the phone questionnaire and by the first wave of the survey.

A first remark concerns the most appreciated aspects (nature, historical and artistic heritage, order and care, public transport, spaces for youth) and sites (square, shopping mall, school, parks, parish, playground or gym) of the life place: It can be noted that foreign boys and girls in general attribute higher positive grades than their Italian peers; that is, they seem to judge the characteristics of their place of life more favourably.

In light of the data collected by the questionnaire, it can be noted that the young foreigners believe they live in the 'centre' and in 'town' much more than their Italian peers do; this means that they actually attribute the meaning of 'centre' to places that the Italians consider 'periphery' and, similarly, they consider 'town' places that for Italians are 'countryside'. This result is to be considered relevant with reference to the different models of mental representation of the territorial categories: For the Italian youngsters the word 'centre' recalls perhaps the historical antique centre of many Italian towns, whereas this does not happen with foreigners; or, by the same token, foreigners may refer the idea of 'countryside' to markedly rural contexts, possibly in generally backward conditions, a model which does not apply to the urbanized countryside common to Northern Italy. It is also relevant that the young foreigners are actually attracted by 'town' and 'centre' more than their Italian peers.

Some questions of the telephone questionnaire openly landscape-related were meant to pick the signals of possible discomfort due to the young immigrants' disorientation or, on the contrary, the well-being tied to a clear identity reference (the fact of 'feeling at home') for those who are Italian or have lived in Italy for a long time. Yet, the responses to those questions do not provide clear confirmation of such an assumption: on the contrary, somehow they deconstruct it and re-open the issue in wider terms. Only 6 per cent of the Italians inquired and 9 per cent of the foreigners feel discomfort in the place where they live, and only less than 10 per cent of the Italians and slightly more than 15 per cent of the foreigners consider it little or not at all nice. Hence, no remarkable or diffuse discomfort emerges. It is thereby worth noting the high percentage (nearly 50 per cent) of Italian youngsters who evaluate the place they live in with indifference. If the limited awareness of and the scarce interest in landscape is a datum already elicited in other surveys carried out at a local scale (Castiglioni and Ferrario 2007), it strikes us here that this datum appears generalized and, as will be shown, to be partially confirmed by the results of the qualitative survey.

The comparison between Italians and foreigners is also very interesting in particular when relating the answers to the young immigrants' length of stay in Italy.

Different stages of the integration process can in fact be highlighted, corresponding to the different perceptions of the place people live in. In the first stage, referred to the sample of girls and boys who have been in Italy for under a year, a sort of shock occurs, wherein opinions are largely clear-cut, either negative or positive (more often positive); when just arrived in a new place, it seems necessary to look around with curiosity: What is considered ugly (possibly because it is different from the familiar place) still does not create particularly relevant discomfort. In the immediately following stage, even though indifference starts to increase, more positive data are recognizable: here we have the higher percentage of 'very nice' and the lower percentage of 'I feel discomfort'. Instead, among those who have been living in Italy for 3-10 years we can clearly see an increase in indifference and the percentage of opinions such as 'quite nice' is maximum: Apparently, in this phase foreign youths look like their Italian peers.

The 'assimilation' process, however, does not proceed further; in fact, young foreigners living in Italy from their first years of life (for over 10 years) or who are born in Italy show quite different results from those of their Italian peers: In these phases we have the higher percentages of feelings of discomfort, and high percentages of opinions such as 'not so nice' or 'not nice at all' referring to the place of life. The relationship with a landscape which should become increasingly familiar does not seem to foster any positive course of territorial integration. A sort of discomfort emerges, resulting perhaps from the frustration of not being able to look completely like their Italian peers. Or, also, such discomfort may arise from a phase where the questioning of one's own identity (Am I really Italian?) is completely open and causes general insecurity. The correlation with other quantitative data has revealed how higher familiarity with the new context is not always a symptom of successful social inclusion, and how the weaving of social and territorial integration is complex and anything but linear.

The inquiry on local scale: First results of a qualitative analysis

In the qualitative stage of the survey, some third-class students of Montebelluna and Crespano del Grappa junior high schools, in the province of Treviso, were involved. In this part of the study we aimed not so much at statistic relevance as at 'retrieving the irreducible originality of single individuals and single situations' (Mantovani 1995: 27), trying to understand how young people perceive their everyday context. In order to pursue this aim, different activities were carried out: students were given the assignment of producing a piece of writing concerning the characteristics of their place of life and drawing a map of it; in addition, a quantitative questionnaire and an open-answer qualitative questionnaire were handed out. In Montebelluna, we conducted 18 in-depth interviews, while a whole class was involved in a collective activity, following to the focus group technique: the students were divided into small groups, each group being required to reflect upon a different issue (the evolution of one's own territory over time, the factors held to be important in order to live well in a place, the importance of the cultural sharing of places like the park and the library – frequently mentioned by students in all the activities carried out – and, finally, the knowledge of the place of living and the ability to orientate themselves in it).

From the collected materials it appears that foreigners do not seem at all 'lost' in the new context of life, even if their references within the territory are partly different from those of their Italian peers: for the immigrants, in facts, the school is a reference place *par excellence*, followed by parks and sport structures. The Italians, on the other hand, assign great importance first of all to the quantity and quality of shops, perceived not only as places where to buy something, but also as meeting points with friends; only after that does the square come and, for them as well, the gym or the sports ground.

The 'territorial competence' shown by foreign minors – higher than that of native minors – clearly emerges from the analysis of mental maps. They tend to develop a better capacity for recognizing the environment – at least on an instrumental level – in order to move about and to use the facilities they have access to. This is probably due to the fact that, coming into a new and unknown place, foreigners felt the need to create a new 'home away from home' (Kershen 2004: 261) and this is why they observed and 'studied' their surroundings much more than their Italian peers, and thus end up knowing it better than them, and developing higher orientation skills.

Among the questions of the qualitative questionnaire, particularly interesting outcomes resulted from analysing the question: 'Me and Montebelluna/Crespano are like …'. Here the youths' fantasy gives voice to a variety of different experiences that would barely have been possible to gather using quantitative techniques. Thus, as far as the place of life is concerned, if one has been in Italy for a short time, s/he might feel 'like the parallels' (from a Rumanian boy in Italy for under a year), but also like 'brothers, which means that I'm fine with the place I live in' (from a Macedonian boy, living in Italy for under two years), just to give two examples. As time goes by, acquaintance with the place increases and this is reflected in such statements as, for instance, '[Me and Montebelluna are like] a bee and a flower' (from a Chinese girl in Italy for five years), 'two good friends and an apple with its own seed' (from a girl born in Columbia, living in Italy for over five years), or simply '[Me and Crespano are like] a normal boy and a town' (from a Macedonian boy, living in Italy for over five years). As for Italians, feelings that are sometimes very different were highlighted, ranging from refusal (['we are like] two magnets: we repel each other!!'; '[we are like] salt and sugar'), to indifference or boredom ('[we are like] a fish and an aquarium, the fish inside an aquarium has quite a dull life'), to affection and friendship ('[we are like] childhood mates, so it's like a friend whom I shared most of my life with'; '[we are like] two things grown up together'; '[we are like] friends that meet everyday and hang out together'), to even a sense of deep belonging ('[we are like] a bio-community: neither of us could survive without the other, since I wouldn't be what I am if I were not born in Crespano'). The time spent in a place is generally an important factor in generating a sense of attachment to it, and this feeling is even stronger when people live in the place they were born. However, as quantitative data confirm, this process is not linear, nor is it always sufficient to ensure good social integration.

The survey highlighted that social factors – for instance, the presence of family and friends – are apparently more decisive than environmental and landscape factors in generating well-being and a sense of belonging to a place. Many boys say they feel fine in the place where they live because here they have a close-knit family and can count on good relationships with their peers. A boy's statement, currently living in Montebelluna but born in Catania from Mauritian parents, is significant: '[…] if I could choose I would stay six months here [in Montebelluna] and six months there [in Mauritius]: here because I have my friends, there because I have all my family'. This twofold tension towards the family's place of origin and towards the place where important interpersonal relationships have been established illustrates well the importance of such factors in the construction of territorial connections.

The answers to the question 'What does the word landscape suggest to you?' allow us to understand what this term evokes in girls and boys. Thus, it came out that it is particularly associated with green and nature. The anthropic element is often mentioned, but mostly as a negative factor that, albeit part of the landscape, 'spoils' it. In their imagination, landscape is and remains a 'beautiful landscape', something which is worth gazing upon, and that sometimes takes almost ideal or idyllic features. Similarly interesting is the fact that sometimes immigrant kids associated the word 'landscape' with their country of origin: for instance, a Rumanian girl living in Italy for under one year says that the word landscape reminds her of 'the town where I used to live before coming to Italy, with its skyscrapers, the park where I went every weekend, my friends, my old school, my neighbours, the streets, the youth centres'. But also a Macedonian boy who has been living here for seven years associates the word 'landscape' with 'my village of origin in Macedonia and not only this but also the moments when I go there for the holidays'. A similar tendency has also been found among the Italians living in Veneto, who were born or have lived for some time in other regions of Italy. Consequently, in all these cases landscape is not only something beautiful, but also something that is loved and remembered with affection.

This part of the research proved fruitful, not only because it allowed an investigation of some problematic issues highlighted by the quantitative survey, but also because it was an occasion for 'cultural mediation' among Italian and foreign students. Indeed, the activities stimulated the youngsters' curiosity, activating a dialogue among them and providing them with the chance to reflect upon issues which they normally do not think about, as for example their relationship with the context of life, and the negative and positive characteristics of such a place, but also the experiences lived by their school peers, with their diversity.

Images of landscape in a group activity: The park and the library as spaces of intersubjective comparison between Italians and foreigners

The qualitative research carried out in Montebelluna has allowed us to investigate the great potentialities of landscape, meant as a relational and emotional dimension, participative of the life context. If this applies to the didactic activities involving the individual students of the class, it is mostly in the group work that the 'question of/on landscape' has shown its capacity to activate paths of experiential sharing of places. Indeed, landscape functioned as grounds for reciprocal exchange of visions, ideas, images of Self, Other and the inhabited territory.

According to the recent geographic approach to 'child studies' (Gagen 2008), which recognizes children's agency and creativity in the cultural construction of places and the consequent need for participatory methodologies while researching on and with them, a specific visual 'action-activity' was carried out in order to investigate the negotiation of everyday landscapes and their meanings between young Italians and foreign immigrants.

From a first activity of free expression (a composition that invited children to describe Montebelluna's characteristics) some territorial contexts spontaneously emerged, towards which the students gave recurring attention. It was then decided to design a group activity around these landscape 'focuses' identified by the kids themselves: the public park and the municipal library. Two very different places of daily routine are contrasted here, the one pivoting on the 'natural' element, the other on the 'cultural' element.

In planning the group-work activities, the photo-elicitation technique was preferred; this is a method drawing on anthropology and visual sociology whereby interviewees interact through the use of photo images which work as stimulus (Harper 2002; Rose 2007; Gold 2004). The use of photo-interview is advised in multicultural contexts, for the great accessibility of the photographic language, and it seems particularly suitable for investigations on landscape, which constantly refer to the visual dimension.

Regarding the park, an ambivalence of attitudes towards the wide area of urban public green can be clearly noted: considered an attractive place by the foreign youths and their families, it is instead perceived as unsafe and disquieting by the Italians, precisely because of the presence of a generic 'danger' associated with the foreign park-goers (on ethno-racially differentiated park use, see Byrne and Wolch 2009). Hence the first discussion among the students in the group activity: the foreign students, upset by the stereotyped vision of their peers, promptly point out, very spontaneously, that we need to refrain from generalizations, and claim that common language intrinsically carries prejudice. The Italians discover 'other' foreign users of the park, and the power of public (and adult) discourse on dangerous places. The focused use of the images helps the group to reflect on the real function of the park: not only as a place of nature but also as a place for sociality.

**Figure 5.1 and Figure 5.2 The park, not only 'natural' space,
but a place for socialization**

Space of mistrust, the park should instead be a place for dialogue. What solutions? The kids are encouraged to take an interest in it, and to highlight possible virtuous processes to fully enjoy the park. It then emerges that only by actually going to the park more assiduously, instead of escaping it, could its positive sociality be ensured and the relationships between Italians and foreigners improve.

As for the library, it must be said that it is a place that has become central for Montebelluna: a high quality architectural project, designed according to modern multifunctional criteria, able to attract people of all ages and to stimulate processes of citizen participation. The library has so become a symbolic place, the lively hub of the town, with an equally relevant physical presence, that it stands out in the urban landscape for its innovative character.

During our on-the-spot 'participant observation' we found, for instance, a group of foreign children of different nationalities that meet there for their afternoon homework, while the library assistants explain to us that social services expressly advise immigrant families to use the library. One of the Chinese kids involved in the activity clearly identified the library as his favourite place in Montebelluna, after his home. In facts, in the group dynamics the foreign children show great acquaintance with this place, perceived as extremely welcoming. The assiduous presence of foreign youths is underlined also by the Italians, who appreciate the library but do not actually see in it that surplus of opportunities that its services offer to the foreign minor.

The library is also the context in which a reconciling of visions is made possible: it is a reassuring place where the foreign presence nourishes the sense of a collective cultural growth. The youngsters recognize that such a perception relies on the quality of place, particularly on its function, inherently cultural. Together with the youths we draw the conclusions that it is precisely through the cultural sharing of places (the same that is practised in school) that it is possible to implement the social integration of foreign citizens.

Figure 5.3 and Figure 5.4 The public library, architectural quality functional to cultural interaction (a group of foreign children doing their homework)

Conclusion

Places and landscapes are what Italians and foreigners daily share in the difficult construction of a plural society: the activities carried out show how urgent it is to engage the foreign component in reflecting and recovering our awareness of how we look at, use and construct places.

The results coming from different research methods show the high potential of dealing with landscape in multicultural groups, considering landscape itself as a powerful medium for eliciting and comparing different feelings, emotions, and conscious and unconscious behaviour patterns. They also highlight that the relationship between teenagers and their place of life represents a very complex issue, probably influenced by several different factors.

Surely, it will be necessary to further explore the achieved outcomes and to implement further surveys on specific aspects, in order to better clarify the role of landscape in teenagers' development of place attachment, both for Italian and for migrating children.

Acknowledgements

The authors are grateful to Eros Moretti and Gianpiero Dalla Zuanna, who coordinated the SIOI project and discussed the results; to Nicola Barban, who did the quantitative analyses of data; to the headmaster, teachers and pupils of the Montebelluna and Crespano schools, for their cooperative participation; to Paola Tasca, for the language advice.

The present work is the result of the authors' close collaboration and content discussion. As for the compilation, B. Castiglioni edited, in particular, the presentation of the quantitative analysis; T. Rossetto edited the Introduction, the paragraph on the group activity on the images of landscape and the final revision; A. De Nardi edited the presentation of the qualitative analysis.

References

Agosti, A. 1996. Geografia tra interpretazione personale e progetto condiviso: Per un'educazione interculturale attraverso la realtà geografica, in *Intercultura e insegnamento. Aspetti teorici e metodologici*, edited by A. Agosti. Torino: SEI, 130-139.

Aime, M. 2004. *Eccessi di culture*. Torino: Einaudi.

Brandalise, A. 2007. L'immagine del territorio e i processi migratori, in *Territorialità. Necessità di regole condivise e nuovi vissuti territoriali*, edited by M. Bertoncin and A. Pase. Milano: Franco Angeli, 32-39.

Byrne, J. and Wolch, J. 2009. Nature, race, and parks: Past research and future directions for geographic research. *Progress in Human Geography*, 33, 743-765.

Castiglioni, B. 2009. *Education on landscape for children*. Strasbourg: Council of Europe Report.

Castiglioni, B. and Ferrario, V. 2007. Dove non c'è paesaggio. Indagini nella città diffusa veneta e questioni aperte. *Rivista Geografica Italiana*, 3, 397-425.

Corna Pellegrini, G. 1999. La riflessione geografica per prepararsi al multiculturalismo, in *Italia crocevia di genti. Immigrazione al positivo: La nascita di una cultura multietnica*, edited by G. Arena, A. Riggio and P. Visocchi. Perugia: Rux, 23-26.

Dalla Zuanna, G., Farina, P. and Strozza, S. 2009. *Nuovi italiani: I giovani immigrati cambieranno il nostro paese?* Bologna: Il Mulino.

Gabb, D. 2006. Transcultural Dynamics in the Classroom. *Journal of Studies in International Education*, 4, 357-368.

Gagen, E.A. 2008. Landscapes of Childhood and Youth, in *A Companion to Cultural Geography*, edited by J.S. Duncan, N.C. Johnson and R.H. Schein. Oxford: Blackwell, 404-419.

Giusti, M. 2008. *Immigrati e tempo libero. Comunicazione e formazione interculturale a cielo aperto*. Torino: UTET.

Gold, S.J. 2004. Using Photography in Studies of Immigrant Communities. *American Behavioral Scientist*, 10, 1-21.

Hannerz, U. 1991. *Cultural complexity: Studies in the social organization of meaning*. New York: Columbia University Press.

Harper, D. 2002. Talking about pictures: A case for photo elicitation. *Visual Studies*, 1, 13-26.

Holloway, S.L. and Valentine, G. (eds) 2000. *Children's geographies: Playing, Living, Learning*. London: Routledge.

Kershen, A.J. 2004. The Construction of Home in a Spitalfields Landscape. *Landscape Research*, 3, 261-275.

Lüginbuhl, Y. 2006. Landscape and individual and social well-being, in *Landscape and sustainable development. Challenges of the European Landscape Convention*. Strasbourg: Council of Europe Publishing, 29-51.

Mantovani, S. (ed.) 1995. *La ricerca sul campo in educazione: I metodi qualitativi.* Milano: Bruno Mondadori.

Minca, C. 2007. Humboldt's compromise, or the forgotten geographies of landscape. *Progress in Human Geography*, 2, 179-193.

Papastergiadis, N. 2005. Hybridity and Ambivalence: Places and Flows in Contemporary Art and Culture. *Theory, Culture & Society*, 22, 39-64.

Papotti, D. 2008. L'approccio geografico al paesaggio: Una rilettura del rapporto fra natura e cultura alla luce della Convenzione Europea del Paesaggio, in *Riconquistare il paesaggio*, edited by C. Teofili and R. Clarino. Roma: WWF Italia – Ministero dell'Istruzione, dell'Università e della Ricerca, 124-138.

Pedroli, B. and Van Mansvelt, J.D. 2006. Landscape and awareness-raising, training and education, in *Landscape and sustainable development. Challenges of the European Landscape Convention.* Strasbourg: Council of Europe Publishing, 119-141.

Pedroli, B., Pinto-Correira, T. and Cornish, P. 2006. Landscape – what's in it? Trends in European landscape science and priority themes for concerted research. *Landscape Ecology*, 21, 421-430.

Rose, G. 2007. *Visual Methodologies: An Introduction to the Interpretation of Visual Materials.* London: Sage.

Rossetto, T. 2008. Nuovi sguardi sul paesaggio. L'immigrazione straniera nei canali prealpini/Ein neuer Blick auf die Landschaft: Die Zuwanderung von Ausländern in die Voralpentäler, in *Le Alpi che cambiano. Nuovi abitanti, nuove culture, nuovi paesaggi/Die Alpen im wandel. Neue Bewohner, neue Kulturen, neue Landschaften*, edited by M. Pascolini. Udine: Forum, 111-125.

Semprini, A. 2003. *La società di flusso. Senso e identità nella società contemporanea.* Milano: Franco Angeli.

Tosi, A. 2002. *Immigrati stranieri in Italia: Dall'accoglienza alla casa.* Bruxelles: European Observatory on Homelessness – FEANTSA.

Zanato Orlandini, O. 2007. Lo sguardo sul paesaggio da una prospettiva pedagogico-ambientale, in *Il Paesaggio Vicino a Noi. Educazione Consapevolezza Responsabilità*, edited by B. Castiglioni, M. Celi and E. Gamberoni. Montebelluna: Museo di Storia Naturale e Archeologia, 39-50.

Zanetto, G., Vallerani, F. and Soriani, S. 1996. *Nature, Environment, Landscape: European attitudes and discourses in the modern period. The Italian case 1920-1970.* Quaderno del Dipartimento di Geografia 18. Padova: Università degli Studi.

Zerbi, M.C. 1993. *Paesaggi della geografia.* Torino: Giappichelli.

Chapter 6

Cultural Hybridism, Identitary Anthropophagy and Transterritoriality

Rogério Haesbaert

> Today's self-proclaimed mobile and multiple identities may be a marker not of contemporary social fluidity and dispossession but of a new stability, self-assurance and quietism. Fixity of identity is only sought in situations of instability and disruption, of conflict and change. [...] The fixity of identity is only sought in times of instability and disruption, of conflict and change. [...] heterogeneity, cultural interchange and diversity have now become the self-conscious identity of modern society. (Young 1995: 4)

Introduction

We live in a time of paradox and perplexity revealed in the most diverse spheres, from the economic to the political, from the cultural to the environmental. For instance, culturalist speeches – and practices – overstate the 'identity vector' in understanding and/or in the very production of the main contemporary dilemmas, while others, sometimes with economistic overtones, diffuse the idea of a widespread capitalist (neo)liberalism (at least prior to the current financial crisis), as if we had no other choice but to accept an inexorable globalized market-cultural homogenization.

In the midst of everything, there appear intermediate or literally 'borderline' 'ways out' such as those which, in the name of the domain of multiplicity and mobility, promote the idea of a kind of 'life on the edge' or 'threshold' in a world of merged, 'mixed', 'hybrid' or 'transcultured' identities – a world in which this discourse of hybridism has a clear positive sense, of a position to be defended and encouraged.

Young, in the introductory quote, reveals this apparent contradiction, arguing for a new form of identity construction moulded on a mobile and multiple character, which would not be in itself representative of crisis and rupture, but rather of 'stability, self-assurance and quietism' towards a new 'self-conscious identity' of contemporary society. 'Particularizing globalization' or 'glocalization', 'stability in fluidity and multiplicity' – these are the ambivalent interplays we face when examining current processes of identity construction.

Imagine now a geographical view on this issue, introducing the spatial, or, from another perspective, territorial dimension into the identity debate. If territories/territorialities (or even landscapes) are suffering as the exclusive basis to identities, likewise considerable changes are evident in how identities are being structured: how does this immanent relationship between territories/territorialities and identity construction take place? In what sense is this ambivalence fuelled when the identification of social groups is related particularly to a spatial/territorial referential regarded as increasingly mobile? How should we express the interplay among differentiation, hybridization, the multiple and 'threshold' character of culture and the multiple territories or multi/transterritoriality in which we live today?

This article aims to discuss the related concepts of hybridism/transculturation and multi/transterritoriality a little further. With that aim in mind, we will start from some general assumptions. The first and most obvious is that our identities are not being diluted by globalization – on the contrary, they may even be strengthened, in forms considered re-essentialized or, as focused here, recreated by mobility into more hybrid ones, emphasizing especially their multiple and threshold/transborder character. Another assumption is that our territories/territorialities, considered (arguably) increasingly unstable, mobile, do not offer, as they did in the past, stable references for the construction of our social/territorial identities – in this case one must ask: is it not possible (as previously indicated in Haesbaert 2004) to territorialize ourselves in and through movement?

We therefore question the contemporary territorial-identity construction from the various manifestations of so-called cultural hybridism (especially that of a Latin American matrix, associated with the concept of 'transculturation'), based on an unequal combination of forces, from a more positive hybridization we propose to call 'anthropophagic' (cannibalistic), a movement consciously assumed by subordinate groups, in their favour, to a more negative hybridization, mainly for the benefit of hegemonic groups' interests (or pre-hegemonic – if we believe in the potential construction, through hybridism, of new power relations from subordination).

Regarding these multiple processes of hybridization, we will focus on their territorial counterparts: a multi- or transterritoriality, also in its various manifestations. This diversity of features of hybridization and (multi- or trans-) territorialization processes, one must not forget, is also inextricably linked to implicated power relations and, within those, to the increasingly market-oriented character, the increasing 'exchange value', embedded in our processes of social/territorial identification.

Hybridism: Ambivalence and anthropophagy

When we refer to the process of 'cultural hybridism' as a major indicator of contemporary globalization or, on the contrary, of identities essentialized by fundamentalist groups, we must be very careful. We must better historicize/geographize our concept of hybridism – or hybridization, to value it more as a process – and recognize, above all, the different subjects that produce it and the geopolitical contexts in which it takes place and where its debate circulates, a bit like the 'geometries of power' proposed by Massey (1994) to render more complex the relations in which the 'time-space compression' and the accessibilities/paces of our time are produced.

It would be interesting, first, to locate hybridism within its Latin American basis, as Latin America is often seen as the 'hybrid continent' *par excellence* – if not in practice, at least broadly in the field of discourse. It does not come as a surprise that we have some of the main representatives of this debate, particularly at the interface between cultural and literary studies in the strict sense. For example, Walter Mignolo (2003), in a post-colonial discourse, advocated an episteme or '*threshold* gnosis', Fernando Ortiz (1995 [1940]) and Angel Rama (1982) developed the concept of transculturation, Édouard Glissant (2005 [1996]) from a Caribbean point of view, advocated an American (and world) 'creolization' and Nestor Canclini (1997 [1989]) focused on our 'hybrid cultures', not to mention the broader work of classics such as Gilberto Freyre and Darcy Ribeiro.

According to Young (1995: 26):

> Hybridism turns [...] difference into sameness and sameness into difference, but in such a way that sameness is no longer the same and difference is no longer simply the different. [...] to break apart and to assemble at the same time and in the same place: difference and sameness in an apparently impossible simultaneity.

Ambivalence is thus a characteristic immanent to the processes of hybridization. But it is Young himself who, despite proposing a more 'dialogical', post-structuralist rather than dialectical reading of the topic, refers to hybridism both as an 'amalgam' and a 'dialectical articulation'. This 'doubled hybridity', he says, commenting on Rushdie, 'has been distinguished as a model that can be used to account for the form of syncretism that characterizes all post-colonial literatures and cultures' (1995: 24). One can say that hybridism operates simultaneously in a double way, 'organically', hegemonizing, creating new spaces, structures, scenes, and 'intentionally', diasporizing, intervening as a form of subversion, translation, transformation (Young 1995: 25).

For Stam, in turn, the dominant discourse on hybridism 'fails in terms of discriminating among several *types of hybridism*, such as colonial impositions [...] or other interactions such as compulsory assimilation, political cooptation, cultural mimicry, economic exploitation, top-down appropriation, bottom-up subversion' (Stam 1999: 60, emphasis added).

As an example, the hybridism of social identities in a (post)colonial context culturally as rich and full of nuances as that of Latin America is not only a tool to break with the cultural 'unit' of the colonizer, deterritorializing both hegemonic (at a more attenuated level) and subaltern (at a much more violent level) groups, but also represents a form of resistance/reterritorialization, sometimes quite rich, creating, through the mixture, new forms of territorial-identity construction.

Some societies and spaces live hybridism more deeply, or are more open and/or are forced to much more intense cultural exchanges. The cultural historian Peter Burke (2003), in a work on hybridism, states that, when imposed, it may represent important cultural losses. But as it should always be considered a two-way street, hybridization can also become an instrument of innovation and/or resistance, as was clearly advocated by the 'anthropophagic' approach of the Brazilian modernist writer Oswald de Andrade.

We intend, therefore, to go a little deeper into the interpretation of what we propose to call 'anthropophagic hybridism',[1] a hybridism of Brazilian-Latin American contextualization with a clearly positive sense, proposed on a pioneer approach by the literary-philosophical reading of Oswald de Andrade. For Oswald, in his *Anthropophagic Manifesto*, 'only anthropophagy unites us' (Andrade 1995: 47), as much from social as from economic and philosophical perspectives. Unlike the view of the colonizers, with their 'immoral and materialistic interpretation' of anthropophagy, this is for him a vision of the world, a *Weltanschauung* resulting from a certain 'primitive stage' of humanity and its rich spiritual world:

> It is in opposition, in its harmonic and communal meaning, to the cannibalism that is anthropophagy driven by gluttony and also anthropophagy driven by famine, known from chronicles of besieged cities and lost travellers. The metaphysical operation linked to the anthropophagic rite is the transformation of taboo into totem. Of opposing value into favourable value. Life is pure devouring. In this devouring that threatens human existence every minute, it is up to man to make taboo into totem. What is a taboo if not the untouchable, the limit? (1995: 101)

Oswald de Andrade's proposal is, in short, to break the 'messianic culture' of the colonizer with the threshold, 'anthropophagic', the so-called 'wild' culture of the native peoples. This has no preconceived purpose at all, no teleology towards divine redemption, but the constant remaking of the Other – and oneself – through 'pure devouring'. Another kind of 'creative destruction' of those that feed themselves constantly swallowing the very force of the Other. In other words, hybridism as force, anthropophagy as a weapon: devouring is instigating constant re-creation, the surfacing of a mythic-poetic thought unconquerable by utilitarianism and the domestication of Euro-colonialist identities and thought. As Maltz states:

> To destroy is to build upon. To swallow in order to, by seizing the "enemy's" tools, be able to combat and overcome it. To swallow the old knowledge, transforming it into raw material for the new. [...] the counterpart of this attitude of ideological and cultural inertia, of brutal assimilation that legitimized the foreign influence, would be the anthropophagic attitude of "swallowing" the

---

1   An interesting exercise, which time constraints do not allow us to perform here, would discuss this 'anthropophagic' vision in light of the concept of 'wild hybridism' of Homi Bhabha, resumed as 'cultural hybridism' by Alberto Moreiras 2001.

> European knowledge, "devouring" it no more to incorporate it mechanically
> but to absorb it dialectically in an attempt to "Brazilianize" our culture, giving
> it an identity. [...] to desecrate the cultural heritage of the colonizer so as to
> inaugurate a new tradition. (1993: 11)

This is why some authors, such as Helena, define anthropophagy as an 'ethos of Brazilian culture' (1983: 91). Thus anthropophagy constitutes the face of hybridism's 'positivity' – which, while on the one hand may represent destruction and impoverishment of past cultures, on the other hand can rejuvenate them and push them towards the new, which may also be richer. Turning taboo into totem, the anthropophagic society violates the untouchable, break the limits (or lives within the limits ...), de-reterritorializes itself in a space where multiplicity is not only a nuisance or a remnant, but a condition of existence and non-stabilizing creation of the new.

The anthropophagic 'philosophy' undoubtedly anticipated post-colonial thought, concerned first and foremost with the contextualization of epistemes by the geo-historical nature of their production (about this, see particularly Mignolo 2003). This cultural reading of hybridization/'anthropophagization' processes may also prompt us to think, at a more concrete level, how to articulate spaces that can reproduce and/or lead to this subverting condition. Thus we propose to rework our concept of 'multi' or 'transterritoriality' (Haesbaert 1996, 2001, 2004, 2007) in light of hybridization processes.

Hybridism and multiterritoriality

Space and territory, as we know, are more than mental ('absolute') references for our location in the world or mere material objects in relation to our surroundings. They are constitutive of our very existence, both in its physical-biological dimension (as 'bodies' – which for some would be our 'first territory') and in its socio-symbolical dimension. Thus, if we say that man is not only a territorial animal but also a 'multiterritorial animal', experiencing several territories concomitantly, this means that the spatial dimension is not a mere stage or appendix of the human condition, but one of its fundamental constituent dimensions.

It is important, at this point, to present, albeit very briefly, the different sides of multiterritoriality. In a less fragmented concept of territory, not defining concepts simply through an empirical outline, a 'portion of the real', but through an issue and a way to approach it, i.e., through a certain gaze and in it a 'focus' on reality, we sustain that territory is the geographic space seen from the 'focus' on relations of power, either in their more strictly material effects of a political-economic nature, or in their more symbolic articulation. Thus, we propose to approach territory through its inherent power relations – from a more 'traditional' state-administrative nature, to its more symbolic configuration, where the very identity construction is seen, first and foremost, as an instrument of power (or, to use a rather controversial term, of 'empowerment') of groups and/or social classes.

Thus there would be (multi)territorialities of a higher functional charge and others of a higher symbolic charge, according to the force of the functions and meanings they are given by different social subjects in (territorializ)action. In the approach here stressed we are more interested in those territories strongly endowed with meaning, which may be involved in a more explicit 'hybridization' in cultural terms. We cannot ignore in this context the strong relationship between political forms of territorial management and the implications in terms of the propensity they carry to the attainment of more (or less) hybrid cultural forms of social identification.

Producing hybrid cultures, therefore, as stated by Canclini (1997 [1989]), also means building somehow 'hybrid', threshold or 'transfrontier' spaces (as in Tijuana, on the Mexico-United States border, which he analysed); multiple territories whose design interferes directly in our conceptualizations of the world, in the construction of our social identities. Contemporary multiterritoriality itself can therefore facilitate processes of hybridization, either by increasing our physical mobility, articulating more than one territory as occurs with migrants in a diaspora, or by the *in situ* territorial diversity itself, as occurs especially in the cosmopolitanism of large global cities. In these, relatively restricted territories, sometimes in the same neighbourhood, an increasingly culturally diverse number of people live.

We can say that multiterritoriality manifests itself in two general forms: one of a more simple nature, which can be called successive or '*lato sensu* multiterritoriality', which involves the linking of multiple (zonal) territories articulated into a network demanding a certain degree of physical mobility for social groups; and another one, of a more specific nature, which could be called simultaneous or '*stricto sensu* multiterritoriality', which involves territories themselves hybrid and/or allows the simultaneous conjunction with other territories (through informational or virtual control mechanisms). In the former case, our hybridization is, say, induced by mobility – it is essential that we experience physical displacement in order to have our multiterritorial experience – and 'control' – with the important caveat that, of course, not everyone moving through different spaces necessarily lives a multiterritoriality beyond its purely functional character.

It is not simply because we have more mobility or inhabit increasingly hybrid territories (or places, in the broad approach to 'place' of Anglo-Saxon geography) that we will, automatically, experience a greater cultural and identity hybridism. The global bourgeoisie, for example, moves a lot, but almost always going to the same places, ignoring the immense cultural – and territorial – diversity which extends around it. Here and there it may even cross paths with the 'Other', but it is as if the 'Other' had been rendered invisible, without dialogue – or when a mandatory dialogue takes place (as in hotels/restaurants and shopping situations), it is a contact of a purely functional nature.

Likewise, it is also not because we live in spaces that reveal high ethnic diversity that we are automatically living multiterritoriality in a cultural sense. There is an important distinction between its potential and effective nature. We may live in a highly cosmopolitan, multicultural city such as London, yet refuse to enjoy this diversity. Accordingly, we have 'many [types of] territories' without building there an effective multiterritoriality, as the latter implies transiting across and especially experiencing this multiplicity of territories/territorialities.

This means that relationships between cultural hybridism and multiterritoriality are not two-way: we may have, for instance, a functional multiterritoriality where no experiences of clear cultural hybridization take place. It is interesting to note, however, that the increased mobility of our time, affecting both our 'successive' (involving physical displacement) and 'simultaneous' ('*in situ*' – in the sense of a culturally multiple place and/or of a 'virtual mobility' that allows 'controlling' territories at a distance) multiterritoriality, is a – highly – *potentially* favouring element in the processes of hybridization. As we indicated initially, we cannot forget that a form of territorialization also takes place 'in and through movement' – many are those who nowadays identify themselves with this mobility so that for them territory, like their identities, is built by the amalgamation of multiple territorialities, or – more 'radically', by 'moving' or 'transiting across multiple territories', which also leads us to think about something like 'transterritoriality'.

Always open hybridism: From transculturation to transterritoriality

Just as hybridism is not an insulated condition, a 'state' proper, but rather a process in constant coming-and-going – or, in other words, in constant *becoming*, multi- or transterritoriality should also be seen particularly within an in-and-out movement and therefore in a process of *transit* among different territories. What matters most here is the condition of the possibility, always open, of our inclusion in 'alien territory' (which also becomes, thus, ambivalently 'ours'), the opening of these territories, which permanently raises the possibility of entering, exiting and/or transiting through these territorialities – or, should we wish so, this condition of transitoriness (in a broad sense, of event).

It is because it emphasizes this idea of movement and *transit* that maybe the most appropriate term is not constructed by the prefix 'multi', but by the prefix 'trans', as I suggested more than a decade ago (Haesbaert 1996). And since I am emphasizing this cultural dimension of territorialization, it is important to make a link, albeit at the introductory level, with the rather close concept of transculturation.

According to Mignolo, the Cuban sociologist Fernando Ortiz, as early as the 1940s and from the perspective of the Cuban reality, suggested replacing the European term 'acculturation', proposed by Malinowski, with 'transculturation': 'While acculturation pointed to cultural changes in a single direction, the corrective transculturation intended to draw attention to the complex and multidirectional processes of cultural transformation' (2003: 233, author's emphasis).

Referring to Ortiz's words, Mignolo points out that transculturation:

> [...] better represents the different phases in the transitive process from one culture to another, because this process does not only imply the acquisition of culture, as connoted by the Anglo-American term acculturation, but it also necessarily involves the loss or uprooting of one's preceding culture, what one could call a partial deculturation. Moreover, it signifies the subsequent creation of new cultural phenomena that one could call neoculturation [...] descendents always have something of both parents, but are always different from either. (Ortiz, apud Mignolo 2003: 235, my emphasis)

Here one can make, clearly, an analogy between this definition of transculturation and that of (trans)territorialization. While the former is seen as a product of the interplay between deculturation and neoculturation, the geographical processes of (trans)territorialization result from the overlap between territorialization and re- (or 'neo', to be faithful to Ortiz) territorialization. Moreiras, although making an argument based on the concept of hybridism, likewise states:

> The concept of hybridism is complex and particularly suggestive because it can be used to group phenomena that arise from both territorialization and deterritorialization. In the latter, hybridism refers to processes of loss in previously determined positions (i.e., hybridism is increasing today because there is deculturation, and deculturation is a violent, irreparable loss). In the former, hybridism refers to the positivity such loss implies, structurally or constitutively (there is no deculturation without reculturation), and reculturation may even produce – under certain circumstances – a threat to the very economy of the system). Hybrid reterritorialization and hybrid deterritorialization are thus two different sides of the same coin. (2001: 342)[2]

Angel Rama, in the field of literary studies in the 1970s, also conceived a concept of transculturation, emphasizing a different scale, the intra-national. While Ortiz focused on the 'transcultural' formation of the Cuban national society in relation to the context of the colonizer, Europe, Rama emphasizes the intra-national relations – between 'centre' ('capital or port', 'avant garde', facing outwards) and 'periphery'

---

2   Although Moreiras places more emphasis on the idea of a bad 'deterritorialization' and a good 'reterritorialization', obviously this is not always the case. On the criticism regarding the sometimes unilateral association between hybridism and deterritorialization, see, besides Moreiras 2001; Kraniauskas 1992; Tomlinson 1999 and our own work, Haesbaert 2004, particularly the item 'Deterritorialization from a cultural approach', 214-234.

(or 'internal regional culture', more rural, facing inwards).[3] Thus another spatial 'interplay' opens up which one might call the 'interplay of scales' (to paraphrase the title of Revel's 1998 book) within which the processes here called multi- or transterritorialization are equally designed. It is not a simple 'accumulation' or 'transition' from one scale to another, but their concurrent experience in terms that resemble, somewhat, what Yves Lacoste called 'differential spatiality'.

'Differential' here implying, of course, not simply a quantitative, extensive (from a cartographically larger and less important scale to one smaller and more important, for instance), 'discrete' difference, but the effective qualitative, intensive differentiation ('continuous difference') from the new amalgam built there. Some 'multiterritorialities', in this sense, represent nothing but a discrete difference, as in the multiterritorial organization of Nation-states, structured so as to 'fit' multiple scales within a single political-territorial order, which extends, for instance, from the territory of the municipality (or county) to the state (or province), from there to the Nation-state and nowadays to political-economic blocs, particularly in the case of the European Union.

The alliance, say, between hybridism or transculturation and multi- or transterritoriality in fact only occurs when a change of territory/territoriality effectively implies a change of behaviour and a cultural mix. It is important here, finally, not to see the space/territory as a mere reflection of these processes of hybridization, but as one of their key constituent elements. It does not come as a surprise that transfrontier spaces, for instance, have become paradigmatic, since they are much more susceptible to processes of hybridization – both through, say, a more 'spontaneous' dynamics, and an 'obligation' or 'necessity', since to merge (national, for example) identities is, in this case, also a survival strategy.[4]

Hybridism and (trans)territorialization: Political implications

Territorial openness and mobility, which some people, mistakenly, associate strictly with processes of deterritorialization, are relevant to stimulate cultural exchanges and processes of hybridization and/or transculturation. Whether these movements are politically and socially positive or negative, however, is a different story – albeit not negligible at all. Beyond the debate about their heuristic, conceptual value, hybridism and multi or transterritoriality have often become a kind of political program. As Nestor Garcia Canclini stated:

---

3   For an analysis of Rama's perspective in the literary field and its contemporary perspectives, see Aguiar and Vasconcelos 2004. Although the authors do not mention it, we can associate, in a more political approach, Rama's 'transculturation' with the concept of 'internal colonialism' developed by González-Casanova (1965).

4   For Moreiras (2001) it would be more of a tactic than a strategy. On this debate, involving Gayatri Spivak's concept of 'strategic essentialism', see especially pages 336-337.

> A policy is democratic both by building spaces for collective recognition and development and by raising the reflexive, critical and sensitive conditions necessary for thinking about what raises obstacles to such recognition. Perhaps the central theme of cultural policies today is how to build societies as democratic projects shared by all without making all equal, in which disaggregation is elevated to diversity and inequalities (among classes and ethnic or other groups) are reduced to differences. (Canclini 1998: 157)

What is, then, the kind of hybridism (or, if you like, transculturation) that we want to promote? What kind of multi/transterritoriality allows the stimulation of these 'positive' (or in other words, as has been previously stated here, 'anthropophagic') processes of hybridization?

We cannot forget that (the discourse of) hybridism is also fashionable – and thus, like everything in fashion, has a high 'exchange value'. As Moreiras stated:

> [...] hybridism can now almost be, in its performative [not merely nominal] aspect, a kind of ideological disguise for capitalist reterritorialization. [...] Arguing in favour of hybridism against the reification of cultural identities as a kind of requirement for a perpetual flexibility is an exaggeration of its usefulness. (2001: 316)

It is 'good' to be hybrid, 'mestizo', créole, because it 'sells' – and it sells because we are told that it is good to mix, circulate through several territorialities, in sum, to consume the *world hybrid* – perhaps a nomenclature that can summarize this more trade-oriented and globalized aspect of hybridism. We are effectively 'global' only if we are 'hybrid'. Here is another key element in this debate, one to which few seem to pay attention: depending on the scale at which it is addressed, hybridism acquires different features and political implications. Speaking of 'local', 'regional' or 'national' hybridism (or transculturation) is not the same as speaking of continental (Latin American), let alone global hybridism – as one of the basic features of a world identity.

As part of the 'flexible' logic of capitalism, only that which is mobile, ephemeral, open to constant change and mixing (to be consumed again) would be good. But, as the great 'thinker of speed' (or of 'dromology') Paul Virilio warned us, in a broader sense:

> [...] we always say that the primordial freedom is the freedom of movement. Yes, but it is not so regarding speed. When you go too fast, you are entirely stripped of yourself, becoming totally alienated. It is possible, therefore, to have a dictatorship of motion [to which we could add: a kind of alienation of territory]. (Virilio 1984: 65)

Complementing that which was indicated earlier, what matters is not *just* 'being in motion' (although some groups over-emphasize this condition), but the possibility we have of triggering this movement when we need to – or, more freely, when we wish to – because the fact that 'the primordial freedom' is, as Virilio suggests, 'freedom of movement', does not mean we have to be necessarily always in motion. It is as if we defended mobility, hybridism and multiterritoriality in such way that they therefore became compulsory. 'Doomed' to hybridism and/or transit among territories, we could, at the extreme, lose any stable point of reference, indispensable to some extent to our human condition – humans who do not have the 'obligation' of the same social-territorial behaviour all the time, because the overlap between motion and rest is a permanent presence in our lives.

Even in a world in which the metaphor of nomadism has become almost commonplace, great mobility and cultural hybridization, of course, do not govern everybody's lives – on the contrary, what we see today is even, in some ways, a reverse process, with a growing set of restrictions on mobility, especially regarding people's mobility, which tends to become even stronger in face of the current global economic crisis.

Relative opening to change, and thus strengthening our autonomy, is very different from permanent, almost complete opening to not developing ties, in a kind of unremitting nomadism. Similarly, closing up (tactically) to resist is very different from closing up – indefinitely – to isolate oneself and/or as a basic form of life. The big issue is not choosing opening up to transculturation, 'anthropophagic' hybridism, multi/transterritoriality in the face of identity closure, mono-culture (!) and uni-territoriality, even because isolated cultures and completely closed territories have, in fact, never existed. Cultures completely open to exchange, to hybridization, have likewise never been produced – that would have been tantamount to ordering, in its extreme, their own disappearance. However, very different forms live side by side in this interplay of opening and (relative) closure, hybridization and (alleged) essentialization.

It is not all about, therefore, opening *or* closure, hybridism *or* essentialization. In the vast array of geo-historical situations and contexts, there is always the possibility of the multiple – not only in the sense of 'living on the edge', through/on the borders, but also of the always open possibility of transiting across different cultures, landscapes and territories. Politically speaking, more important than conceiving of our lives and our identities as inherently 'hybrid' and 'multiterritorial', is the certainty that, if and when we decide, we have the alternative of changing territories, trying other forms of cultural identification, exchanging values – and that no one will force us into permanent hybridization, nor into constant mobility in the vast territorial multiplicity of our times.

That is why we must speak of a space-time always 'alternative' – not only in the sense of representing an alternative, the creation of the new, but also of allowing alternation – between the more and the less hybrid, between the more and the less open, that is, a space-time that combines interchange, extroversion and mobility with equally vital withdrawal, introspection and rest. And the alleged 'balance' between these dimensions can only be assessed through rigorous empirical work and consideration of the multiple needs and interests at stake for each individual, group and/or social class.

References

Aguiar, F. and Vasconcelos, S. 2004. O conceito de Transculturação na obra de Angel Rama, in *Margens da Cultura: Mestiçagem, Hibridismo & outras misturas*, edited by B. Abdala Jr. São Paulo: Boitempo.

Andrade, O. 1995. *A Utopia Antropofágica.* 2nd ed. São Paulo: Globo.

Bayart, J.-F. 1996. *L'Illusion Identitaire.* Paris: Fayard.

Bhabha, H. *O local da cultura.* Belo Horizonte: EdUFMG.

Bourdieu, P. 1989. *O Poder Simbólico.* Lisboa: Difel; Rio de Janeiro: Bertrand Brasil.

Burke, P. 2003. *Hibridismo Cultural.* São Leopoldo: Editora da Unisinos.

Canclini, N. 1997 [1989]. *Culturas Híbridas.* São Paulo: Editora da USP.

Castoriadis, C. 1997. *World in fragments: Writings on politics, society, psychoanalysis, and the imagination.* Stanford: Stanford University Press.

Gatens, M. and Lloyd, G. 1995. *Collective Imaginings: Spinoza, Past and Present.* London and New York: Routledge.

Glissant, E. 2005 [1996]. *Introdução a uma poética da diversidade.* Juiz de Fora: Editora da UFJF.

González-Casanova, P. 1965. Internal Colonialism and National Development. *Studies in Comparative International Development*, 1(4), 27-37.

Haesbaert, R. 1996. O binômio território-rede e seu significado político-cultural, in *A Geografia e as transformações globais: Conceitos e temas para o ensino (Anais do Encontro 'O ensino da Geografia de 1° e 2° Graus frente às transformações globais').* Rio de Janeiro: UFRJ.

Haesbaert, R. 2001. Da desterritorialização à Multiterritorialidade, in *Anais do IX Encontro Nacional da ANPUR.* Rio de Janeiro: ANPUR.

Haesbaert, R. 2002. *Territórios Alternativos.* Campinas: Contexto; Niterói: EdUFF.

Haesbaert, R. 2004. *O Mito da Desterritorialização.* Rio de Janeiro: Bertrand Brasil.

Haesbaert, R. 2007. Território e Multiterritorialidade: Um debate. *GEOgraphia*, 17. Niterói: Programa de Pós-Graduação em Geografia.

Helena, L. 1983. *Uma literatura antropofágica.* Fortaleza: UFC.

Kraniauskas, J. 1992. Hybridism and Reterritorialization. *Travesía*, 1(2), 143-151.

Lévy-Strauss, C. 1995 [1977]. *L'Identité.* 3rd ed. Paris: Presses Universitaires de France.

Maltz, B. 1993. Antropofagia: Rito, metáfora e pau-brasil, in *Antropofagia e Tropicalismo*, written by Maltz, B., Teixeira, J.E. Ferreira, S. Porto Alegre. Editora da UFRGS.

Massey, D. 1994. A global sense of place. *Space, Place and Gender.* Minneapolis: University of Minnesota Press, 24-29.

Mignolo, W. 2003. *Histórias locais/Projetos Globais: Colonialidade, saberes subalternos e pensamento liminar.* Belo Horizonte: Editora da UFMG.

Moreiras, A. 2001. *A exaustão da diferença: A política dos estudos culturais latino-americanos.* Belo Horizonte: EdUFMG.

Rama, A. 1982. Literatura y cultura, in *Transculturación narrativa en América Latina.* México: Siglo Veintiuno.

Revel, J. 1998. *Jogos de Escalas.* Rio de Janeiro: Fundação Getúlio Vargas.

Spivak, G. 1988. Subaltern Studies: Deconstructing Historiography, in *Selected Subaltern Studies*, edited by R. Guha and G. Spivak. New York: Oxford University Press.

Stam, R. 1999. Palimpsestic Aesthetics: A meditation on hybridity and garbage, in *Performing Hybridity*, written by J. May, J.and J. Tink. Minneapolis/London: University of Minnesota Press.

Tomlinson, J. 1999. *Globalisation and Culture.* Chicago: Chicago University Press.

Virilio, P. 1984. *Guerra Pura.* São Paulo: Brasiliense.

Young, R. 1995. *Colonial Desire.* Oxon and New York: Routledge.

Chapter 7

From Landscaping to 'Terraforming': Gulf Mega-Projects, Cartographic Visions and Urban Imaginaries

Mark Jackson and Veronica della Dora

Introduction

'The World' is a 9 x 7 km archipelago of 300 private artificial islands arranged in the shape of a world map, each of them available at a price. It is one of several large-scale artificial island projects planned and constructed in Dubai and the wider Gulf region. You cannot see 'The World' from the ground. Like any map, The World is visible only from above. A cartographic, rather than landscape feature, The World is the product of a process its makers call 'terraforming'. This chapter explores terraforming and its new terrestrial logic.

   Much attention has been paid to the 'production of landscape': as 'social formation' (Cosgrove 1985); as a physical and semiotic creation in the hands of the powerful to naturalize dominant discourses; and as a screen, decoded to reveal unequal power-relationships (Duncan 1990; Cosgrove and Daniels 1988).[1] From this last critical perspective, the aesthetic dimension of landscape has been charged with duplicity. Don Mitchell's *The Lie of the Land*, for example, offers a materialist interpretation of 'landscape production' in the context of early twentieth-century California, which aims to uncover the 'other side of the California Dream' (1996: 14). Idylls of California's agricultural scenery hide inequities of capitalist agriculture and migrant labour exploitation. Mitchell's work is echoed by the art historian W.J.T. Mitchell, who characterized landscape as 'the dreamwork of imperialism' and 'the medium by which this evil [colonialist exploitation] is veiled and naturalized' (2002: 10 and 30). Similarly, Kenneth Olwig's *The Body Politic* (2002), part of a larger project aimed at rescuing 'landscape's substantive nature', illustrates how scenic landscape was used as a technological device by

---

   1   For example, by deconstructing 'landscape's capacity to 'naturalize' social or environmental inequities through an aesthetics of visual harmony, geographers and art historians have long recognized that [...] landscapes [as ...] paradigms of [...] social and environmental order [are] often painstakingly constructed by rapacious landowners in the course of destroying more communal but less profitable fields, farms and dwellings' (Cosgrove 2006: 51).

early seventeenth-century Stuarts to unify Britain under the 'natural' authority of divine right.[2]

In what follows, we aim to read the terraforms emergent in the Gulf from a background which treats their built forms as objects that both emerge from particular landscape and visual materialities as indicative of emergent social orders, but which are at the same time quite different from traditional landscape forms. While troubling traditional landscape aesthetics and the moral connotation of 'looking at landscape' (Lowenthal 2007), Dubai's 'The World' archipelago might *also* quite comfortably be interpreted as the ultimate 'veil' concealing the unequal power-relationships of global capitalism (Davis and Monk 2007). From a sense of upsetting the established discursive frames of landscape, while at the same time drawing from its premises of production and social ordering, we situate our interpretation of these artificial islands, and similar developments, as evidence of a contemporary shift from landscaping to terraforming. The Gulf's artificial islands and related engineered geographical forms are products of an iconography of the imagination predicated by, and made possible through, the assemblage of particularly recent social practices: digital visualization technologies, earth engineering capacities, development driven market agendas, energy application, and neo-liberalizing legitimations. This is not to say that these developments do not have historical precursors or family resemblances, but rather that the emergent terraforms and their situated practises participate through specific, and perhaps new, materialities we find significant for thinking about the relationships between landscape and development within and beyond Europe.

Extraordinary visions in the form of fancy-shaped archipelagos, massive water-based canal developments, and landmark buildings, all characterize how expanding Persian Gulf cities seek to attract foreign capital and, in the process, configure new imaginaries of the urban. These imaginaries and very real, material forms are built on deeply unjust labour and environmental practices (Davis 2007; Caplin 2009). However, unveiling these inequities is not the primary goal of our contribution. Our focus, rather, is on 'terraforming'.

Terraforming is a new phenomenon similar to, and yet, different from, the landscapes and processes of 'landscape production' traditionally explored by cultural geographers. We argue that present land and earth forming practices in places like Dubai combine large-scale geo-engineering techniques for re-shaping the Earth into new physical and biotic formations which order markets, 'lifestyles' and spatial imaginaries. These imaginaries constitute powerful visualizations of social promise and security.

The chapter falls into four sections. First, we define 'terraforming' and its historical contexts, and compare those to 'landscaping'. Second, we focus on the terraforming of 'The World', in the context of Dubai (and the Gulf) as a model for

---

2 More recently, Olwig has continued to bridge the gap between landscape as a material phenomenon and landscape as a representation through concepts of alienation, commodification, and reification (Olwig 2005).

new intriguing forms of neo-liberalizing development around the world. Third, we discuss the shape of The World and the cartographic logics underpinning its iconic visuality. The last section explores the links between terraforming, 'world image', and mapping, by focusing on the parallel emergence of virtual Google Earth communities and terraformed utopian communities.

Shaping the land, 'forming' the Earth

> Arabian Canal's masterplan [...] creat[es] organic diversity through a rich array of distinct developments and landforms ... . Water enables life along the Canal and [...] create[s] a comfortable environment. Buildings are designed and grouped to provide shade to keep temperatures cool [... .] Arabian Canal sets an important new benchmark for [...] development both in Dubai and worldwide. (Arabian Canal's 'Masterplan Overview' and 'Design Philosophy' www.arabiancanal.com)

'Landscaping' and 'terraforming' are transformative processes at once physical and imaginative. 'Landscaping' refers to any activity that modifies the visible features of an existing area of land such that it may conceal or embellish a specifically built space so as to render it part of a continuous or harmonious landscape. 'Landscaping' can be paraphrased, philologically, as 'land-shaping', wherein *Land* designates a bounded, cultivated area (Olwig 1996: 633). But landscaping means more than 'cultivating': it is deemed both science *and* art, and requires deft observation and design skills. It serves functional ends of human habitation, and at the same time, creates forms of symbolic human feeling; it is inherently aesthetic. As planned 'staging', landscaping can be traced back to Renaissance gardening and the engineering of Palladian landscapes for the visual consumption of Venetian patricians, who gazed at their land properties from the balconies or porches of their villas (Cosgrove 1993, 1985). Interestingly, 'terraforming' is the term used by the developers of the artificial islands being fabricated in the Persian Gulf (and indeed around the world) for a new class of extra-wealthy 'global patricians'.

Landscape first emerged 'as a way of seeing the external world, in the fifteenth and sixteenth centuries' (Cosgrove 1985: 46). It was, and remains, 'a visual term, that arose initially out of Renaissance humanism and its particular constructs of space [and] was, over much of its history, closely bound up with the practical appropriation of space' – Palladian villas in the Veneto region, for example, were built on land reclaimed through complex hydrological engineering (ibid. 2008: 60). Renaissance landscaping symbolized an achieved harmony between human life and the hidden order of creation (Cosgrove 1988). As such, it abhorred bold enterprises, like natural gigantism, as monstrosities. One of the more despised ancient myths was that of Dinocrates, Alexander the Great's architect who planned to carve the mountain-peninsula of Athos into a colossal statue of his patron holding in his left hand the city of Alexandria (della Dora 2005). Fifteenth-century humanists took the legend as a negative model and a threat to an ideal combination of balance, harmony and practicality. A commentator like Buonaccorso Ghiberti was so embarrassed by the legend that he represented Dinocrates as withdrawing the whole idea after second thoughts and offering elaborate explanations to Alexander of its impracticality (Schama 1995).

Figure 7.1 and Figure 7.2 Francesco di Giorgio Martini, Representation of Dinocrates offering his project to Alexander the Great, 1476 and Thomas Reis, Portrait of Sheykh Muhammad al-Makhtoum offering the Burj al-Arab to The World

Source: Figure 7.1 courtesy of Biblioteca Centrale Nazionale di Firenze; Figure 7.2 courtesy of the artist.

In the Gulf today, however, gigantism and bold projects have become normative terra-forms. Everything in Dubai, for instance, is marked by superlatives (the largest mall, the tallest tower, the first desert ski resort, the most luxurious hotel, etc.). But terraforming goes beyond mere superlatives. Like landscaping, terraforming suggests the modification of the natural environment. It suggests domestication, but with very different rhetorical and political strategies than those celebrated in Renaissance Italian reclamations – and on a much larger scale. If sixteenth-century Venetian productions of landscape closely involved land drainage and reclamation

(Cosgrove 1988), the production of terraforms in contemporary Dubai implies the *creation*, rather than simple transformation, of land.

Terraforming, literally 'Earth-shaping' is a neologism that comes from science fiction. First coined in 1950 by American writer Jack Williamson (Boston et al. 2004: 976), its cultural history is far more recent than that of 'landscaping'. Terraforming is the hypothetical process of deliberately modifying a planet's atmosphere, temperature, surface topography and ecology so as to allow the habitation of terrestrial life forms (McKay 1990; McKay 1982). In 1961, astronomer Carl Sagan proposed the terraforming of Venus. He imagined seeding the atmosphere of the planet with algae to remove carbon dioxide and reduce the greenhouse effect until surface temperatures dropped to 'comfortable' levels. In 1973, Sagan's attention turned to Mars. Three years later, NASA officially addressed the issue of terraforming (or planetary modelling) and concluded that it was possible for Mars to be made into a habitable planet. In 1976 the first conference session on terraforming was organized, and in 1982 terraforming entered the academic lexicon with the famed paper 'Terraforming Mars' (McKay 1982).[3] Urban and offshore development projects in the Gulf have adapted the rhetoric and logic of terraforming to fully terrestrial means.

Figure 7.3 Daein Billard, Terraforming Mars, 2006
Source: www.wikimedia.org.

Although 'terraforming' commonly refers to extra-terrestrial, planetary scale environmental engineering, it is important to recognize four key aspects to terraforming which contextualize the developmental logics underpinning 'terrestrial terraforming'. The first is that terraforms aim to alter environmental conditions previously unsuitable to terrestrial, and particularly, human life. Thus seeding clouds, melting stored water on frozen Mars, introducing bacteria to infertile soils,

---

3   Mars continues to be the paradigm planet for exploring the imaginary possibilities of global environmental engineering and extra-terrestrial colonization (see Daly and Frodeman 2008; Cockell and Gerda 2004; Zubrin 2002; Haynes 1990), with arguments today defending Martian terraforming as conducive to understanding more fully terrestrial eco-systems (Haynes 1990; Zubrin 2002).

or introducing supportive soil structures where previously there were none. In the context of terraforms in the Gulf, the development of new landforms and city spaces from desert and sea are indicative of an ideological shift towards creating new, and supposedly superior, built spaces, *ex nihilo*. Importantly, this creation symbolically reifies the continental or 'pre-terraformed' as unsuitable to particular expectations of human life.

Second, terraformed change is engineered. Central to terraforming is the notion that rationalized conditions for life yield instrumental controls from which large-scale complex systems might emerge (Luke 1999). Technical capacities and their application make terraforming thinkable. Indeed, the imaginaries made possible by new technologies are often *only* imagined as possibilities. Implementing terraforms in any extant way is very often long subsequent to an imagined technical materiality. Yet, while terraforming Mars has long been a dream of speculative scientists, in the terrestrial context, terraforming new city spaces from the sea and desert is marketed as imminent reality – emerging artificial mega canals and islands well illustrate the point. Indeed, in most cases, 'terrestrial terraforming' is often presaged by highly affective imaginary productions of the proposed developments through computer generated imaging. Advertising videos, on-line interactive environments and satellite mapping technologies are integral to the speculative economies and developmental lure of these promissory spaces. Terraforming is both a promissory and a speculative practice of futural anticipation; in the case of extra-planetary formation it is a distant futurity; in the case of terrestrial development it is very contemporary, and central to how the spaces are, literally, materialized.

Terraforming, thirdly, aims at producing stability. Terraformed environments promise to support life over such a term that, ideally, complex self-ordering systems emerge from seed events. 'Start with a lifeless planet. Add a few Earth organisms and stir. Wait about 10,000 to 100,000 years. Voila! A living, terraformed planet will emerge' (Boston et al. 2004: 975). New emergences of social, technical and market possibility, within the context of the Gulf cities, terraformed islands are imagined, marketed and sold as stable and timeless (but also modern) places beyond the social and environmental turmoil of the global present. The world is running out of places where it can start over: 'sand and sea along the Persian Gulf [...] provide the final *tabula rasa* on which new identities can be inscribed: palms, world maps, cultural capitals and financial centres' (Koolhaas 2007: n.p.). These emergent places are envisioned without crime or want, and benefit only from the best the cosmopolitan urban has to offer. Stability *and* urbanity are their – impossible – allure.

Finally, terraforming happens on a large scale. Whole planetary environments are changed, with extensive habitats grown where previously absent. Technology and innovation intervene to make new life possible. So large is the scale of extra-planetary terraforming, Boston, Todd and McMillen speak in terms of tens of thousands of years for life to emerge. Although not planetary, the formations in the Gulf are also very large indeed. Dubai's island developments promise themselves as 'epic wonders, Odyssean creations of world scale possibility: "A vision made real. The World is today's great development epic. An engineering odyssey to create an island paradise of sea, sand, and sky, a destination has arrived that allows investors to chart their own course and to make the world their own"' (www.theworld.ae).

The chief designer to the Arabian Canal, advertised as one of the largest and most complex engineering mega projects in the world, and a corollary development to The World, boasts that the Canal '[...] will create life in the desert [and ...] be a globally-recognised landmark destination' (www.arabiancanal.com/media-centre/press-room/1007/unearths/). Engineered by a company called 'Limitless', the canal project, a 75-kilometre long waterfront housing development through the desert, claims its engineering works, 'defy description', and will 'cool temperatures' so as to provide 'varied and distinctive topograph[ies]' (www.arabiancanal.com/engineering/applied-science/). The first phase of the project to move one billion cubic metres of earth has been completed with construction begun on Canal Front, an inaugural community with 'an arresting new landscape [...] teeming with life, energy and activity' (www.arabiancanal.com/canalfront).

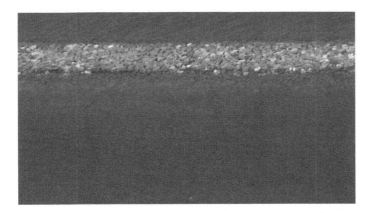

Figure 7.4 The World's breakwater, 1 May 2007
Source: Image courtesy of Imre Scott/wikimedia.

Venus and Mars have not been terraformed, but the Arabian Sea has – and in the most literal senses. Water has been turned, not simply into earth, but into '*The World*'. 'Life' is being brought to the desert. Unlike landscaping, 'terraforming' in Dubai denotes the creation of 'your own private island-world' *from scratch*, from a sea surface, or from a metonymically similar arid desert surface, each marketed as a *tabula rasa* waiting to be inscribed by human imagination and action, according, of course, to certain economic principles and opportunities. '[T]he sea has become, like the desert, a developer's ... canvas for unlimited exploitation' (AMO 2007: 262).

The World and the Gulf

> [...] Welcome to your very own blank canvas in the azure waters of the Arabian
> Gulf. Where orchestrating your own versions of paradise [...] is a much needed
> inoculation against the ordinary [...] The World really can revolve around you.
> (From 'Bespoke', Principle I of 'The World's' Principles, www.theworld.ae)

While artificial islands are, themselves, nothing new, unique to the significance of
The World is the way the development promises investors 'the chance [...] to try
terraforming their territorial waters' (AMO 2007: 262). Each island in The World
is of varying size and shape, and is identified with a city or area corresponding to
a planetary cartographic location ('Kathmandu', 'Peru', 'Angola', 'Los Angeles',
etc.). In the best mapping tradition, the entire development is surrounded by a 27-
kilometre oval breakwater, and is thus a self-enclosed cartographic world set aside
from the surrounding world: a figurative *and* literal islanding (Gillis 2004).

Figure 7.5 The World on 18 October 2007
Source: Image courtesy of Imre Scott/wikimedia.

The shape of The World is the product of a cartographic vision, and a bold
imaginative act. Unlike most maps, however, The World archipelago is 'a vision
made real': its islands are physically built using sand dredged from the bottom of
the sea. Thirty-four million tonnes of rock and thirty-two million tonnes of sand
were engineered into place with satellite mapping and geo-systems science, and
settled with new compaction technologies by an army of foreign labourers and
expertise from around the world. The resulting 'world' forms an iconic destination
where 'a shoreline of systematic multiplication and ad-infinitum stretching'
is revealed from 'a previously fraught and barren landscape' (www.theworld.
ae). Both the possibility for newly built space, but, perhaps more importantly,

'epical proportions of land' created 'to generate terrific revenues' (ibid.) are the end result of the islands' terraforming. This is not landscaping. It does not intend to edify with the aim of bringing to the fore, or enhancing, latent symmetries, inherent rationalities, or aesthetic integrities. It is the creation of new imaginary possibilities in the refusal of the already existing continental landscape. The continental is explicitly cast aside as 'ordinary' or even as waste; it needs to be re-imagined and re-created – re-developed – in order to make it both extraordinary, and fundamentally, profitable. Islands figure, classically, as imaginative loci for extending the remit of experiential possibility; but, with the earth as it is – full, spent and anxious – artificial island developments terraform ostensibly new worlds. The seventh principle of 'The World' states the difference explicitly: 'No-where but in a place like this can the world be so richly *re*-imagined' (www. theworld.ae/world-principles.html, emphasis added).

The cultural iconicity of the island imaginary is absolutely central to the refinement of these new liveable spaces, and to the attraction of the millions to whom the new spaces are targeted. Islands function, Baldacchino writes, as 'sites of innovative conceptualizations, whether of nature or human enterprise, whether virtual or real' (2006: 3). Between the initial speculative plans for the developments of Dubai's three Palm islands, The World archipelago, the new city development called Waterfront City, and the Arabian Canal, some four million people were, originally, to populate the emergent city spaces of Dubai and its surrounds. These numbers and projects have been significantly scaled back due to the global economic crisis of 2008 and 2009. Significant, however, is less the fact that fewer people than expected will be housed and entertained on the island terraforms, and more that the virtual imaginaries of these material cartographic objects are predicated on consumptive and appropriative discourses of island-ness, isolated remove, individualized self formation, and marketized space.[4]

Excluding the ordinary to the periphery, terraformed 'island-ing' works to enclose an imaginary space as a globally attractive centre for re-defining urban possibility. These possibilities are imagined and visualized for the consumer through hypothetical, almost science fiction means. For example, the openness of lifestyle possibilities are evidenced to consumers and interested investors by the fact that, in the case of 'The World', islands are sold bare, as 'natural' terraforms waiting to be colonized. Possibilities for these barren sites are offered up to the consumer through the media of computer generated graphical images of water wonderlands, luxurious oases of cosmopolitan participation, technologically

---

4 As terra forms entirely dependent on technologically sophisticated construction techniques, these are spaces enabled almost exclusively for, and by, 'an electronic capitalism [which] enables the most successful to secede from the rest of society' (Reich as quoted in Spivak 2008: 161); they are thus concentrations of marketed opportunity which materialize the 'exclusion of a margin that is rejected' as ordinary (Brown 2003: n.p.; Raffoul and Pettigrew 2007: 3).

advanced mobility, and meteorologic calm. But, it is up to the purchaser – within the confines of agreed to guidelines – to make of their island what they will.

Choice and self-fashioning are delimited through pre-determined private development relationships. Expression and individuality are legitimated as long as they are disciplined through specific, privatized relationships which in turn reinforce the stability and security of the development as a whole. Diversity is welcomed so long as it benefits the maintenance of the socio-technical system of the project.[5] As exclusive spaces, terraformed islands, canal communities, and similar geo-engineered environments actively prohibit. Gated community paradigms reinforce securitized access with panopticist surveillance apparatuses, walled compounds, and vetted participation. Housing and building opportunities are conceived as value adding strategies for investment and consumerist attraction. Whereas Palladian patricians might have gazed out on their shaped landscapes in reflective, if also self assured, repose, the landscape of the present Gulf terraforms is one framed through a gaze of speculative opportunity. Gone is an assumption of developing, however stratified, an integrated and functioning *res publica* in harmony with a symbolic landscape (or a productive fertile land); island terraforms celebrate individualized consumption, investment and self-fashioning, only tangentially are they cosmopolitan, aesthetic or edifying spaces.[6]

These developmental terraforms, unlike planetary earth systems, are private spaces which acculturate themselves to the imperatives of the commodity form (Abbas 2008). Through them, Dubai seeks to create itself and the Arabian Gulf as an iconic landscape for the development of global urban imaginaries whose interactive possibility is determined through terraforms as transactable objects. To communicate the possibility of appropriating such spaces as totalizable and containable, to give them a sense of place, but of a place beyond history, the visual tropes of the terraforms need be both recognizable, and, at the same time, able to deliver a particular, if also indefinite, promise. It is for this reason that the most stricking of the terraforms, The World, appears as a map. A giant commodity logo

---

5   As evidence of this constraining of possibility, when it first advertised itself to the market, The World's island terraforms were available on an invitation-only basis (AMO 2007: 266).

6   Adding water-based frontage to potential housing and property is the key reason for the Gulf's island developments such as The World. Waterfronted properties are more valuable than non-waterfronted properties. In a desert environment like certain Gulf States, the presence of, and access to, water is a defining characteristic for determining terraform principles of emergent, new life. Dubai's real-estate market depends, almost entirely, on its willingness and capacity to add value. The city-region's property development market thus legitimizes its often exorbitant exchange-values precisely through the construction of ever more spectacular and enticing additions. If this exchange value can appeal to imaginative geographies which speak to long held social and private desires for exclusivity, remove, isolation and control, while at the same time offering access, amenity and cosmopolitan luxury through a nearby city, then so much the better.

made of sand and water, The World evocatively encapsulates both its global scope *and* terraforming's 'planetary' connotations through its cartographic form.

Visualizing The World as a map

> [...] Layered with meaning and taking inspiration from all corners of the globe, here is where all the good things come together in a luminous display of humanity and community, progress and momentum. Nowhere but in a place like this can the world be so richly re-imagined. (From 'Legacy', Principle VII of 'The World's' Principles, www.theworld.ae)

In his last book, *The Invention of the Earth*, Italian geographer Franco Farinelli argued that 'every map is an island' (2007: 105). Maps are self-enclosed miniature worlds floating in our everyday lives. Maps represent a piece of the Earth's surface separated and on its own, or the flattened surface of the entire Earth as viewed from above. The legacy between islands and cartographic images of the Earth is profound and longstanding. In ancient Greece the world itself was imagined as an island surrounded by an ocean of chaos. This conception endured throughout the Middle Ages on *mappae mundi,* Christianized world images featuring a circular 'world island' composed of three continents and surrounded by the river Ocean (Gillis 2004; Cosgrove 2001). The idea of 'Spaceship Earth' with human inhabitants as 'riders on an isolated, self-sustaining organism' (Von Bertalanffy 1968) and appropriations of the iconic NASA AS17-148-22727 (1972) picture by Environmentalist groups uncannily echoes this world-island imagery: spaceship Earth is nothing but a beautiful, fragile island floating in a dark, unknown universe (Gillis 2004: 167-68). From the early fifteenth century to the end of the seventeenth century, the earth's surface was narrated through maps of islands in the so called *isolarii* (or island books), specialized atlases featuring one or more islands on each page, accompanied by notes of historical, geographical and anthropological character, as well as by legends and other curiosities (Tolias 2007; della Dora 2008). Imaginary islands on the map often anticipated great geographical discoveries: Christopher Columbus and his followers did not venture into an empty Atlantic, but rather in a sea full of islands – real and imagined (Gillis 2004).

As with mapping and islanding, terraforming also comes from above. An epigram soon to surround the Palm Jebel Ali, an island development neighbouring The World will read 'it takes a man of vision to write on water'. Vision, however, is not a mere equivalent of sight. Vision incorporates both the ocular act of registering the external world, and a more abstract and imaginative sense of creating and projecting images (Nuti 1999). Vision also implies futurity. The world is a vision, for we can only access it through cartographic representation and cultural mediation (Cosgrove 2001). *The World in Dubai* – still mostly barren – is also a vision: it requires imagination both on the producers and consumers side; its futurity is anticipated through a faith in images and promotional videos more than in the 'tangible reality'.

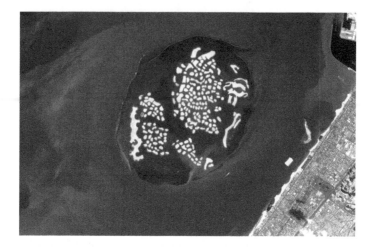

Figure 7.6 The World Archipelago, Persian Gulf
Source: Image courtesy of NASA/wikimedia.

Denis Cosgrove (1985) and Ken Olwig (2002) have shown how, in the different contexts of Renaissance Italy and Georgian England, landscape was consistently reformulated as an 'idea' that bounded territorialized spaces onto *paesi* or *lands* by means of such mathematical and graphic techniques as perspective, projection, geometry and trigonometry – in other words, through a process of mapping, or, in Gillis' words, of 'islanding' (2004).

> In painting and garden design landscape achieved visually and ideologically what survey, map making and ordnance charting achieved practically: the control and domination over space as an absolute, objective entity, its transformation into the property of the individual or state. And landscape achieved these ends by use of the same techniques as the practical sciences, principally by applying Euclidean geometry as the guarantor of certainty in spatial conception. (Cosgrove 1985: 46)

Landscaping, or land-shaping was thus a cartographic process; the map preceded the territory. Today, in the Gulf and in other parts of the world (Jackson and della Dora 2009), territory is no longer simply modified, or land-shaped. It is *a* map. Unlike scenic Palladian or Georgian landscapes, the new terraforms are maps that can be viewed only through the mediation of technology – the same technologies through which they have been produced. Dredgers do not see what they are doing. They blindly follow their GPS; nor are estate owners given a totalizing view of The World from the balcony of their villas: The World is invisible from ground (or sea) level, and Dubai is invisible from much of the more exclusive parts of The World. Exclusivity demands a visible separation, or at least the illusion of

Figure 7.7 Artificial Archipelagos, Dubai
Source: Image courtesy of NASA/wikimedia.

separation, and the ultimate backdrop to the separatist lifestyle promoted through a terraformed development is one constructed without reference to an extant world; the terraformed deliberately evokes a worlding beyond terrestrial historicity. This exceptionalism is reinforced through the hyper-visual and technologically mediated apparatus of our interactions with the terraforms.

Conceived so as to be visible only 'from above' Gulf terraforms require a new type of God-like viewer: the millionaire enjoying the view of The World from his private helicopter or from Dubai's seven-star hotel, the tourist approaching to Dubai's International Airport, or, more commonly, and, more significantly, an increasing global population of Google Earth armchair travellers after 'new wonders'. Visibility is produced through new economies and technologies of space. As a terraformed cultural icon, The World has put Dubai on the map. And new mapping technologies (such as Google Earth) bring The World (and many other yet-to-be island worlds) to life.

World does not equal Earth (and The World does not equal Google Earth)

Terraforming embodies both the abstraction of future visions and the materiality of the earthly element (terra). It shapes the earth and creates worlds. World, however, does not equal Earth. And *The* World does not equal Google Earth. Earth denotes an organic rootedness and nurture. 'World', by contrast, has a socio-spatial meaning. It signifies cognition and agency. Earth implies attachment and habitation, the ground from which life springs, is lived, and returns at death. World is a semiotic creation. We are born into the world, may engage it or retreat from it. Worlds may coincide or collide. We may imagine past or better worlds. In Greek and Latin 'earth', environmental and elemental, is gendered female. It is soil, especially fertile soil ploughed for cultivation. When applied to the planet, it attaches through its agrarian connotation to the sense of locale actually opposed to the global. World, by contrast, is gendered male. Spatial, it implies mobility and communication across a global surface. While 'earth' calls for contemplative distancing, 'world' calls for action (Haraway 1997; Cosgrove 2001).

Differences between the various 'worlds' that terraforms produce, and earlier landscape reclamations, are significant. Palladian landscape reclamations transformed God's creation with the aim of allowing nature's perfection to be revealed. Today's terraforms are not worried to justify their moral implications by relating divine and human action. Many terraformed island developments implicitly avow appeals to moral justification by rendering their developments conducive to economically configured identities. Thus, the relation of polis to nature through improvement rationales becomes re-configured and re-spatialized as lifestyle consumption. This is not to say, of course, that terraforming is apolitical – quite the opposite. If Earth is intrinsically elemental and poetical, then world is intrinsically political, or rather *geo*-political. The point can be illustrated, once again, cartographically.

When we open Google Earth we are welcomed by a friendly orthographic rendering of the terrestrial globe which we can spin as we like (Crampton 2008). The globe is democratic and 'politically correct'. It is a physical globe floating in the dark universe; 'Spaceship Earth' calling for global unity, a globe without political boundaries (unless we choose to visualize them). Unlike world maps, the globe is not ethnocentric, for all the points on its surface are equally distant from the centre. In fact, you can choose to centre Google Earth 'on your own place' and to continuously re-centre it. The message is clear: Google Earth offers itself as a safe space for free-speech, or what Crampton (2003) calls *parrhēsìa*, open to 'everyone' – ideally. By contrast, The World in Dubai is a flat map. Like any flat map, The World has a projection and a centre. The World is centred on Dubai; as a prominent slogan in the city reads, 'The Earth Has a New Centre'.

Like any map, The World in Dubai is simply *a* vision of the world. It too is about inclusions and exclusions (Wood 1992; Harley 1988). For example, it is not quite worldly enough to include Israel in its island countries available for purchase. The World re-inscribes traditional Western cartographies. While several countries are excluded from the map of 'The World', other geographical regions are included. You can buy 'Texas', but not Nebraska; you can buy 'New Delhi' and 'New York', but not Kolkata or Manila. In other words, a representational geo-politics is at play in how the new terraforms construct representational knowledge apparatuses.

At the same time, the cartographic image, thanks to its common sense perception as a 'scientific' instrument (rather than as a cultural construction), is invested with disinterested 'neutrality'. Another terraform example, this time in the Black Sea, best illustrates the point. Federation Island is a proposed 250-hectare island shaped in the form of a map of Russia (to be realized for the 2014 winter Olympic Games) deliberately inspired by The World in Dubai. Its location near the shores of Abkhazia invests this terraform as a 'stamp' of Russian territorial hegemony in the troubled secessionist region. Yet, the cartographic shape is perceived by its designers as 'neutral' enough to elude such a contentious political discourse.[7]

What is fascinating about Federation Island is the fact that we can interact with it through the visual media of Google Earth, despite the fact that *it does not yet exist*. The visual materiality of the promised terraform in the context of Google Earth produces a geographic imaginary which is rendered uncritical through the medium, and its technical claims, to cartographic truth. We use Google Earth as a visual reference to legitimate spatial claims. We also use Google Earth and similar mapping technologies to find our way to new and unfamiliar places. But the presence of *promissory* terraforms, whether they be Russia's Federation Island, or Dubai's still incomplete The World, within such a cartographic space, legitimates and naturalizes their determinately political presence as a normalized geographies.

If The World is a cartographic vision and the product of a cartographic logic, it is also a map made of earth (or sand). It is, as we said, a 'vision made real', a terraformed vision. But the paradox is that as new 'democratic mappings' such as Google Earth claim to bring ordinary people together in virtual communities, 'real communities' recede to create their own worlds: miniatures marked by luxury and globality, by exclusivity and removal – removal from the continental chaos of what Mike Davis called our 'planet of slums' (2006). As The World's developers eschatalogically claim, 'there is nothing after *The* World'.

Herein terraforms assume their significance and power. Terraforms do not exist as isolated objects. They circulate through complex visual, cartographic, and political horizons of possibility which shape geographical imaginaries of development, landscape and futurity. As part of these diverse networks of association, they work to re-define the urban imaginaries of development in ever more spectacular ways, yet also in ways which veil their processes of production and social ordering. In many ways, these terraforms – artificial islands, waterways, promissory cities – are landscapes just as Cosgrove, Daniels, Mitchell, and Olwig define earlier such geographical features. But, for us, these terraforms extend the scalar and normative exercises of landscaping to create both material and visual promises for re-imagining what it means to inhabit the Earth. They world Earth differently. Bringing almost extra-terrestrial and spectacular vision in the form of technical formations, they promise a beyond to development and human habitation,

---

7 As Dutch architect Erick van Egeraat commented, 'within the constraints of iconography an alternative to representing the map could be to represent a national icon. Canada plans a floating maple leaf, Switzerland has built an embassy in the form of their white cross, and for Holland we [...] consider building a tulip. For Russia though, what can we think of? A giant hammer and sickle? Haha, that's no virtue. Compared to that, a map is safely neutral' (eikonographia.com/?p=1851).

one which is not limited by the vagaries, muddles, failures and disappointments of history, but one which is 'Limitless'.

Conclusion

Landscape has never been simply scenery; it is always already substantively political (Olwig 2005). But whereas landscape distances the perceiver by highlighting absence and presence, ordering, etc., and, through this framing, alienates prior social relationships to land, in the case of terraforms, their power and uniqueness lies in the fact that they do not alienate viewers or participants from an experiential past or commons. Terraforms idealize an image of material space, and naturalize latent expectations of development by *creating* already enclosed land, or, at least, the conditions for their creation 'ex nihilo'. These are places whose appeal lies precisely in the fact that they are, arguably, without history. Thus, the key signifiers within these sites are 'sea', 'sun', 'sand', and 'sky'. In these post-disenchantment spaces continental modernity is accepted as unsalvageable. Post-landscapes, they invoke affectual triggers that inhere and attract through terraforming promise. The World's Principle IV, 'Visceral', sums up this post-landscape affectual shift to the terraform: 'The thrill is back. And this time, it's here to stay. Your pulse races. Your heart pounds. A once-in-a-lifetime experience has begun. A whirlwind of sights, sounds, and emotions [...] casts life in a brilliant new light. You may have discovered your perfect soul mate. Sand, sea, sky. Yours' (www.theworld. ae/world-principles.html). History is eradicated.

We have indicated a shift in geographical imaginaries from landscaping to terraforming in the context of Dubai. These projects are not isolated. The island paradigm travels, literally and figuratively. Terraforms are part of a material circulation whose geographic imaginaries are a product of developmental logics and visual technologies. Representational apparatuses, these terraforms also veil specific political, social and aesthetic effects. They work to actively exclude histories and practices associated with limited and failed continental ordinariness. The world is begun anew and re-imagined through spectacular developmental strategies of marketized space. Normalized as fantastic, the expectant promise of future stability is envisioned as a terraformed extraordinary.

There is nothing after The World. (www.theworld.ae)

References

Abbas, A. 2008. Faking Globalization, in *Other cities, other worlds: Urban imaginaries in a globalizing age*, edited by A. Huyssen. Durham: Duke University Press, 243-266.

Bagaeen S. 2007. Brand Dubai: The instant city; or the instantly recognizable city. *International Planning Studies*, 12, 173-197.

Baldacchino, G. 2006. Islands, island studies, island studies journal. *Island Studies Journal*, 1, 1, 3-18.

Bertalanffy, L. von 1968. *General system theory: Foundations, development, applications.* New York: George Braziller.

Boston, P.J., Todd, P. and McMillen, K.R. 2004. Robotic lunar ecopoiesis test bed: Bringing the experimental method to terraforming, in *Space Technology and Applications International Forum*, edited by M.S. El Genk. Melville, NY: American Institute of Physics, 975-985.

Brown, W. 2003. Neo-liberalism and the end of liberal democracy. *Theory and Event.* 7, 1, n.p.

Caplin, J. 2009. Mirage in the desert oasis. *Harvard International Review*, 30, 4, 28-32.

Cockell, C. and Horneck, G. 2004. A planetary park system for Mars. *Space Policy*, 20, 291-95.

Cosgrove, D. 1985. *Social formation and symbolic landscape.* Totowa, NJ: Barnes and Noble Books.

Cosgrove, D. 1988. The geometry of landscape: Practical and speculative arts in sixteenth-century Venetian territories, in *The iconography of landscape*, edited by D. Cosgrove and S. Daniels. Cambridge: Cambridge University Press, 254-76.

Cosgrove, D. 1993. *The Palladian landscape: Geographical change and its cultural representations in sixteenth-century Italy.* University Park, PA: Penn State University Press.

Cosgrove, D. 2001. *Apollo's Eye.* Baltimore, MD: Johns Hopkins University Press.

Cosgrove, D. 2006. Modernity, community and the landscape idea. *Journal of Material Culture*, 11, 49-66.

Cosgrove, D. 2008. *Geography and vision.* London: IB Tauris.

Cosgrove, D. and Daniels, S. (eds) 1988. *The iconography of landscape.* Cambridge: Cambridge University Press.

Crampton, J. 2003. *The political mapping of cyberspace.* Edinburgh: Edinburgh University Press.

Crampton, J. 2008. Keyhole, Google Earth, and 3D worlds: An interview with Avi Bar-Zeev. *Cartographica*, 43, 85-93.

Daly, E.M. and Frodeman, R. 2008. Separated at birth, signs of rapprochement: Environmental ethics and space exploration. *Ethics & the environment*, 13, 1, 135-51.

Davis, M. 2006. *Planet of slums.* London: Verso.

Davis, M. 2007. Sand, fear and money in Dubai, in *Evil Paradises: Dreamworlds of neoliberalism*, edited by M. Davis and D. Bertrand Monk. New York: The New Press, 48-68.

Davis, M. and Monk, D.B. 2007. Introduction, in *Evil Paradises: Dreamworlds of neoliberalism*, edited by M. Davis and D. Bertrand Monk. New York: The New Press, ix-xvi.

della Dora, V. 2005. Alexander the Great's mountain. *Geographical Review*, 95, 489-516.

della Dora, V. 2008. Mapping a holy quasi-island: Mount Athos on early Renaissance *isolarii*. *Imago Mundi*, 60, 139-65.

Duncan, J. 1990. *The City as Text: The Politics of Landscape Interpretation in the Kandyan Kingdom*. Cambridge: Cambridge University Press.

Farinelli, F. 2007. *L'Invenzione della Terra*. Palermo: Sellerio.

Gillis, J.R. 2004. *Islands of the mind: How the human imagination created the Atlantic world*. New York: Palgrave MacMillan.

Haraway, D. 1991. *Simians, cyborgs and women: The reinvention of nature*. New York and London: Routledge.

Harley, B. 1988. Maps, knowledge and power, in *The Iconography of Landscape*, edited by D. Cosgrove and S. Daniels. Cambridge: Cambridge University Press, 277-312.

Haynes, R.H. 1990. Ecce ecopoiesis: Playing God on Mars, in *Moral Expertise: Studies in practical and professional ethics*, edited by D. MacNiven. London and New York: Routledge, 184-197.

Jackson, M. and della Dora, V. 2009. 'Dreams so big only the sea can hold them': Man-made islands as anxious spaces, cultural icons, and travelling visions. *Environment and Planning A*, 41(9), 2086-2104, DOI: 10.1068/a41237.

Landis, G.A. 2001. Searching for life: The case for halobacteria on Mars. *American Institute of Physics Conference Proceedings*, 552, 39-96.

Lowenthal, D. 2007. Living with and looking at landscape. *Landscape Research*, 32, 635-56.

Luke, T. 1999. Eco-managerialism: Environmental studies as a power/knowledge formation, in *Living with nature. Environmental politics as cultural discourse*, edited by F. Fisher and M.A. Hajer. Oxford and New York: Oxford University Press, 103-120.

McKay, C.P. 1982. Terraforming Mars. *Journal of the British Interplanetary Society*, 35, 427-33.

McKay, C.P. 1990. Does Mars have rights? An approach to the environmental ethics of planetary engineering, in *Moral Expertise*, edited by D. MacNiven. Routledge, New York, 184-197.

Mitchell, D. 1996. *The lie of the land: Migrant workers and the California landscape*. Minneapolis, MN: University of Minnesota Press.

Mitchell, W.T.J. (ed.) 2002 [1994]. *Landscape and power*. Chicago: The University of Chicago Press.

Nuti, L. 1999. Mapping places: Chorography and vision in the Renaissance, in *Mappings*, edited by D. Cosgrove. London: Reaktion Books, 90-108.

Olwig, K. 1996. Recovering the substantive nature of landscape. *Annals of the Association of the American Geographers*, 86, 630-53.

Olwig. K. 2002. *Landscape nature and the body politics*. Madison, WI: University of Wisconsin Press.

Olwig, K. 2005. Representation and alienation in the political land-scape. *Cultural Geographies*, 12, 19-40.

OMA-AMO. 2007. *The gulf*. Baden: Lars Müller.

OMA-AMO. 2007a. Terraforming, in *Al Manakh*, edited by O. Bouman, M. Khoubrou, R. Koolhaas. Amsterdam: Stichting Archis, 262-265.

Raffoul, F. and Pettigrew, D. 2007. Introduction, in J.-L. Nancy. *The Creation of the World, or Globalization*. Oxford: Blackwell, 1-26.

Schama, S. 1995. *Landscape and memory*. New York: Vintage Books.

Spivak, G.C. 2008. Megacity – 1997: Testing theory in cities, in *Other Asias*. Oxford: Blackwell, 161-174.

Tolias, G. 2007. Isolarii, fifteenth to seventeenth century, in *The History of Cartography*, edited by D. Woodward. Vol. III. Chicago: University of Chicago Press.

Wood, D. 1992. *The power of maps*. New York: Guildford Press.

Zubrin, R. and McKay, C. 1997. Technological requirements for terraforming Mars. *Journal of the British Interplanetary Society*, 50, 83-92. [Online] Available at www.users.globalnet.co.uk/~mfogg/zubrin.htm [accessed 15 February 2008].

Zubrin, R. 2002. The case for terraforming Mars, in *On to Mars: Colonizing a new world*, edited by R. Zubrin and F. Crossman. Ontario, CA: Collector's Guide Publishing, 179-80.

Websites consulted as primary sources

www.arabiancanal.com
http://eikonographia.com
www.thepalm.ae
www.theworld.ae

PART II
Landscape History, Heritage and Social Change

Chapter 8
European Landscapes: Continuity and Change

Johannes Renes

Introduction

During the last decades historical research has become ever less important in landscape studies. This is also true for the Permanent European Conference for the Study of the Rural Landscape (PECSRL), which started during the 1950s as a forum for historic, particularly morphogenetic, research into European landscapes. During their first decades, these conferences focused on the history of village types and field-patterns. During the 1980s and 1990s, the conferences became more oriented towards planning, by presenting historic landscapes as part of our heritage. Since the 1990s, the emphasis on modern developments grew, when growing numbers of ecologists and planners visited the conferences. One of the side-effects of the more interdisciplinary character of the conferences was a growing confusion regarding the term 'landscape'.

During recent conferences, many papers were presented on historic and heritage landscapes, but only a small minority was based on research that went farther back than the nineteenth century. This chapter's main aim is to stress the importance of research into the long-term history of landscapes. After an introduction, I will present some general information on long-term landscape change. The next section will deal more specifically with changes in European landscapes during the Early Modern Period. The chapter will conclude with some remarks on the consequences for heritage planning.

Landscape

Since the 1980s, there is an ongoing discussion on landscape definitions. To summarize this discussion as much as possible, two meanings are essential. The first is the medieval meaning of landscape as a territory, including the institutions that govern and manage it. This meaning can still be found in, for example, the Dutch region of Drenthe, which is called 'the old landscape'. Also a number of territories in the former Dutch East Indies, which were subjected to indirect rule, were known as 'self-governing landscapes'. These territorial definitions of landscape can be traced through the old German *Landschaftsgeographie* to modern landscape ecology. Landscapes according to this definition are of course subjective, but at the same time they can be investigated and mapped by fieldwork and archival study.

The second meaning developed when painters started to make pictures of rural scenes and called them 'landscapes'. In due course, not just the paintings, but also their object itself became known as landscape. Dutch painters re-introduced the word landscape in the English language, where the word therefore had a more visual meaning than on the Continent. These visual definitions turn landscape into a composition that is made within one's mind. According to these definitions, without observers there is no landscape.

There is common ground, as the majority of landscape researchers is aware of the subjectivity and of the constructed character of our ways of looking at landscapes. Also, following publications by Olwig (1996), there seems to be a consensus that there is also a 'substantive' landscape. Still, the two types of definitions lead to different types of landscape research. On the one hand, much research focuses on the visual aspect of landscape and is usually more concerned with the perceiver than with the perceived landscape. On the other hand, there is still much research on what can be called the 'physical topography of an area' (Olwig 1993). In this chapter, I will stay closer to this latter approach and will look primarily at the concrete landscapes of settlements, field-patterns and agrarian land-use types.

It is good to realize that, in the shadow of the academic discussions on landscape concepts and meanings, also the more traditional research into landscape history has made progress. Much of it stresses the dynamics of landscape development. Although landscapes do show continuities, they are also subject to transformations. Compared to some decades ago, the history of the cultural landscape is now seen as more dynamic and complex.

Moreover, in landscape planning as well as in landscape ecology it is still too often taken for granted that the recent and present transformations of European landscapes are more or less unique. Recent dynamics are presented as opposed to 'traditional' landscapes. The Belgian geographer Marc Antrop (1997) defined traditional landscapes as

> those landscapes having a distinct and recognisable structure which reflects clear relations between the composing elements and having a significance for natural, cultural or aesthetical values. [...] They refer to these landscapes with a long history, which evolved slowly and where it took centuries to form a characteristic structure reflecting a harmonious integration of abiotic, biotic and cultural elements.

Another, somewhat shorter, definition is used in a recent research project in Galicia (Calvo-Iglesias et al. 2009): 'Traditional agrarian landscapes are those prior to the industrialization period that still have preserved characteristic features from older times'.

Implicitly, and sometimes explicitly, these landscapes are described as not just more stable but also less troubled than the present ones. Figure 8.1 shows a recent example of this way of thinking of a more or less linear development, in which human influence grew and landscape diversity as well as biodiversity grew with it, reaching the peak around 1900. Among Dutch ecologists, there is a long tradition behind this graph. Already during the 1930s, the ecologist Victor Westhoff, a key figure in the history of Dutch nature conservation, made a distinction between the earlier human activities that enriched nature, as opposed to the recent human influences that have diminished diversity. Also in discussions on landscape as heritage, often the twentieth century is seen as fundamentally different from earlier periods. The main aim of landscape preservation is then the safeguarding of those landscapes that have 'survived' the assaults of twentieth-century transformation. In an influential paper on this topic, the English landscape archaeologist Christopher Taylor (1972) distinguished between Zones of Survival and Zones of Destruction.

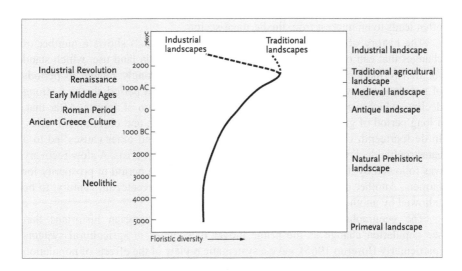

Figure 8.1 The traditional landscape model, showing agriculture and landscape development and the effects on floristic diversity through time

Source: Paracchini et al. 2007: 22.

In my opinion, the distinction between modern, dynamic cultural landscapes on the one hand and 'traditional', relatively stable landscapes on the other, is a-historical or, at best, an example of the writing of desired history. In this chapter, I will stress the importance of a more historic approach to landscape studies. Thus, I will return to some of the themes and approaches of the early PECSRL-conferences, which have almost been forgotten but nevertheless provide valuable insights for modern landscape research and planning.

In the following sections, three aspects of European landscape history will be discussed, each confronting one aspect of the 'traditional landscapes'-model. Firstly, 'evolution and transformation' will oppose the static character of past landscapes by showing their dynamic histories. Secondly, 'integration' will oppose the idea of more or less self-sustained local and regional landscapes. Thirdly, the section on 'people and landscapes' will argue that even in landscapes that seem to show a large degree of continuity, it is often difficult to talk of continuity in population groups and their relation to their surrounding landscapes.

Evolution and transformation

Many landscapes have undergone a number of transformations during the last millennia. Between such dynamic periods, there have been periods of relative stability, in which landscapes could become 'old', which in the present period often leads to an interest from the heritage-sector.

The history of land use in Central Europe (Figure 8.2) shows a number of changes that can only be described as radical. Changes in land use, which stand for transformations of the European landscape, are strongly related to periods of population growth or decline. The end of the Roman period shows a strong decline, which went together with desertions and a return of forests. After that, a long period of growth started, accelerating in the tenth and eleventh centuries. In de fourteenth century a combination of epidemics and other causes led to a catastrophic decline in population and, hence, an agrarian crisis. A slow recovery was followed by the so-called 'long sixteenth century', a period of prosperity for farmers. Another crisis followed in the middle of the seventeenth century, to be followed by another period of growth a century later.

The geographer David Grigg (1980), following agrarian historians that see population change as the prime mover of changes in agricultural systems (particularly Boserup 1965), gave a systematic review of the effects of population pressure. In his vision, population growth in a pre-industrial society will mean a growth of agrarian production by reclamations, specialization, new varieties, more productive crops and reduction of fallow. Another effect is the growth of non-agrarian occupations. Migration and birth-control are mechanisms to diminish population pressure. It is interesting to note that a number of these developments, in particular specialization, the growth of non-agrarian occupations and migration to towns, point to a stronger economic integration during these periods. As opposed to the effects of growth, periods of crisis were characterized by extensification, but also by de-integration.

One of the growth-periods, from the tenth to fourteenth centuries (the High Middle Ages), was probably the main formative period in the history of European landscapes. North-western Europe more or less exploded and people from this

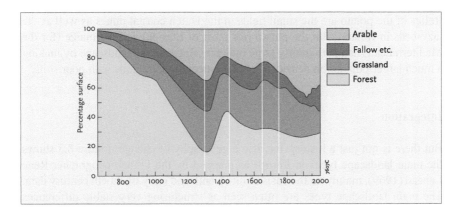

**Figure 8.2 The development of land use in Germany since the
seventh century AD as reconstructed by Bork et al.**

Source: Bork et al. 1998: 221.

region colonized regions that were, at least in most cases, thinly populated. From England, a first wave of settlers moved into Ireland, from France people moved southward to fill the open spaces in the lands that were conquered from the Muslims and from the German lands people moved eastward, following routes through the coastal marshes and through the mountain ranges and avoiding the much more populated lowlands. The westward and southward moving colonists quickly merged with the existing population, but many of the ethnic Germans in Eastern Europe kept their own language and identity and added to the ethnic mosaic that characterized Eastern Europe until the Second World War and in some regions even longer. In most of Central Europe, colonization, reclamation and urbanization led to a situation around 1300 in which almost all present towns and villages already existed and in which some regions were more populated than even today.

This period was followed by the crisis-period of the fourteenth and early fifteenth centuries. Arable was converted to grassland, fish-ponds and forest and thousands of settlements were deserted. On the remaining arable, the emphasis on grain gave way to a broader spectre of crops (Thirsk 1997). The remaining farmers usually had more land than before. Such was the degree of economic integration in Europe that these processes took place almost everywhere on the continent.

Later growth-periods were the 'long sixteenth century' and the period from the second half of the eighteenth century onwards. Again, new lands were reclaimed, in less densely populated parts of Europe, sometimes by immigrants from the core regions (see for example Klassen 2009). Other changes included the introduction of new techniques and new crops, the diminishing of fallow etc. Many of these changes occurred within an existing landscape structure. Some new crops brought direct as well as indirect changes. The potato had some direct effects on the landscape, as it was grown partly on places where no other crops survived.

Relics of the potato are the small fields in the Dutch coastal dunes as well as the lazy-beds in Ireland. But the potato (for most of Europe) and the maize (for the Mediterranean) had particularly large indirect effects on the landscape by making a much larger population density possible, especially in regions with poor soils.

Integration

But there is not just a history, but also a geography of change. Figure 8.3 shows the main landscape types in Europe as mapped by the French geographer René Lebeau (1969), mainly on the basis of nineteenth and early twentieth century data. The main landscape types are often seen as structuring very stable differences within the European landscape. In the large majority of landscape histories, it is taken for granted that most of our landscapes were founded in a certain period and, after that, remained more or less intact until the nineteenth or early twentieth centuries.

It is certainly true that many landscapes give an image of stability and may even look timeless. Everyone who takes the fast train from Paris to the west of France is struck by the almost sudden change of the landscape, from the huge grain fields of the Paris Basin, the open fields with their large nucleated villages hidden in the stream valleys, to the small-scale enclosed '*bocage*' landscape of mixed-farming in the West of France. This is perhaps the most fascinating landscape divide in Europe; yet, even this divide is perhaps much stronger now than in medieval periods. In the early Middle Ages these landscapes must have looked less different from each other, each consisting of small arable fields surrounded by large areas of pasture and common waste.

The map of landscape types does not show a stable jigsaw of the continent in regions with typical, traditional ways of life. All landscape types have their own history, which is often related to the functioning of regions within larger economic and cultural developments. The open fields started as small areas of permanent arable, probably around the ninth or tenth century. During the following centuries they developed into the agricultural core regions of Europe, from the twelfth century onwards supplying the growing towns. As such, they were already part of an integrated economic system. The bocage-landscapes were less specialized but were also, as suppliers of particularly animal products, integrated in the larger economy. These landscapes show a gradual development, in which ever more parts of the commons were enclosed and used for arable or intensive pasture.

The open fields had their heyday in the period around 1300, which was characterized by population pressure and an emphasis on the production of grain. The open field regions were the bread baskets of Europe at the time. Grain farmers grew rich and the central towns in the grain-producing regions, such as the Paris Basin, could afford to build huge cathedrals.

Later periods brought different developments that were again related to economic processes on a larger scale. From the fourteenth century onwards,

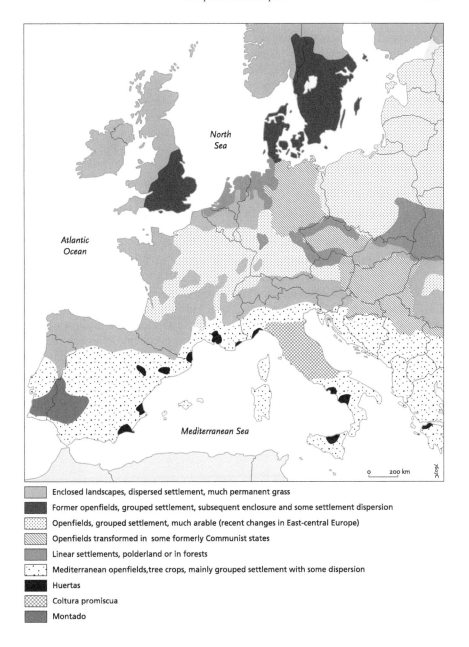

Enclosed landscapes, dispersed settlement, much permanent grass

Former openfields, grouped settlement, subsequent enclosure and some settlement dispersion

Openfields, grouped settlement, much arable (recent changes in East-central Europe)

Openfields transformed in some formerly Communist states

Linear settlements, polderland or in forests

Mediterranean openfields, tree crops, mainly grouped settlement with some dispersion

Huertas

Coltura promiscua

Montado

Figure 8.3 Landscape types after René Lebeau (1969)

Note: English version by H. Clout 1998.

Source: Reproduced by permission of Edward Arnold (Publishers) Ltd.

English open fields were turned into sheep pasture, supplying wool for the growing textile industry. Landowners who succeeded in buying most of an open field then evicted the tenants and 'enclosed' the whole village territory into one large sheep farm. When the demand for grain grew again during the sixteenth century, the production of grain moved eastward. During this period, open fields with a three-field system were introduced in parts of Russia, including parts of the Baltic States. These regions became integrated in the European economy as suppliers of grain and other products (wood) to the core regions in north-western Europe.

One typical aspect of European landscape research is the underestimated dynamics of the Early Modern Period. But when we look at the main landscape types, they all show changes in function, structure and location during the Early Modern Period. During this period, in particular the emerging European world-system, as described by Wallerstein, Braudel and others, was a strong force that led to a reconstruction of the European countryside. It was the late Hans-Jürgen Nitz, who succeeded in connecting the economic theories of Wallerstein and Braudel with developments in European landscapes (Nitz 1993).

Within Europe, the core regions show a surprising degree of continuity (Figure 8.4). There is a direct line from the dual-core Europe from the late Middle Ages to the present so-called Blue Banana. Nitz (Figure 8.5) succeeded in showing that already during the sixteenth and seventeenth centuries, many European regions were oriented on the then core-region in north-western Europe. It must be possible to make similar maps for Mediterranean Europe, focusing on Northern Italy and the town of Istanbul.

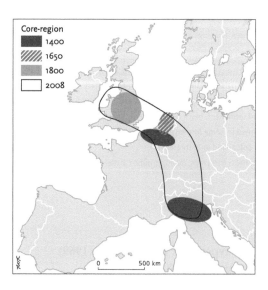

Figure 8.4 Core-regions in Europe since the Middle Ages

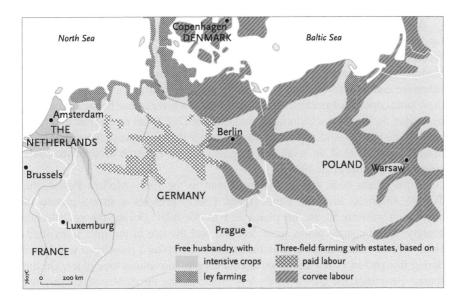

North Sea

Copenhagen
DENMARK

Baltic Sea

Amsterdam
THE
NETHERLANDS

Berlin

Brussels

POLAND Warsaw

GERMANY

Luxemburg

Prague

FRANCE

Free husbandry, with
intensive crops

ley farming

Three-field farming with estates, based on
paid labour

corvee labour

0   200 km

**Figure 8.5 Land use around the North Sea and the Baltic during
the seventeenth century**

Source: Redrawn from Von Thünen 'Zones in early modern northern Central Europe', in
Nitz, *The Early Modern World-System in Goegraphical Perspective*, Franz Steiner Verlag,
Stuttgart 1993.

In the northern half of Europe, the map shows regions with the most intensive agriculture in the core region. Here, a substantial part of the population lived in towns. Somewhere during the end of the sixteenth century, the county of Holland must have passed the point where more than half the population lived in towns of at least 2,500 inhabitants. Here, we can already speak of an urbanized landscape, with farmers working for an urban market.

But even for many regions in the periphery of Europe, the integration into the larger economic system meant a stronger market orientation and taking a role in the new system. For the eastern part of the Baltic, it meant the already mentioned specialization in grain production, whereby these regions could only compete in the European market by cutting costs. The same system that brought prosperity to family farms in the coastal plains around the North Sea stimulated the emergence of large estates with forced labour in parts of the Baltic. During the eighteenth and ninteenth centuries, the English enclosures became a more general procedure that was also used for arable lands, as part of the making of more efficient arable and mixed farming landscapes. The English example was followed by other countries, particularly Denmark (Figure 8.6), Sweden and some parts of Germany (Nitz 1992).

Also other landscapes showed changes. Recent research in the bocage-landscapes rewrote the history of these seemingly timeless landscapes. Early medieval sources give little evidence of hedges and give the impression of landscapes with small settlements, surrounded by some fields but mainly by extensive common forests and grazings. In the course of the last thousand years, these landscapes have gradually been filled with newly reclaimed fields, surrounded by hedges. In France the historian Annie Antoine (2002) has made clear that large parts of the bocage landscapes are much younger than we always thought and are in fact post-medieval. In southwest England, recent research brought the same conclusion (Turner et al. 2006).

In his recent book on the Mediterranean in the Early Modern Period, the Turkish/American historian Faruk Tabak (2008) presents a strong picture of alternating movements to the plains and to the hills. During the seventeenth century, many coastal plains were deserted and the use of the hills intensified. Here, typical mixed-farming systems such as the *coltura promiscua* developed during this period. A few centuries later, a reverse movement took place. On the Iberian Peninsula, the growth of the wine trade during the eighteenth century led to a strong growth of the planting of cork oaks, which were combined with pasture in the savanna-like landscapes that were called *montados* (Portugal) or *dehesa* (Spain). Many of the *dehesas* are now thought to date from the eighteenth to the early twentieth century (Plieninger 2004).

Even landlocked regions such as Hungary played a role in the Early Modern integrated Europe by exporting horses and cattle overland to the Rhineland. It is therefore good to realize that changes were not only taking place in the core regions. Partly even the opposite is true: the economies in the European core regions were characterized by a varied economic basis which, just like a rainforest in ecological terms, had a certain flexibility and therefore stability or, to use a better term, resilience. Peripheral regions, on the other hand, were often dependent on a single product, making them vulnerable to change. Small changes in the core regions could lead to fundamental changes in the periphery.

People and landscapes

Some developments, such as reclamation of new arable lands, lead to changes in the morphology of the landscape. Other developments, however, do not necessarily bring changes in landscape morphology, but nevertheless transform the functioning of landscapes and the lives of those in it.

A series of maps of a village in Mecklenburg-Vorpommern, northeast-Germany (Figure 8.7), shows the complexities of local historical developments. A medieval village with a three-field system developed into a landed estate with the estate farm exploiting the complete former village territory. The former free farmers were reduced to the status of farm labourers. A short-lived land-reform in 1946, which divided the estate, was quickly followed by collectivization. The collective

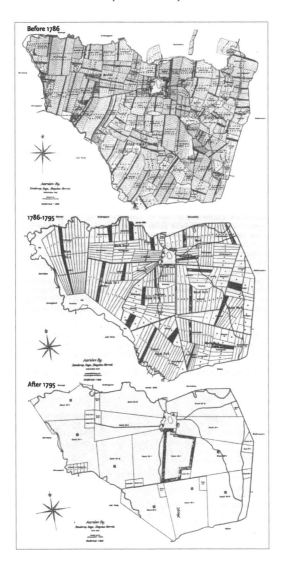

**Figure 8.6 The village of Aarslev in Denmark shows some of the
landscape changes in southern Scandinavia during the
Early Modern Period**

Note: The first map shows the village of Aarslev in 1786, with an open-field system. One farm owned 78 small strips. The reallotment of 1786, connected to a change in agrarian system (the introduction of the so-called *Koppelwirtschaft*), diminished the number of parcels; the same farm now had 21 strips, dispersed over the nine fields. Only nine years later, the common-field system was abolished and every farm received a single, enclosed field. On the edges of the field the local cottagers received small plots.

Source: Frandsen 1992: 195-196.

farm shows certain continuities with the old estate farm, especially where it used the same buildings. The manor house usually became the office of the local party-leader. After 1989 many of the collective farms were privatized, collapsed and were taken over by private persons, often from Western Germany. They now run the former village as a large private farm.

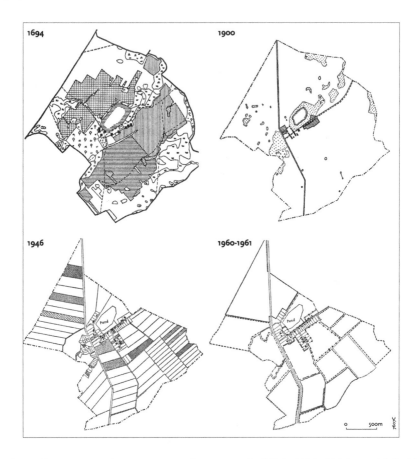

Figure 8.7 Stresow (Vorpommern, Germany) in 1694, 1900, 1946 and 1960-61
Source: Mayhew 1973: 196-197 (after Bethien).

So, the first transformation did transform the landscape. But since then, although ownership patterns, economic systems and political systems showed drastic changes, in the landscape structure much continuity can be distinguished, showing a degree of path dependency.

This brings me to another aspect: the relation between landscapes, landscape change and the inhabitants of that landscape. In old history writing and in modern planning, there is much nostalgic thinking about local people being heavily connected with their surrounding landscapes and therefore being the best suited for shaping the future of their landscapes. This idea of a long-standing local population as partner is, at least partly, romantic wishful thinking. The example of the estate shows that the connection of people with their landscapes can change quickly: a farm labourer could become a free farmer in 1946 only to find himself in the position of a labourer on a state farm a few years later. People living in a landscape they shaped themselves are the exception, not the rule.

Even more traumatic were the relations between people and their landscapes in many parts of Central and Eastern Europe as a result of the huge twentieth century population movements. Figure 8.8 shows the small town of Slavonice in the southern part of the Czech Republic.[1] For many centuries the small town was named Zlabings and was populated mainly by ethnic Germans. In 1938, after the Munich Agreement, the region was handed over to Nazi Germany. The Jewish population and the ethnic Czechs were evicted and the 'Unterer Platz' or Lower Square was renamed 'Adolf-Hitler-Platz'. In 1945 the German population in turn was evicted and the town was repopulated with Czechs, partly from nearby, partly from further away (part of the new empty spaces in the Czech mountains were repopulated with so-called Wolga Czech, the descendents of people who had migrated to Russia during the nineteenth century). The communists, who came to power in 1948, renamed the square 'Namesti Miru' or Peace Square. In 1953, after the Russians withdrew from Austria, the town became part of the border zone, just behind the Iron Curtain, and was more or less isolated. A few years later, the government decided that the built heritage was threatened by crisis and depopulation, and the town was taken out of the border zone. It more or less survived the rest of the communist period. Since the fall of communism, the town was discovered by artists, who settled there in some numbers. Now the town is a relatively quiet tourist attraction near a minor border crossing.

Stories like this can be told in many places in Central and Eastern Europe. The town of Gdánsk was rebuilt after the Second World War by and for a recently settled Polish population and the former Polish town of Lwów is now the Ukrainian town of Lviv and is inhabited and maintained by Ukrainians. These new inhabitants prove to be able to develop strong ties with a town that was built by other ethnic groups or even former enemies. This example shows that sometimes behind a heritage landscape lies a very troubled history.

This is not the only example of a traumatic history. Does Figure 8.9 show a traditional landscape? The west of Ireland is one of the iconic heritage landscapes of Europe, full of relics from every period in history and prehistory and even inhabited with some of the last people who still speak Irish. But again, it is a landscape with a troubled history. Two centuries ago, this landscape was full of

---

1   The example of Slavonice is based on general literature on the postwar history of the region (particularly Antikomplex 2007), extended with information from the local museum.

Figure 8.8 The town of Slavonice (Czech Republic)

Figure 8.9 Ruined church in Co. Connemara (Ireland)

people. Around 1840 Ireland had some 8 million inhabitants, as many as England. Now Ireland has five million, a tenth of the English population. The potato-crisis, but also the failure of nineteenth-century Ireland to industrialize, made this poor and overcrowded landscape into a poor and empty landscape. The ruined church is a relic of this history.

Research into long-term landscape history

To gain insights into these processes, the long-term history of the European landscape has to be studied. This type of research is time consuming and is only possible by interdisciplinary collaboration. There are precedents of such research. A landmark study was the Swedish Ystad project, where a region in South Sweden was the subject of an interdisciplinary project with a time scope of 6000 years (Berglund 1991).

In the Netherlands, during the last ten years the so-called 'biography'-concept (Roymans et al. 2009; Kolen 2005) was developed. The term 'biography of landscape' is used for the first time by the American geographer Marwyn Samuels in the well-known book *The Interpretation of Ordinary Landscapes* (1979). But Samuels used the term to apply for more research on agency, on the individuals who made the landscape. The true inspiration came from anthropologists, such as Kopytoff (Kopytoff 1986; Kolen 2005), who wrote on the biography of things, making clear that objects have their own life histories. Objects are handed over from one person to the other and from one generation to the next. In this process they are damaged and repaired, are lost and refound and thereby get ever different meanings. An object can start its life as a religious token, can then become a holiday souvenir, then become recognized as a piece of art and in the course of time can become a heritage object.

In the same way, landscapes are handed over from one generation to the next, being transformed and gaining as well as losing meanings. The researchers from different disciplines who work under the banner of 'biography of landscape' look at landscapes in this way. This means research that focuses on the long-term history of the landscape, sometimes covering thousands of years. It also means that attention is being given to the many different stories and narratives that different people and groups tell about the same landscape. It leads to the collection of oral history, to find out these different stories. It also means a new beginning for the study of field-names. These names have always been used to give insights in past landscapes and in landscape change. Now they are re-interpreted as sources for past perceptions of landscapes.

The long-term history of the landscape also includes the changing meaning of objects from the past. As the American geographer Donald Meinig once made clear: 'one aspect [of landscape] is so pervasive as to be easily overlooked: the powerful fact that life must be lived amidst that which was made before. Every landscape is an accumulation' (Meinig 1979: 44). Anyone who looks for it can find many examples of landscape features that have been re-used and re-interpreted by later generations.

The biographic approach may become an important tool, not only by its long-term perspective, but also because it has the potential to combine traditional research into landscape changes with new research questions from cultural geography and social sciences.

Disadvantages of the 'traditional landscapes model' for planning

So, the idea of traditional landscapes has disadvantages for historical research. It gives stories that are too simple, too one-dimensional and with a false impression of stability, and by doing that it fails to show the complex histories and the dark sides of landscape histories. Many landscapes have troubled, not to say traumatic histories.

But what does this mean for landscape planning and landscape management? One recent Dutch example may illustrate how chances are being missed when a simplistic vision of landscape history is applied. Recently the provincial government developed plans to build 3000 houses in Waterland, one of the recently designated National Landscapes. The new houses were to be built in a way that saves the rural landscape as much as possible (LA4SALE 2004). The landscape architect who was in charge of designing the possibilities started with a typology of settlement types, based on nineteenth century maps, and designed an extension plan for each of the types. In the example of the settlement of Ransdorp, recent topographical maps show a nucleated village and two rows of outlying farms. These farms were recently relocated from the village; therefore nucleated settlement was seen as characteristic.

The problem is the lack of historical perspective, let alone research. The whole plan completely fails to recognize the historic layers of the landscape. In this region one of the first large-scale landscape archaeological projects in the Netherlands took place during the 1980s (Bos 1986, 1988). It resulted in the discovery of a completely different settlement pattern (Figure 8.10). During the first reclamation phase, somewhere during the tenth or eleventh century, the strip-field pattern was laid out and each farmer built a house on a small dwelling mound within his field. The farms were standing more or less in a row.

In the course of time, agriculture became ever more difficult as the peatland subsided because of oxidation. The land gradually became too wet for arable and even animal husbandry became difficult. During the fourteenth and fifteenth centuries, a mixed economy existed, with the men working part of the year as sailors or fisherman. This developed into the main source of income and ever more people left the fields and settled near the church, where a nucleated village developed. The economic change was successful, and the village started to build a huge church. Then, at the end of the sixteenth century, the town of Amsterdam urged the sailors working for Amsterdam merchants to live in the town. The village shrank and the large church was left unfinished.

Figure 8.10 The village of Ransdorp on a topographical map of c. 1900, with the sites of medieval farms* and the location of new farms built since the 1960s

Note: * Bos 1986.

Gradually, drainage was improved and agriculture became more important again. During the second half of the twentieth century, a land consolidation project made it possible for farmers to build a new farm in the fields, away from the crowded village. By returning to the land, they in fact restored the medieval settlement-pattern.

A better historic analysis in cases like this does not automatically lead to a better plan. But it does lead to a more sophisticated discussion and it can bring original solutions that add new interesting stories to an already rich landscape.

Discussion

The history of European landscapes is characterized by periods of transformations and of relative stability. These developments were powered by economic and demographic, but also by political and cultural factors. Therefore, research in landscape history needs to be interdisciplinary. To get a real understanding of the processes behind landscape change, this type of research needs to take the long term into account.

A better knowledge of long-term landscape change is also necessary for landscape and heritage planning. A simple distinction between 'traditional' and 'modern' landscapes can only be used to protect the former and to develop the latter. But it is not enough to protect. We need more insights into the processes of landscape change and into the resilience of different landscape features. It is not the primary task of landscape historians to make landscape museums, although incidentally there is nothing wrong with a museum. But our primary task is to help in developing concepts for ways of modernization that combine a living landscape, a prosperous population, a high biodiversity and an interesting heritage.

References

Antikomplex. 2007. *Zmizelé Sudety/Das verschwundene Sudetenland.* 5th ed. Domažlice: Nakladatelství Českého Iesa.

Antoine, A. 2002. *Le paysage de l'historien; archéologie des bocages de l'ouest de la France à l'époque moderne.* Rennes: Presses Universitaires de Rennes.

Antrop, M. 1997. The concept of traditional landscapes as a base for landscape evaluation and planning. The example of Flanders Region. *Landscape and Urban Planning,* 38, 105-117.

Berglund, B.E. (ed.) 1991. *The cultural landscape during 6000 years in southern Sweden.* Copenhagen: Munksgaard (Ecological Bulletins 41).

Bork, H.-R., Bork, H., Dalchow, C., Faust, B., Piorr, H.-P. and Schatz, T. 1998. *Landschaftsentwicklung in Mitteleuropa.* Gotha/Stuttgart: Klett-Perthes.

Bos, J.M. 1986. Ransdorp in Waterland; de ruimtelijke ontwikkeling van een veennederzetting. *Historisch-Geografisch Tijdschrift,* 4, 1-5.

Bos, J.M. 1988. *Landinrichting en archeologie: Het bodemarchief van Waterland,* Amersfoort: ROB.

Boserup, E. 1965. *The conditions of agricultural growth; the economics of agrarian change under population pressure.* London: Allen & Unwin.

Calvo-Iglesias, M.S., Fra-Paleo, U. and Diaz-Varela, R.A. 2009. Changes in farming systems and population as drivers of land cover and landscape dynamics: The case of enclosed and semi-openfield systems in Northern Galicia (Spain). *Landscape and Urban Planning,* 90, 168-177.

Clout, H.D. 1998. The European countryside: Contested space, in *Modern Europe; place, culture and identity,* edited by B. Graham. London, Sydney, Auckland: Arnold, 287-309.

Frandsen, K.-E. 1992. When the land was sold; the sale of the Crown Estates in Denmark 1764-1774 and the impact of the sale on the rural landscape, in *The transition of the European rural landscape: Methodological issues and agrarian change 1770-1914,* edited by A. Verhoeve and J.A.J. Vervloet. Brussels: NFWO-FNRS, 190-202.

Grigg, D.B. 1980. *Population growth and agrarian change; an historical perspective.* Cambridge: Cambridge University Press (Cambridge Geographical Studies 13).

Grove, A.T. and Rackham, O. 2001. *The nature of Mediterranean Europe; an ecological history.* New Haven and London: Yale University Press.

Klassen, P.J. 2009. *Mennonites in Early Modern Poland and Prussia.* Baltimore: Johns Hopkins University Press.

Kolen, J. 2005. De biografie van het landschap; drie essays over landschap, geschiedenis en erfgoed. PhD thesis, Amsterdam: VU University.

Kopytoff, I. 1986. The cultural biography of things: Commodization as process, in *The social life of things; commodities in cultural perspective*, edited by A. Appadurai. Cambridge: Cambridge University Press, 64-91.

LA4SALE 2004. *Bouwen voor Waterland; verkenning naar de invulling van het regionaal woningbouwprogramma vanuit een landschappelijk en cultuurhistorisch perspectief.* Haarlem: Provincie Noord-Holland.

Lebeau, R. 1969. *Les grands types de structures agraires dans le monde.* Paris: Masson.

Mayhew, A. 1973. *Rural settlement and farming in Germany.* London: Batsford.

Meinig, D.W. 1979. The beholding eye; ten versions of the same scene, in *The interpretation of ordinary landscapes; geographical essays*, edited by D.W. Meinig. New York, Oxford: Oxford University Press, 33-48.

Nitz, H.-J. 1992. The temporal and spatial pattern of field reorganization in Europe (eighteenth and nineteenth centuries); a comparative overview, in *The transition of the European rural landscape: Methodological issues and agrarian change 1770-1914*, edited by A. Verhoeve and J.A.J. Vervloet. Brussels: NFWO-FNRS, 146-158.

Nitz, H.-J. 1993. The European world-system: A Von Thünen interpretation of its eastern continental sector, in *The early-modern world-system in geographical perspective*, edited by H.-J. Nitz. Stuttgart: Steiner (Erdkundliches Wissen 110), 62-83.

Olwig, K.R. 1993. Sexual cosmology: Nation and landscape at the conceptual interstices of nature and culture; or, what does landscape really mean?, in *Landscape; politics and perspectives*, edited by B. Bender. Providence, Oxford: Berg, 307-343.

Olwig, K.R. 1996. Recovering the Substantive Nature of Landscape. *Annals of the Association of American Geographers*, 86, 630-653.

Paracchini, M.L., Terres, J.-M., Petersen, J.-E. and Hoogeveen, Y. 2007. High nature value farmland and traditional agricultural landscapes; open opportunities in the development of rural areas, in *Europe's living landscapes; essays exploring our identity in the countryside*, edited by B. Pedroli et al. *Landscape Europe*, Zeist: Wageningen/KNNV.

Plieninger, T. 2004. Built to last? The continuity of holm oak (*quercus ilex*) regeneration in a traditional agroforestry system in Spain, in *Weidewälder, Wytweiden, Wässerwiesen – traditionelle Kulturlandschaft in Europa*, edited by W. Konold, A. Reinbolz and A. Yasui. Freiburg (Culterra; Schriftenreihe des Instituts für Landespflege 39), 5-62.

Roymans, N., Gerritsen, F., van der Heijden, C., Bosma, K. and Kolen, J. 2009. Landscape biography as research strategy: The case of the South Netherlands Project. *Landscape Research*, 34, 337-359.

Samuels, M.S. 1979. The biography of landscape; cause and culpability, in *The interpretation of ordinary landscapes: Geographical essays*, edited by D.W. Meinig. New York/Oxford: Oxford University Press, 51-88.

Tabak, F. 2008. *The waning of the Mediterranean, 1550-1870: A geohistorical approach*. Baltimore: Johns Hopkins University Press.

Taylor, C.C. 1972. The study of settlement patterns in pre-Saxon England, in *Man, settlement and urbanism*, edited by P.J. Ucko, R. Tringham and G.W. Dimbleby. London: Duckworth, 109-113.

Thirsk, J. 1997. *Alternative agriculture; a history from the Black Death to the present day*. Oxford: Oxford University Press.

Turner, S. (ed.) 2006. *Medieval Devon and Cornwall: Shaping an ancient countryside*. Macclesfield: Windgather (Landscapes of Britain).

Chapter 9
The Brioni Archipelago:
Functional Identity of a Historical Landscape

Ivančica Schrunk and Vlasta Begović

Introduction

Brioni (Brijuni in Croatian) is a small archipelago of 14 islands and islets in the northern Adriatic, along the western coast of Istria, outside the city of Pula, in Croatia. The entire archipelago has been a national park since 1983. The two largest islands, Veli Brijun and Mali Brijun, are the most 'valuable landscapes' for their cultural-historical and natural values, but also for the recreational and aesthetic values.

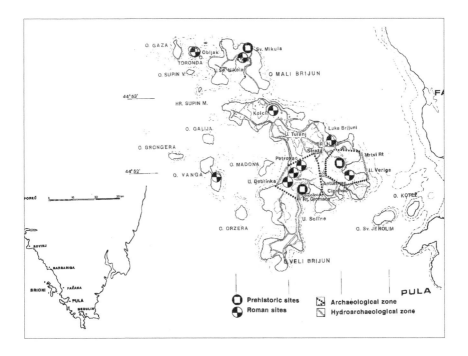

Figure 9.1 The Brioni archipelago with archaeological sites

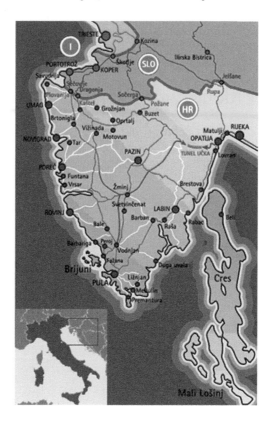

Figure 9.2 Map of the northern Adriatic region

There is also a particular spacial quality of Brioni's location between the eastern and western coast of the Adriatic, as well as on the Mediterranean edge of the sub-Alpine region. The network of these geographical relations shaped the Brioni landscape and seascape as a socio-cultural space throughout history. The archipelago has been on the major route for cultural and political interactions from prehistory to modern times.

The Brioni islands can be described in an ecological-geographical sense as a combination of two Braudelian categories of Mediterranean landscapes: low, marshy plain and islands. Both categories, as Paul Claval noted for such landscapes, historically had periods of prosperity and periods of decadence and isolation (Claval 2007a). For most of their history the largest two islands were rural landscapes with primarily agricultural character, except for the last 100 years. There is no vital agricultural production in recent times, considering the period since the Second World War.

The publication of the papers presented at the 21st PECSRL conference in 2004 (Roca et al. 2007) inspired us to examine Brioni's biography in the context

Figure 9.3 The Brioni archipelago, aerial view
Source: Photo by R. Kosinožić.

of this PECSRL session's themes of historical landscapes and identity. The story is one of shaping and reshaping of this landscape and seascape, as the relations between nature and society changed in cycles around the permanent axis of its geography. The permanence of the geographical axis is understood here only in a geo-ecological sense. Meanings and perception change axes, just as they change landscapes, when we take the semiotic approach to their study (Cosgrove 2003). We ask: which of these cycles – if not all of them – can provide an identity that may function in the present context of the national park? We do not have a definite answer, but will attempt to suggest some possibilities. By examining the long-term history of the cultural landscape, we first interpret how it was constructed and how identities were formed in reciprocal processes of change. We then attempt to apply these concepts to the practice of heritage management by experientially connecting past and present through the meanings and ideas of the landscape.

Cultural-historical landscape

Landscapes are, in their most basic definition, humanized and therefore perceived realities, if we may borrow P. Claval's expression. In geographical studies, landscapes synthesize geo-ecological factors, spatial patterns, scenic and aesthetical qualities, and social-cultural traditions. Archaeological and anthropological inquiries utilize a wide range of definitions for the concept of landscape. Considering the complexity of diachronic perceptions, which are conditioned at community or individual scale, archaeological landscapes may be interpreted as constructed, conceptualized and ideational (Knapp and Ashmore 1999). In a phenomenology approach, as C. Tilley (1994: 37) states, the landscape is an 'unstable' concept and human perception keeps redefining and recreating it. In this chapter our approach is integrated and our concern is with the issues of cultural heritage identity and management, which are best addressed with a guiding concept of cultural landscape. The concept of cultural landscape is very complex, as it depends on context and observer, and has many scholarly and working interpretations (comprehensive summary in Jones 2003; working definition in the World Heritage Convention, 2009 documents). Michael Jones provides a comprehensive summary of its usage and how it 'reflects different methodologies and serves different research agendas and power interests' (2003). Also in Palang and Fry's study (2003), cultural landscape is not a single definition but 'a set of interfaces between different approaches and understandings' (Semm and Palang 2004: 50, with other bibliography). From our point of interest in this chapter, the concept of cultural landscape has an important role 'in cultivating narratives of identity' (Moore and Whelen 2007: xi) and 'in understanding social consciousness through time' (Claval 2007b: 92).

The archaeological landscape of Brioni we have investigated is historically significant, and a unifying concept of cultural landscape is best suited for our positions as archaeologist and architectural historian. We use the term historical landscape in the title of this chapter in order to emphasize the significance of material remains of various periods. Therefore, here we do not distinguish conceptually between historic and cultural landscapes, although we acknowledge the difference, as Julie Riesenweber expressed it: that the former 'through the high degree of material integrity evoke some period or event in the past', while the latter are 'significant places in which some traces of the past endure yet undergo constant change' (Riesenweber 2008: 29). However, both of these definitions are equally applicable to Brioni. The issue of constant change is central to our discussion of the function and identity of a historical landscape. We do understand 'that landscapes are dynamic constructions that have been almost constantly modernized and adapted to new functions' (Vervloet 2007: 436). In our interpretation 'identity' is associated with function, which in turn is closely related to people's lifestyles and mentalities.

Long-term history and landscape transformations

The relationship between people and their environment was always dynamic and changing. The farther the historical view stretches back in time, the better we understand the social and cognitive/mental processes of cultural and natural land change over time. Our own study of the Brioni archipelago from prehistory to the present time has shown that the dynamics of change in the past have been underestimated in comparison to those in more recent times (Begović and Schrunk 2007). We fully agree with J. Renes' view of the 'false impression of stability' with regard to the landscapes of the past. There is indeed no justification in seeing modern cultural landscapes as dynamic, but 'traditional' ones, or archaeological ones for that matter, as stable (Renes 2008: 2). Long-term history as a research strategy in landscape studies, for both archaeology and geography, uncovers the true dynamics of material, functional and symbolic change.

The oldest visible element of the geo-ecological time depth of this landscape goes very far back. Literally, the first footprint was imprinted in the Mesozoic Era. A therapod or two lived in the space of the archipelago, which took the present form in the postglacial period. The footprint can be seen in the rocky shore of the Barban peninsula. The first evidence of human impact dates to the middle Neolithic, c. 4000 BC, when we would expect that the practice of agriculture, and probably salt collection, introduced the first significant modification of the landscape. Traces of huts were found beside the salt marsh in the bay of Saline. Nothing permanent is left in the landscape, only a handful of stone tools and fragments of pottery, which now dwell in the museumscape in Pula.

The 5000-year history of cultural layers begins with the Bronze Age and Iron Age – thirteenth – second century BC The first permanent markers still visible in the landscape were stone-built hill-forts of the Bronze and Iron Age inhabitants, known as Histrioi or Histri in the Greek and Roman sources. Their name was recorded only because they were pirates and difficult opponents. The peninsular region still bears the name Istria-Istra. These settlements have not been researched archaeologically, but there are many known parallels from the Istrian coastal territories (Suić 2003).

The first historical transformation – when events are known from texts – began with the Roman expansion in the early second century BC. Cultural and environmental consequences of the Roman conquest and settlement were profound and the changes long lasting. For the first time, there was no clear division between natural and cultural landscape in the archaeological record, so much was the natural landscape conditioned by human actions. Material culture dominated natural landscape. We may also see this change as the first modern transformation – in regard to the role that economy, lifestyle and mentality had in forming the landscape. Cash crops (olives and grapes) and intensive exploitation of natural resources (stone and salt) created wealth, which was used for extensive building in stone. The economic and social power permitted and demanded a luxury lifestyle and leisure. In this landscape of production and leisure, a Roman senatorial family built one of the most luxurious maritime villas in the Roman world in the Verige bay on the largest island (Schrunk and Begović 2000).

The established economic and political power structure continued through the Late Roman, Byzantine and early medieval periods, but lifestyle and mentalities changed with Christianity and militarization. New spirituality, insecurity, and the need for defence and fortifications added a new layer of material culture onto the landscape. An agglomerated, walled settlement – a Byzantine *castellum* in Madona Bay – housed the island population from the fifth to the sixteenth century. In Late Antiquity and the early Middle Ages Brioni also became a significant seascape marker of navigational routes in the Adriatic, the safest European communication network from the sixth to the ninth century. The prominent role of the seascape also reflected in the changed perception of the island's geo-ecological character.

Figure 9.4 Roman maritime villa, ideal reconstruction
Source: V. Begović and Z. Gregl.

Figure 9.5 Roman maritime villa, peristyle

The name of the largest island, and probably the settlement as well, changed into Brevona. The source of the meaning is the Latin word *brevis/breve*, presumably as the reference to the shallow sea in Madona Bay.

The Venetian rule begun in 1331 and for a century and a half continued the course of the lifestyles established in Late Antiquity. The new regime only built updated defences. The period of decadence and isolation, and of plagues and malaria, set in with the sixteenth century. In that period the *castellum* was abandoned and the population transferred to the north-eastern side, facing the Istrian coast. The coastal network, rather than the trans-Adriatic navigation, became more important. Spiritual sites were also moved to the newly developed settlement area. The first early Christian churches of St. Mary and St. Peter were abandoned and even new saints, St. Roch (protector against plagues) and St. German, became patrons of the new churches. The social process of the *castellum* site abandonment is archaeologically and historically poorly documented and still unclear. Bio-environmental reasons, namely the unhealthy marshy areas on the south side of the island, seemed to have caused the 'emptying' of the landscape and internal migration (Gazin-Schwartz 2008). Renes (2008) observed that the Mediterranean coastal wetlands become inhabitable when infested with malaria and other diseases, as the result of poor maintenance of their drainage systems.

The period of disease and isolation lasted for three centuries. The Austrian rule in the nineteenth century brought back Brioni's strategic role. Between the mid- and late nineteenth century two forts were built on Mali Brijun and five on Veli Brijun, one of which was one of the biggest forts in the Mediterranean. The Austrian forts were built on hill tops, like those in the Bronze and Iron Age. The regional warfare brought a change in the network of social relations, which in turn brought prosperity to Brioni, when it again became included in the system of regional defences.

The last major transformation of the Brioni landscape – which lasts to the present day – began in 1893, when an Austrian steel magnate, Paul Kupelwieser, bought the concession to develop the islands as an elite tourist resort. The general European enthusiasm for the Mediterranean, ancient/classical Mediterranean in particular, governed him in his projects. Brioni was an ideal Mediterranean leisure landscape, almost in the heartland of Austria. Leisure had to be coupled with usefulness and the land's aesthetic appeal, as in Roman times. There were also strong physical and mental connections with the English aristocratic landowners of the seventeenth-eighteenth century, who 'reaffirmed through the cultural achievements, best exemplified by the huge tracts of land they had beautified, the economic and social responsibilities they had' (Claval 2004: 30). Kupelwieser eliminated malarial marshes, renewed the old vineyards and olive groves and created an English landscape park. An Istrian forester composed the landscape with parks, forests and meadows. He even implanted tropical plants and animals. The building of elite villas and hotels followed.

Ancient ruins of Roman villas and small temples, explored by Anton Gnirs, an Austrian archaeologist, completed the Mediterranean character of a mythological landscape. This period lasts to the First Word War. Then, the landscape of leisure and luxury was once again, as in late Roman times, adapted to warfare efforts. The old Austrian forts received soldiers again and the hotels and villas accommodated officers and navy commanders. At the end of the war, Austria lost and Italy gained Brioni. Elite tourism resumed, more villas were built, and tennis courts and a golf course were added. Italian archaeologists carried on research of Roman sites.

In the post-Second World War settlement, Brioni went to Yugoslavia. The exclusive and safe islands offered an ideal setting for a presidential retreat and a state residence. Like the Roman senator in Antiquity, the Yugoslav president

Figure 9.6 Early twentieth-century hotels in Brioni harbor

Tito created his landscape of power and production with a secluded residence, stately villas, and with vineyards and tangerine groves. New tropical animals and plants were brought in, many as presidential gifts, to enhance or to naturalize the landscape. Even a new archaeological site – the abandoned Byzantine/ medieval *castellum* – was completely excavated next to the presidential residence. This political landscape was not a new phenomenon for Brioni, but the radical demographic change was the first in its history. All the remaining inhabitants on the big island were relocated to the coastal town of Fažana. This change is still in force to this day and there are no permanent residents on the islands. Those employed in the tourism economy or in the national park offices live in the coastal communities of Fažana and Pula. (Such a demographic collapse may have occurred 2,000 years earlier as the consequence of the Roman conquest, but we lack historical or archaeological evidence of it.)

After Tito's death in 1980, the political character of Brioni's cultural and natural landscape, then imbued with the memory of the charismatic leader, needed a new public identity. The solution was a National Park, created in 1983. National parks must have an appearance of 'wilderness' and the old political and economic causes of this 'landscape of clearance' justified the new reality for the government. The largest island opened to ordinary and elite tourism. It became a popular destination of daily tourist excursions and an occasional venue for state or scientific conferences. However, it also remained an exclusive summer resort for government officials, and a high-end vacation place with accommodation in old villas and two luxury hotels. Croatian independence brought no change, only legal protection from any development under the jurisdiction of four state ministries, because of Brioni's competing significance for natural and cultural heritage, tourism and state affairs. Interestingly, this administrative arrangement is evocative of Paul

Opdam's four mutually dependent 'constructs of the landscape': the ecophysical, social, economic and decision-making landscape (Opdam 2006: 54). Recently, a part of the historic landscape became an open-air theatre – literally. Theatrical performances are held in the nineteenth-century Austrian fort on Mali Brijun. This seasonal tourist attraction in the northern Adriatic parallels summer performances in two other World Heritage sites in Croatia, in the central and southern Adriatic, respectively Diocletian's palace in Split and in the old city of Dubrovnik.

Functional identity

In the biography of Brioni's landscape three key functions, which are also closely related to its identity, dominate throughout history: (1) productive; (2) defensive/ security; and (3) leisure and entertainment. In some periods – notably the Roman period – we find the combination of two or all three of them and see that those functions are interdependent. The exception is the period of the last 100 years when the third function predominates (although for President Tito security was also a concern). The productivity function has been completely neglected in recent times, as land use shifted from agriculture to recreation. This shift is not surprising in view of the general 'post-productivity phase' in the Mediterranean today, especially on the islands. Brioni is entirely a 'place of consumption' today (Pinto Correia 2008).

The Brioni islands are presented to the public today as a recreational resource in which cultural heritage plays an important role. The managed landscape consists of a collection of disconnected, individual monuments and natural attractions. Visitors cherish the aesthetic value of the nature, but do not understand the dynamics of cultural landscape change. It is not at all clear to them how and why the individual monuments are part of a spatial/historical framework. How do we integrate natural and cultural themes and offer the public an understanding of Brioni's story of changing relations between nature and society? How do we use the narrative capacities of the landscape? The past must have some meaning in our contemporary lives, if historic preservation is successful (Melnick 2008).

Assessment of historical/cultural landscapes should be the first step in defining both function and identity for the management objective, and therefore for the meaning in the present time. In assessing Brioni as a cultural landscape with a number of historical periods and narratives, we choose Jelier Vervloet's 'Atlantis' model of cultural stratification – multi-layered landscape. In his words, 'Atlantis refers to the situation in which a lot of elements of different periods are still visible at the surface in their original spacial distribution' (Vervloet 2007: 435). The idea of Atlantis does evoke a reality that has disappeared and surfaces only in the minds of the storytellers and listeners/observers. This parallel may further be extended to the roles of researchers/heritage managers and consumers/tourists.

Vervloet has recognized that the cultural landscape is a complex and changing reality. Considering the special complexity in assessing an Atlantis model of cultural landscape, he asked 'what could be the leading philosophy in this case: the age of the different relics, their variety, or the story these collections of material are telling us about the shaping and reshaping of a landscape?' (Vervloet 2007: 435). He seems to favour the third option and we find it convincing and applicable to Brioni. Vervloet is also concerned with temporal and spatial coherence. The respective goals, in his words, are 'to construct a continuous story of the transformation processes in the past, leading to the landscape in which we are living today' (Vervloet 2007: 435).

On Brioni, an important factor and signifier of identity in connecting the past and living landscape is lacking. A permanent population is not present in the contemporary landscape. We understand that 'population is a constitutive element of the landscape and thus of territorial identity', and that 'permanent population acts as an agent of change in landscapes and lifestyles' (Roca and Leitão 2008). The status of a national park institutionalized the 'emptied' landscape. Nature returned and overcame vineyards, olive groves and quarries. Visitors perceive the natural scenic areas on Brioni as original nature and generally do not know that the 'wilderness' is a recent ideological and political creation. National parks are often such landscapes, but mostly in colonial political contexts (Gazin-Schwartz 2008; Guernsey 2008). In the continuous story of change, we must 'try to understand the dialectic between the cultural and the natural world' (Stepenoff 2008). Today, the agents of change in Brioni's landscape and lifestyle are short-term visitors/ consumers and the government stakeholders. If we do not want to make Brioni an open-air museum of cultural and natural history, how do we then find a functional identity for it in the living context of the national park?

The Brioni archipelago is a complex heritage resource, but today it is treated as a material thing. The approach to its management is only structural, in which the archaeological sites and historic buildings, as well as nature, are interpreted and presented as physical features. They individually provide goods and services, but the present-day landscape as a whole lacks a defining identity to which people could relate the complexity of social experience in the past. Claval (2008) saw 'the problem of identities as a consequence of an underevaluation of the symbolic services provided by landscapes'. A phenomenological approach to archaeological landscapes, 'in which the past is understood and interpreted from a sensuous human scale', 'seems to be of most direct relevance for conceptualizing the complex lived experience of place and landscape in the past and the present' (Tilley 2004: xiii and 2). Such a humanistic interpretative approach, which focuses on the existential meaning of landscape, may in practice enable Brioni's historical landscape to provide symbolic services to the visitors.

Brioni is a unique place where we have this continuity of meaning and where visitors could have an experiential engagement with the landscape. The experience starts with present-day landscape. Brioni's current functional identity as recreational, leisure landscape preserves the idea, and even some forms, of the elitist landscapes of the past. The late nineteenth and early twentieth century elitist landscape is almost entirely visible and many of its amenities are currently in use. Understanding the 'lived experience' in Roman times may be achieved by observation and the encounter with the architecture of the maritime villa, still visible in its original spatial setting. Modern technology would have to be used to communicate reconstructions of the invisible architecture. However, historical identities were multiple and non-elites constructed landscapes as well. To experience the whole story, we may suggest a thematic presentation of the historical landscape in relation to all three functional identities we presented earlier. Paths would link the sites representative of the historical mentalities and lifestyles. These 'places' and 'spaces' tell us the story of human experience in constructing and reconstructing the landscape and the seascape until our own time.

This is our suggestion for a starting point to move the assessment and management of Brioni as a national heritage resource in a new direction. Both should be approached in an interdisciplinary way and at landscape scale, and should be based on the understanding and interpretation of the long-term history of the cultural landscape and of multiple identities. In presenting the story of reciprocal transformations of the landscape and identities, past and present may be connected experientially through the meanings and ideas of the landscape.

References

Ashmore, W. and Knapp, A.B. (eds) 1999. *Archaeologies of Landscape. Contemporary Perspectives*. Oxford: Blackwell.

Begović, V. and Schrunk, I. 2007. *The Brioni Islands. Past, Architecture and Cultural Heritage*. Zagreb: Golden Marketing-Tehnička Knjiga.

Claval, P. 2004. The Languages of rural landscapes, in *European Rural Landscapes: Persistence and Change in a Globalising Environment*, edited by H. Palang et al. Dordrecht: Kluwer Academic Publishers, 11-39.

Claval, P. 2007a. About Rural Landscapes: The Invention of the Mediterranean and the French School of Geography, in *European Landscapes and Lifestyles: The Mediterranean and Beyond*, edited by Z. Roca et al. Lisbon: Edições Universitárias Lusófonas, 15-31.

Claval, P. 2007b. Changing Conceptions of Heritage and Landscape, in *Heritage, Memory and the Politics of Identity: New Perspectives on the Cultural Landscape*, edited by N. Moore and Y. Whelan. Aldershot: Ashgate, 85-93.

Claval, P. 2008. Concluding keynote lecture, PECSRL 23rd session, paper, Obidos, 5 September 2008. [Online] Available at http://tercud.ulusofona.pt/PECSRL/Presentations/Paul%20Claval.pdf [accessed: 15 February 2010].

Cosgrove, D. 2003. Landscape: Ecology and Semiosis, in *Landscape Interfaces: Cultural Heritage in Changing Landscapes*, edited by H. Palang and G. Fry Dordrecht: Kluwer Academic Publishers, 15-21.

Gazin-Schwartz, A. 2008. Abandoned, Avoided, Expelled: The Creation of 'Empty' Landscapes, in *Landscapes of Clearance: Archaeological and Anthropological*

Perspectives, edited by A. Smith and A. Gazin-Schwartz. Walnut Creek, CA: Left Coast Press, 25-45.

Guernsey, B. 2008. Constructing the Wilderness and Clearing the Landscape: A Legacy of Colonialism in Northern British Columbia, in *Landscapes of Clearance: Archaeological and Anthropological Perspectives*, edited by A. Smith and A. Gazin-Schwartz. Walnut Creek, CA: Left Coast Press, 112-123.

Jones, M. 2003. The concept of cultural landscape: Discourse and narratives, in *Landscape Interfaces: Cultural Heritage in Changing Landscapes*, edited by H. Palang, and G. Fry. Dordrecht: Kluwer Academic Publishers, 21-51.

Knapp, A.B. and Ashmore, W. 1999. Archaeological Landscapes: Constructed, Conceptualized, Ideational, in *Archaeologies of Landscape. Contemporary Perspectives*, edited by W. Ashmore and A.B. Knapp. Oxford: Blackwell, 1-30.

Longstreth, R. (ed.) 2008. *Cultural Landscapes: Balancing Nature and Heritage in Preservation Practice*. Minneapolis and London: University of Minnesota Press.

Melnick, R.Z. 2008. Are We There Yet? Travels and Tribulations in the Cultural Landscape, in *Cultural Landscapes: Balancing Nature and Heritage in Preservation Practice*, edited by R. Longstreth. Minneapolis: University of Minnesota Press, 197-209.

Moore, N. and Whelan Y. (eds) 2007. *Heritage, Memory and the Politics of Identity: New Perspectives on the Cultural Landscape*. Farnham: Ashgate.

Opdam, P. 2006. Ecosystem methods: A spatial concept for integrative research and planning of landscapes, in *From Landscape Research to Landscape Planning. Aspects of Integration, Education and Application*, edited by B. Tress et al. New York: Springer, 51-66.

Palang, H. and Fry, G. 2003. Landscape Interfaces, in *Landscape Interfaces: Cultural Heritage in Changing Landscapes*, edited by H. Palang and G. Fry. Dordrecht: Kluwer Academic Publishers, 1-13.

Palang, H., Soovali, M., Antrop, M. and Setten, G. (eds) 2004. *European Rural Landscapes: Persistence and Change in a Globalising Environment*. Dordrecht: Kluwer Academic Publishers.

Pinto Correia, T. 2008. The Specificity of Mediterranean Landscapes Facing the Multifunctionality Challenge. PECSRL, 23rd Session, keynote lecture, Lisbon, 1 September 2008. [Online] Available at http://tercud.ulusofona.pt/PECSRL/Presentations/Teresa%20Pinto%20Correia.pdf [accessed: 15 February 2010].

Renes, J. 2008. European Landscapes: Continuity and change. PECSRL 23rd session, paper, Obidos, 4 September 2008. [Online] Available at http://tercud.ulusofona.pt/PECSRL/Presentations/Johannes%20Renes.pdf [accessed: 21 January 2010].

Riesenweber, J. 2008. Landscape Preservation and Cultural Geography, in *Cultural Landscapes: Balancing Nature and Heritage in Preservation Practice*, edited by R. Longstreth. Minneapolis: University of Minnesota Press, 23-34.

Roca, M.N.O. and Leitão, N.M. 2008. Population as an element of territorial identity. PECSRL 23rd Session, paper, Obidos, 4 September 2008. [Online] Available at http://tercud.ulusofona.pt/PECSRL/Presentations/POPULATION%20AS%20A%20CONSTITUTIVE%20ELEMENT%20OF%20TERRITOR IAL%20IDENTITY.pdf [accessed: 15 February 2010].

Roca Z., Spek, T., Terkenli, T., Plieninger, T. and Hochtl, F. (eds) 2007. *European Landscapes and Lifestyles: The Mediterranean and Beyond*. Lisbon: Edições Universitárias Lusófonas.

Schrunk, I. and Begović, V. 2000. Roman estates on the island of Brioni, Istria. *Journal of Roman Archaeology*, 13, 253-276.

Semm, K. and Palang, H. 2004. Life-ways in the Setu Cultural Landscape. *Pro Ethnologia*, 18, 49-67. [Online] Available at http://www.erm.ee/pdf/pro18/semm&palang.pdf [accessed: August 2009].

Smith, A. and Gazin-Schwartz, A. (eds) 2008. *Landscape of Clearance: Archaeological and Anthropological Perspectives*. Walnut Creek, CA: Left Coast Press.

Stepenoff, B. 2008. Wild Lands and Wonders: Preserving Nature and Culture in National Parks, in *Cultural Landscapes: Balancing Nature and Heritage in Preservation Practice*, edited by R. Longstreth. Minneapolis: University of Minnesota Press, 91-105.

Suić, M. 2003. *Antički grad na istočnom Jadranu*. 2nd revised and updated edition. Zagreb: Golden Marketing-Tehnička Knjiga.

Tilley, C. 1994. *A Phenomenology of Landscape*. Oxford: Berg Publishers.

Tilley, C. 2004. *The Materiality of Stone: Explorations in Landscape Phenomenology*. Oxford: Berg Publishers.

Tress, B., Tress, G., Fry, G. and Opdam, P. (eds) 2006. *From Landscape Research to Landscape Planning. Aspects of Integration, Education and Application*. New York: Springer.

Vervloet, J.A.J. 2007. Some Remarks about the Changing Position of Landscape Assessment, in *European Landscapes and Lifestyles: The Mediterranean and Beyond*, edited by Z. Roca et al. Lisbon: Edições Universitárias Lusófonas, 433-438.

World Heritage Convention, documents. 2009. [Online] Available at http://www.international.icomos.org/centre_documentation/bib/culturallandscapes.pdf [accessed: 14 February 2010].

Chapter 10

Expanding the European Landscape: Aqueducts and the Spanish Usurpation of México

William E. Doolittle

Introduction

From AD 1519 to 1821 that portion of North America extending between 7 and 47 degrees North latitude and between 80 and 125 degrees West latitude was claimed by Spain and considered by the crown as the *Virreinato de Nuevo España* (the Viceroyality of New Spain). Today, only a portion of this territory is considered 'New Spain' (Casagrande 1987). It encompasses that part of present-day México lying between 18 and 22 degrees North latitude (Figure 10.1).

Virreinato de Nueva España ca. 1800
Republic of México since 1848
"New Spain" today

Figure 10.1 Northern and southern limits of New Spain

This area is the most urban and densely populated part of México, containing cities with such Spanish names as Guadalajara, Veracruz, Cuernavaca, Puebla, León, and Aguascalientes. It was the area in which the Spanish crown concentrated its economic and cultural activities during the viceregal period, and where the indigenous civilizations of the Tarascans, the Aztecs, the Tlaxcalatecans, as well as smaller groups of people, such as the Otomi and the Totonoc, systematically lost their lands, identities, and lives, concomitantly with the increase in the number of settlers from Spain. Perhaps surprisingly, Spanish colonial immigrants did not find themselves in a foreign land. They were among foreign people, to be sure, but they were not in unfamiliar environments. With the exception of the seasonal rainfall regime (winter rains in the Mediterranean and summer rains in the New World), central México is environmentally similar to much of Spain, especially central, southern, and eastern Spain whence most of the colonist originated (Boyd-Bowman 1973). Indeed, the late, great anthropologist, Eric Wolf (1959: 199) noted that Spaniards adapted easily to the environments of México because they were 'sons of a dry land themselves'. He went on to claim 'they were master builders of aqueducts', and it is in this regard that they had what is argued in this chapter to be their greatest single and direct impact on transforming the visual appearance of the rural landscape of México, cementing in the eyes of the remaining native inhabitants that the land was no longer theirs.

Theoretical context

A great deal of recent research has focused on the role of landscapes in the construction of identities. As mental constructs, landscapes help people know who they are and from whom they differ (Claval 2007). In most cases, the visual appearance of the biophysical environment itself was the basis of identity. For example, the special qualities of the Alps created an image that German, French, Italian, and Romanche-speaking people could share in the building of the Swiss nation.

Geographers and anthropologists have long been aware of the tenacious links native peoples have with their lands, their environments (Bonnemaison 1981; Raison 1977). Similarly, they are aware of the negative consequences and the loss of identity when people are moved from their land. There are, perhaps, no better examples than British colonists pushing aborigines further and further into the middle of the Australian continent, with people often preferring death over migration (Unaipon 2001), or the North American settlers displacing indigenous people, often forcing them onto reservations (Sturgis 2007). In almost every case such as these, one group's loss of landscape and identity is replaced with another group's newly created identity on the landscape, or a newly created landscape to reflect the interloping group's identity.

Creating an identity on a new landscape, or creating a new landscape to reflect one's identity, typically involves building features that are similar to ones found in the former homeland. The operative here is *build*. When one thinks of built environments, one typically envisages towns or cities, or at least a complex or conglomeration of streets, edifices, and accompanying utilities. Built environments are nearly always urban (Rapoport 1990). Indeed, the term 'built environment' connotes a totality of transformation. Built environments reflect the culture of the builders, provide a tangible manifestation of the culture's presence, and hence contribute to the building of identity.

But this is only half the story in some cases. Built environments can be rural, territorial (Agnew 1999). They can also constitute messages or texts sent to people and cultures other than those responsible for their creation (Duncan 1990). In other words, they can be icons or imprints of authority to legitimate power (Cosgrove 2006) that, conversely, can symbolize loss of landscape identity (Haesbaert 1997). Long overlooked in studies of landscape and identity are surviving remnants of once larger native populations, reduced in number as foreigners intruded on their land but not displaced, and the features built by the culture usurping the landscape, the arena where social forces encounter one another (Mitchell 2000). In a sense, landscape usurpation can be envisaged as a new twist to the old idea of environmental conquest being such a major achievements that it becomes a federative theme (Olwig 1984). An excellent example of such rural landscape usurpation is the construction of aqueducts in México and, hence, the expansion of the European landscape to the New World.

Ancient aqueducts of México

Prior to the arrival of the Spaniards in 1519, aqueducts were constructed in two locales, both within the Basin of México. One aqueduct was built in the 1400s from a spring at the base of Chapultepec Hill across the narrow, western portion of Lake Texcoco to the island city and Aztec capital of Tenochititlán (Doolittle 1990: 120-127). It was built as a series of artificial islands connected with a conduit made of split and hollowed out tree trunks (Figure 10.2).

This aqueduct provided fresh drinking water to inhabitants surrounded by a brackish lake resulting from interior drainage. Rebuilt incorporating a masonry-supported canal on the islands, an earlier construction relying on an earth-supported canal was destroyed by flooding. With plank walkways paralleling the conduit/canal, this aqueduct doubled as a causeway.

The Chapultepec aqueduct was sufficiently important to the Aztecs that it appears on a map drawn in 1522, sent to Spain in the second letter of Hernán Cortés to the king, and published as a woodcut in 1524 (Cortés 2005). Indeed, this aqueduct was so important to the Aztecs that Cortés had it destroyed as part of his razing of the entire city. As is the case with so many conquerors who suffer from consequential myopia, his actions soon came back to haunt him. Deciding to build the capital of New Spain on the ruins of the fallen imperial city, Cortés was immediately faced with building a new water supply system, including an aqueduct. Construction

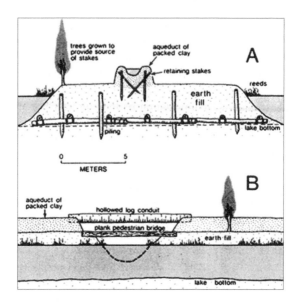

**Figure 10.2 Cross-sectional (A) and profile (B) diagrams of the first
Chapultepec aqueduct built in the 1940s**

Source: Reprinted with permission of the University of Texas Press.

of the aqueduct began in 1555, following the route of its Aztec counterpart, but involving a series of arches, clearly a Romanesque trait introduced from Spain. The spring dried up as the demands of the growing México City increased and the aqueduct eventually fell into disrepair. Today, only a portion of it remains in the median of Avenida Chapultepec, a relic of the colonial era (Figure 10.3).

Unlike this obviously urban construction, the other area in which the Aztecs built aqueducts was definitely rural. Far to the east, on the opposite side of Lake Texcoco, and in the foothills of the surrounding mountain ranges, the Aztecs tapped two springs, and transported water several kilometres via canals to cultivated and terraced hill slopes that constituted a royal pleasure garden. A new way to perceive and imagine landscapes, one that involved 'aesthetic feelings' emerged in the fifteenth century not only in western Europe (Berque 1995) but in the New World as well.

Maintaining the proper gradient of these canals required the construction of four aqueducts, spanning the saddles between adjacent hills (Doolittle 1990: 127-134). These aqueducts were all quite similar. They were 150-400 meters long, trapezoidal in cross-sectional shape, up to 40 meters wide at the base, two to three meters wide at the top, and more than 20 meters high. Construction was rock rubble and earthen fill with no mortar. The canals along the tops were made of a concretion laid atop a one meter high dry masonry rock wall with the chinks filled with irregularly shaped rock fragments (Figure 10.4).

**Figure 10.3 A remnant of the heavily remodelled Chapultepec
aqueduct today**

**Figure 10.4 The upper portion of an Aztec aqueduct built east of
Texcoco in the 1400s**

Given that these structures were not plastered over, vegetation colonized the
sloping sides of the aqueducts. From a distance, these aqueducts look almost
natural today. They blend into the landscape so naturally that only observers with

Figure 10.5 A distant view of the Aztec aqueduct shown in Figure 10.4

Note: The dashed line indicates half of its base, the interface between the aqueduct and the hill on the left.

keenly trained eyes are able to distinguish them from the hills they connect and the saddle they fill (Figure 10.5).

Although the myth of Native Americans making no environmental impacts has been soundly debunked (Denevan 1992), these aqueducts demonstrate that some monumental constructions were quite innocuous.

In addition to rebuilding the Chapultepec aqueduct, which they so stupidly destroyed, the Spaniards embarked on what can only be described as an aqueduct building frenzy that began in 1537. An on-going study has thus far identified 104 aqueducts built during the viceregal period (Table 10.1). Ninety-five, or 91 per cent of these, were located within the area referred to today as 'New Spain' (Figure 10.1).

**Table 10.1 Chronological and regional assessment of aqueduct
              construction in México**

Period of construction	Number	Within New Spain	Beyond New Spain
1537–1600	17	14	3
1601–1700	12	12	0
1701–1810	56	51	5
Yet to be dated	19	18	1
Total	104	95	9

Source: Doolittle 2009.

**Figure 10.6 The largest section of arches in the aqueduct built from
Cempoala to Otumba**

One of the first aqueducts built in México during the viceregal era is also one of the most impressive. Faced with water shortages at his mission in Otumba, 50 kilometres northeast of México City, the Franciscan priest Francisco de Tembleque oversaw the construction of an elaborate water delivery system that began at springs another 45 kilometres further to the northeast. Construction began ca. 1541 and ended *ca.* 1557. It required the building of a few small aqueducts to cross narrow gorges, but in three low-lying areas it involved monumental arcades. The largest of these is more than a kilometre long, 38 meters high, and has 67 arches.

Construction is *mamposteria* (mortared rock rubble) with ashlar quoins. Piers have two distinct levels, a wide lower level and narrower upper level, separated by a bevelled cornice. Spandrels are narrower yet with springers atop another bevelled cornice. The channel is flush with the spandrels, and shallow; the crown measures a mere 1.5 metres. The spans of the arches are six times wider than the thickness of the piers, and combined with the thinness of the crown, results in a most slender and graceful appearance. The remains of adobe centrings are still standing between some of the piers. Adobe is an ideal framing material in this timberless landscape. This aqueduct functioned for more than 200 years. It shows few signs of maintenance or repair which is surprising given its location in a tectonically active area (*Engineering News* 1888, Ramirez de Alba 1991: 26-28, 38-40) (Figure 10.6).

By any standard the construction of this aqueduct is a remarkable accomplishment. Tembleque, who had no training as an architect or engineer, became legendary among his contemporaries almost immediately. He was even brought in as a consultant to the México City council during its planning of the municipal water system (Kubler 1948: 117). And, herein lays a mystery. Where did Tembleque obtain his knowledge?

Born and raised in Castilla-La Mancha, south of Toledo (Valdes 1979: 15-240), Tembleque probably never saw an aqueduct until he entered the Franciscan order. Indeed, although the landscapes of several parts of Spain have remnants of aqueducts built in Roman times – 1,500 years earlier – only one of these aqueducts was still operating during Tembleque's lifetime. That aqueduct is the one in Segovia. Constructed of rusticated ashlar blocks, and having thick piers and narrow spans, this aqueduct is quite different from the one Tembleque designed. Despite these differences, it may have served as model given that it is an arcade and was carrying water at the time.

The architecture of Tembleque's aqueduct might have been Spanish, but the engineering was Aztec. The arch was unknown in the New World prior to the sixteenth century, but *mamposteria* was not. Furthermore, it is well-documented that Tembleque used labourers from Texcoco and Tenochititlán (Kubler 1944: 18), the two places in México where the Aztecs had built aqueducts a century earlier. In other words, the aqueduct credited to Tembleque is actually a hybrid of Old World architecture and New World engineering.

The relationship between native people and this aqueduct is complex. Although their contributions in terms of labor inputs is well known, far less recognized is their intellectual inputs. Upon hundreds of the stones that comprise this feature, and typically high up and under the arches, are plainly visible pictographs that are in effect the signatures of the individuals and groups of artisans who carried out the work (Garabay, K. 1961). These glyphs may not have been sanctioned by Tembleque, and indeed he might not have seen them until after the centrings were removed and it was too late to obliterate them. They remain today as silent reminders that as Spanish as this aqueduct may appear on the landscape, it could not have been constructed without the knowledge and labour of native people who no longer are identified with this land.

This aqueduct became a source of Spanish pride almost immediately after completion. Of all the maps made of different parts of New Spain in the sixteenth century, only one contains the image of an aqueduct, and it is the map included in the *Relación Geográfica de Cempoala* (Figure 10.7).

The aqueduct carrying water to Otumba is clearly visible in the southeast corner. Drawn by a native cartographer who incorporated many native icons to portray the landscape, the aqueduct is strikingly Spanish.

Finally, an image of this aqueduct created quite a sensation in the Mexican media during the summer of 2008. The secretary of tourism for the state of Hidalgo used the aqueduct as background in a promotional image that appeared on billboards, in newspapers, and on national television. Using the slogan, *Hidalgo en la píel* (Hidalgo, in the flesh), the central figure of the image is a very Spanish-looking nude female soap opera star painted in such a way that she blends into the aqueduct (see http://www.cronica.com.mx/nota.php?id_nota=357261). Surprisingly enough, the controversy did not revolve around the image of a nude woman, but rather the

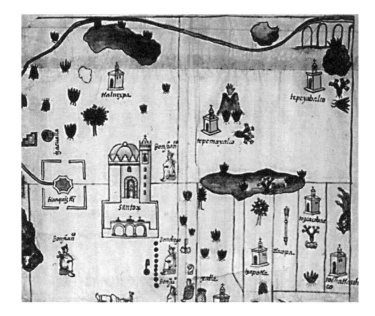

Figure 10.7 SE corner of the 1580 map of Cempoala, Hidalgo, México
Note: North is to the left.
Source: Reprinted with permission of the Nettie Lee Benson Latin American Collection, University of Texas Libraries, Austin.

use of the image of the aqueduct without the permission of the National Institute of Anthropology and History, the government agency charged with overseeing México's cultural heritage. Regardless, the image portrays the cultural landscape of México as anything but native or indigenous. It proudly portrays an image of New Spain and therefore the usurpation of the landscape.

The construction of aqueducts began in the sixteenth century, but it continued in the seventeenth and reached its zenith in the eighteenth century (Table 10.1). All of the seventeenth century constructions were concentrated in the area known today as 'New Spain' (Figure 10.1) and particularly in the vicinities of Cuernavaca, Puebla, and México City. A good example of one built on the rural landscape is that found just north of the city of San Miguel de Allende (Figure 10.8).

Details of construction remain to be unravelled, but a few things are evident by visual inspection. This aqueduct is smaller and less graceful than the one at Otumba. The spans of the arches are only about three times the width of the piers. Although constructed of *mamposteria*, the undersides of the arches are made of bricks set radially. There are small, almost imperceptible cornices at some springers, under the arches but not where the piers meet the spans. Fragments of stucco suggest that it was at one time completely plastered over. Based on appearances alone, one has to conclude that the skill of this aqueduct's designer and builders was not on par

Figure 10.8 Close-up view of the remains of a seventeenth century aqueduct near San Miguel, Guanajuato

with that of Otumba. It was on the far northern frontier of Spanish territory at the time and far from any source of skilled native craftsmen. Exactly when it was built is currently unknown; however, the seventeenth century seems most likely.

San Miguel was founded in 1543 but remained a very small place for a long time. A major irrigation system was constructed south of there some time before 1576, but the nearest masonry aqueduct was not built until 1637, further south yet in the city of Celeya (Murphy 1986: 14, 26). A map drawn in 1723 – curated in México's *Archivo General de la Nación*, section *Tierras*, vol. 258, exp. 4, f. 90 – shows both the town of San Miguel and the aqueduct, thereby documenting that construction was completed by that time (Wright Carr 1996).

Although small, this aqueduct was an important feature in regards to Spanish usurpation of the Mexican landscape. It did not simply supply water to either the town or to fields, but instead it carried water to a *batán* or fulling mill, a completely new form of technology in the New World, and one that accompanied sheep, a recently introduced species that overran the central Mexican countryside in a brief period of time. As evident on the map, the aqueduct was part of a complex system of water control. The system began with a masonry *presa* or dam built across a deeply incised gorge. Prior to the arrival of the Spaniards, their livestock, and their technology, this gorge was used by natives only as a place to fetch drinking water and harvest wild plants. The dam raised and impounded water, creating a reservoir. From there a canal carried water along the west side of the gorge a few hundred meters. In one place along its course water was turned out into a circular penstock that increased water pressure to turn the horizontal turbine of a *molina harina* or flour mill. Water not diverted into the penstock continued down canal another 50 meters. Here the canal turned sharply to the left and water flowed over the arcade to the mill, a large masonry building of which little remains today.

The final two aqueducts to represent usurpation of the Mexican landscape and an expansion of the European landscape are those associated with the Hacienda Xalpa and the community of Guadalupe. As such, the former represents construction on a large, private rural land holding, and the latter represents a public works project on the rural fringe of a city.

Xalpa is located on the far northwest of the Basin of México. The area remains quite rural after nearly 500 years. The main agricultural lands of the *hacienda* were on the basin floor, which is typically arid. Water was brought in through a very long canal system from the surrounding mountains to the west. At least two aqueducts are known to be associated with this system. One is near farm land and is rather small, transporting water over the narrow Río Cuatitlán near the town of Huehuetoca (Romero de Terreros 1949). The other, and by far the more impressive of the two, is known as 'Arcos del Sitio' (Figure 10.9).

Located 22 kilometres northeast of the town and Jesuit College of Tepotzotlán, this aqueduct has arches in four tiers. Construction began in 1764 with the Jesuits overseeing construction of the bottom two tiers (Ramírez de Alba 1991: 29) and completed under the direction of the master mason Librado Rivera (Romero de Terreros 1949: 124). This aqueduct is *mamposteria* construction with ashlar basaltic quoins. The bottom two tiers consist of two arches separated by a thick pier and flanked by wall-like abutments fitted against volcanic outcrops. The third tier has arches two to three times wider than the piers, with the piers being smaller than the top of the tier below. It is possible to walk around the piers on, and across the base of, the third tier. The piers on the third and fourth tiers were built in sections, each being narrower than the one below. Arches are tangent to the third tier, but those on the top tier are set back on cornices at each springer. The channel running along the top of the aqueduct is more than a meter wide and deep. Because of the decreasing width of the pier with height, the aqueduct takes on a more graceful appearance near the top, in contrast to the robust appearance of the base. By any standard, this aqueduct is a magnificent accomplishment in terms of both engineering and architecture. At 60 meters in height everything about it resonates power and authority, with a distinctive European flavor (Baxter 1934: 129).

México City early on became the seat of Spanish authority in New Spain. Surrounded by communities comprised mainly of indigenous people, one such community garnered a great deal of attention due to the alleged apparition known as Our Lady of Guadalupe. As the population continued to increase and the city expanded during the viceregal period, water shortages became greater. The solution to Guadalupe's water problem began in 1743 with the construction of an aqueduct that carried water from the Río Tlalnepantla several kilometers to the northwest. Traversing low-lying basin land, much of which was inundated seasonally with brackish water, the aqueduct is less than two meters high in most places, consists of approximately 2,300 arches, and was completed in 1751 (Romero de Terreros

Figure 10.9 Arcos del Sitio of the aqueduct of the Hacienda Xalpa
Note: The bottom tier, a pair of arches, is obscured by vegetation.

1949: 75-82). Remains of the nonfunctioning aqueduct are preserved over much of its course in the medians of some minor thoroughfares (Figure 10.10).

The longevity of this aqueduct was hindered by subsidence and tectonic activity. In many places it appears to run uphill while in other places it is fractured, dislocated, and shows signs of collapse and rebuilding. These flaws, of course, are a function of neither design nor construction, but external environmental factors. Built under the guidance of the accredited architect Domingo de Trespalacios y Escandón, this aqueduct was described at the time as being 'magnificent' (Romero de Terreros 1949: 77). Construction was principally *mamposteria* but in places where it turned, the junctions were made of elaborately carved ashlar blocks replete with ornamentation (Figure 10.11).

Although the aqueduct is well within the urban area of México City today, it was built in a rural setting. Indeed, the aqueduct is clearly visible in the center of the landscape painting titled *Valle de México desde el cerro del Tenayo* by José María Velasco in 1900, when the area was still very rural.

The Guadalupe aqueduct was built under the authority of the viceroy and had the weight of the Spanish government behind it. In other words, in addition to providing water to an indigenous community, it was a strong visual statement by the Spanish government as to who was in control of the land – the landscape – once the exclusive domain of the native people.

Figure 10.10 The Guadalupe aqueduct as it appears today in the median of a México City avenue

Conclusion

In his seminal comments prior to the 2008 PECSRL meeting, Paul Claval (2008a) noted that landscapes have traditionally been shaped by the oldest and generally still dominant cultures. This is not unlike the doctrine of first effective settlement (Zelinsky 1973). His notion that 'newcomers have no landscape of their own' does not pertain to the Spaniards in Viceregal México. These people encountered a landscape not unlike that of Spain, but one inhabited by vastly different people. Concomitantly with wholesale depopulation (Whitmore 1991), they stamped their own identity onto the rural Mexican landscape with the construction of aqueducts, thereby usurping it while sending a strong signal about power and authority to the surviving natives. In effect, Spaniards expanded their portion of the European landscape to México, and, although they didn't call it 'development', their actions fall firmly under that rubric by today's standards.

In his closing address to the 2008 PECSRL meeting, Professor Claval (2008b) reminded participants that landscapes once conceived as stable realities are now recognized as never having been stable, and that a new landscape perspective is emerging, one in which the identity crisis of modern society is in part linked with the destruction and disruption of traditional landscapes. The rural landscapes of much of our world are currently being transformed, perhaps not for the better. Such transformations, however, have been occurring as long as people have been invading other peoples' territories. Spanish aqueducts in México stand as but one example.

**Figure 10.11 One of the many bends in the long Guadalupe aqueduct
augmented with distinctive Spanish ornamentation**

Acknowledgements

I thank Matt Fry and Matt LaFevor for their assistance during some of the fieldwork
and for reading earlier versions of this manuscript. This chapter is dedicated to the
late Denis Cosgrove, scholar and friend, who never saw a landscape he did not
find interesting.

References

Agnew, J. 1999. Regions on the mind does not equal regions of the mind. *Progress in Human Geography*, 23, 101-110.

Baxter, S. 1934. *La arquitectura hispano colonial en Mexico.* México: Departamento de Bellas Artes.

Berque, A. 1995. *Les raisons du paysage: De la Chine antique aux environnements de synthèse.* Paris: Hazan.

Bonnemaison, J. 1981. Voyage autour du territoire. *L'espace géographique*, 10, 249-262.

Boyd-Bowman, P. 1973. *Patterns of Spanish emigration to the New World (1493-1580).* Buffalo: State University of New York at Buffalo Council on International Studies.

Casagrande, L.B. 1987. The five nations of México. *Focus*, 37, 1, 2-9.

Claval, P. 2007. About rural landscapes: The invention of the Mediterranean and the French school of geography. *Die Erde*, 138, 7-23.

Claval, P. 2008a. The idea of landscape. [Online] Available at http://tercud. ulusofona.pt/PECSRL/Presentations/IDENTERRA_Idea_of_landscape.pdf [accessed: 13 January 2010].

Claval, P. 2008b. Conclusions and Implications of the PECSRL Meeting, Unpublished paper presented at the 2008 PECSRL meeting in Obidos, Portugal.

Cortés, H. 2005. *Five Letters 1519-1526*. London: RoutledgeCurzon.

Cosgrove, D. 2006. *Geographical imagination and the authority of images*. Stuttgart: Franz Steiner.

Denevan, W.M. 1992. The pristine myth: The landscape of the Americas in 1492. *Annals of the Association of American Geographers*, 82, 369-385.

Doolittle, W.E. 1990. *Canal irrigation in prehistoric Mexico: The sequence of technological change*. Austin: University of Texas Press.

Doolittle, W.E. 2009. Cartographic chronology of aqueduct construction in viceregal México. Unpublished paper presented at the annual meeting of the Association of American Geographers, Las Vegas.

Duncan, J. 1990. *The city as a text: The politics of landscape interpretation in the Kandyan Kingdom*. Cambridge: Cambridge University Press.

Engineering News. 1888. The aqueduct of Zempoala, México, 7 July 1888.

Garabay K., A.M. 1961. Glifos en los arcos de Otumba, in *El Padre Tembleque* edited by O. Valdes. México: Bibliotecas Enciclopédica del Estado de México, 171-184.

Haesbaert, R. 1997. *Des-territorialização e identidade: A rede 'Gaúcha' no nordeste*. Niteroi: Editora da Universidade Federal Fluminense.

Kubler, G. 1944. Architects and builders in Mexico, 1521-1550. *Journal of the Warburg and Courtland Institutes*, 7, 7-19.

Kubler, G. 1948. *Mexican architecture of the sixteenth century*. New Haven: Yale University Press.

Mitchell, D. 2000. *Cultural geography: A critical introduction*. Oxford: Blackwell.

Murphy, M.E. 1986. *Irrigation in the Bajío region of colonial Mexico*. Boulder: Westview Press, Dellplain Latin American Studies 19.

Olwig, K. 1984. *Nature's ideological landscape*. London: George Allen and Unwin.

Raison, J.-P. 1977. Perception et réalisation de l'espace dans la société. *Annales E.S.C.*, 3, 412-432.

Ramírez de Alba, H. 1991. *La construcción en el Estado de México: Us studio técnico con referencia histórica*. Toluca: Gobierno del Estado de México, Secretaría de Desarrollo Urbano y Obras Públicas.

Rapoport, A. 1990. *The meaning of the built environment: A nonverbal communication approach*. Tucson: University of Arizona Press.

Romero de Terreros, M. 1949. *Los aqueductos de México en la historia y en la arte*. México: Universidad Nacional Autónoma de México, Instituto de Investigaciones Estéticas.

Sturgis, A.H. 2007. *The trail of tears and Indian removal.* Westport: Greenwood Press.

Unaipon, D. 2001. *Legendary tales of the Australian aborigines.* Victoria: Miegunyah Press.

Valdes, O. 1979. *El Padre Tembleque.* México: Biblioteca Enciclopedica del Estado de México.

Whitmore, T.M. 1991. A simulation of the sixteenth-century population collapse in the Basin of México. *Annals of the Association of American Geographers,* 81, 464-487.

Wolf, E. 1959. *Sons of the shaking earth.* Chicago: University of Chicago Press.

Wright Carr, D.C. 1996. Mapas, planos e ilustraciones de Guanajuato y Querétaro en el Archvo General de la Nación. *Lince,* 4, 3-9.

Zelinsky, W. 1973. *The cultural geography of the United States.* Englewood Cliffs: Prentice-Hall.

Chapter 11

Uncovering a Past Landscape: Rio de Janeiro in the Seventeenth Century[1]

Mauricio de Almeida Abreu

Introduction

Given the importance of sugarcane complex in the first centuries of Brazilian colonization, it is paradoxical that information about it is currently so scarce. The fact is that the data we have – whether about production, prices, uses of the slave and free labour – are very fragmentary and never constitute historically prolonged series. Therefore, those who venture into the study of the old world of sugar are required to be very creative in dealing with these issues. Even the French historian Fréderic Mauro (1989) and the Brazilianist Stuart Schwartz (1988), who studied the precious records of the Jesuit sugarcane plantations in Bahia, in particular the *Engenho* of Sergipe do Conde, were forced to make interpolations and conjectures to fill the gaps of the available documents.[2]

The problem is even greater in Rio de Janeiro. Although the *engenhos* in Rio de Janeiro kept records detailing production, revenues, expenditure and payment of tithes, as shown in some wills and inventories, the truth is that the documentation simply disappeared.[3] This lack of knowledge, in turn, becomes even more serious

1   This research has the support of the Conselho Nacional de Desenvolvimento Científico e Tecnológico-CNPq and of the Fundação Carlos Chagas Filho/Fundação de Amparo à Pesquisa do Estado do Rio de Janeiro-FAPERJ.

2   In Brazil all big sugarcane plantations were called *engenhos*. The word *engenho* in this context actually means mill, but it is also used to refer to the whole complex involving the sugarcane plantation itself, and all lands of an estate with a mill for processing the sugarcane and produce sugar. For this reason we will keep the Portuguese word *engenho* for all estates with plantations and mills. Owners of those *engenhos* were called *senhores de engenho* (masters of mills). Those who used to plant sugarcane but did not process it were called planter, never *senhores*, and could be owners of estates or have only plots of different origins.

3   It is known that until 1644 there was no requirement for written accounting, as that year can be found on the letter in which the *Provedor da Fazenda* (royal authority in charge of the royal financial control) of Rio de Janeiro, Francisco da Costa Barros, suggested to the Crown to require the owners of *engenhos* (*senhores de engenho*) 'to keep production books, with numbered and signed sheets, for greater control of the collection of tithes'. We do not know, however, when this practice actually began. See AHU-RJ, Box 2, No. 42.

as we know, from coeval authors or even modern historians who managed to access the information saved from destruction, that sugarcane production had increasing importance in the economic and social life of the captaincy in the seventeenth century. The Brazilian-born Friar Vicente do Salvador (1564-1635), who wrote the first history of Brazil, reported that Rio de Janeiro, which previously 'dealt more with flour for Angola than with sugar', had, in 1627, 40 *engenhos* (1982: 334). The British historian C.R. Boxer, in turn demonstrated that from 1638 to 1642 an average 20 to 25 caravels laden with sugar left Guanabara Bay heading to Portugal every year, attesting to the full integration of the captaincy into the large sugarcane plantation complex (1973: 173). On praising the greatness of Brazil for its drugs and mines, the Jesuit Antonil, who visited the *Engenho* of Sergipe do Conde, also pointed to the sugar prosperity of Rio de Janeiro by stating that in 1711 the captaincy had 136 *engenhos* (1982: 140).

Figures provided by Friar Vicente do Salvador and Antonil are good indicators of the increasing importance of Rio de Janeiro in the colonial sugar scenario, and are mandatory citations by all those approaching the seventeenth century. The truth, however, is that apart from the obvious acknowledgement of the growth of the sugarcane complex in Rio de Janeiro, very little advance has been made in generating new knowledge on the subject. Certainly the lack of information has contributed to this fact. In 1790 a big fire destroyed the building of the local Municipal Council that used to house the majority of the early documents of the local agencies of royal administration and corresponding responses from local authorities, including the control and commercialization of the sugar production. However, it is also true that information still survives and can therefore shed light on the old world of sugar. It should be acknowledged, however, that access to these records is precarious, either on account of their poor condition or because of their dispersion in different document collections.

The difficulties are similar regarding pictorial representations of sixteenth century Rio: the earliest image we know of the city's landscape was made by French traveller François Froger in 1695; the first urban map is that by Brigadier Massé, from 1713, also a Frenchman. Actually the first trustworthy map came up with the 'Topographic Map of the Captaincy of Rio de Janeiro' by Manoel Vieira Leão in 1767. The latter map is also the earliest cartographic document we have of Rio de Janeiro´s sugarcane complex and indicates the *engenhos* that were in operation at the time. However, it serves us little when we discuss the seventeenth century: there is no guarantee that the *engenhos* represented in 1767 were those that existed in mid-eighteenth century and the map tells nothing about past *engenhos*. Therefore, if we are to advance in the knowledge about the early days of sugar production in Rio, we need not only to be creative in the treatment of the documentation that still exists, but also to produce our own cartographic representations.

In this chapter we intend to shed a little more light on this still poorly known rural Rio de Janeiro which existed before the eighteenth century. Supported by a thorough analysis of sources which included all notaries volumes that survived

Figure 11.1 A conjectural image of an engenho in Rio de Janeiro in the seventeenth century (drawing, 2009)
Source: Reprinted by permission of Francisco Viana.

time, and other documents scattered throughout several archives, we have been able not only to go back to that distant past, but also to identify and locate the *engenhos* that existed then; to name their owners; and to rescue, albeit minimally, the role played by *senhores*, planters, and slaves at the time. We have also been able to track down the trajectory of the *engenhos* through time, thus allowing the incorporation of a diachronic approach into the analysis. We must recognize, however, that little progress has been made regarding the quantitative ascertainment of sugar production.

Due to the limitations of the available sources, we adopted a variety of methodological procedures in the course of the research. The empirical analysis presented here addresses solely the identification and location of the *engenhos* of Rio de Janeiro during the first two centuries of colonization.

A brief context of sixteenth-century economy

Several authors who analysed sixteenth-century Brazil (see Godinho 1953; Ferlini 2003) pointed to the existence of four different economic scenarios. The first of these, very favourable to the development of sugarcane production, began in the mid sixteenth century and reached the third decade of the seventeenth century; it was a time when the price of sugar tended to remain at relatively high levels, which stimulated the growth of sugarcane production in the Brazilian captaincies, in particular Pernambuco and Bahia. This favourable scenario was followed by a time of transition that showed fluctuations in the price of sugar and lasted until the 1650s. Then began a period of great economic difficulties, or even acute crisis, which was mainly characterized by a sharp fall in the price of sugar and by the intensification of taxation of Brazilian captaincies, whose vast contribution was called upon to help meet the commitments made by the Portuguese Crown with England and Holland. This time of difficulty lasted until the early 1690s, which, in turn, marked the beginning of a period of resumption of prices and growth of sugarcane crops reaching as far as the eighteenth century.

While the explanations of these situations vary among authors, there is some agreement linking the instabilities that started in the 1630s to a series of political and economic events that affected both Europe and Brazil. Among these the conflict between Spain and the United Provinces stands out, with the consequent Dutch occupation of Pernambuco (1630-1654), which, although it stimulated sugar production in faraway captaincies such as Rio de Janeiro, led to the loss of many ships carrying sugar to Portugal and required, moreover, a remarkable effort of fortification of the Atlantic coast, which the Portuguese Crown could only afford by imposing new taxes. Secondly, the Portuguese restoration of 1640, with the consequent state of belligerency that developed in the Iberian Peninsula until 1668, not only exhausted the royal treasury but also stopped the profitable illegal trade with Buenos Aires and Potosi; also, it determined that the needs of the colony were to be provided for largely by the colony itself. Thirdly, the taking of Luanda by the Dutch in 1641 cut the supply of African slaves to Brazil, leaving Brazilian captaincies virtually isolated from their main source of labour supply until 1648, when Angola was re-conquered by an expedition led by Salvador Correia de Sá e Benevides, largely funded by Rio de Janeiro's *senhores de engenho*. Fourthly, the entrance of the Antilles into the sugar market starting in 1650 not only led to the loss of important European consumers once supplied by Brazilian sugar but also increased the demand for slave labour and dropped the price of sugar. Finally, in order to ensure income to metropolitan traders and protect the transportation of colonial sugar to Portugal, the Crown issued, at the end of the 1640s, several rules that reorganized the trade with the colony and ultimately increased the latter's difficulties.

There is disagreement in the historiography, however, about the magnitude and extent of the unfavourable circumstances prevailing in the second half of the seventeenth century. Most authors equate them to a period of general crisis in agriculture, with repercussions in Europe and Brazil. Sampaio (2000), however, recently argued that this crisis had been restricted to sugarcane production and had not lasted nearly as long as it is usually claimed. Without denying the importance of export crops for the colonial economy, the author argues that the economic life of the captaincies had some autonomy, which grew over time.[4] In the case of Rio

---

4   The argument that the colonial economy cannot be explained exclusively by external factors is advocated by the Brazilian historians like Joao Fragoso and Manolo Florentino. See Fragoso 1992; Fragoso and Florentino 1993; Florentino 1991.

de Janeiro, this relative autonomy would have started as early as the seventeenth century when a market economy for foodstuff was taking its first steps, easing the effects of the 'great economic crisis of the second half of the seventeenth century', which Sampaio (2000: 23) believes to have affected the colony less than is usually believed; Sampaio also limits its detrimental effects on Rio de Janeiro to the 1660s and 1670s.

Questions that remain unanswered

Many questions need further clarification if we are to better understand how these old sugarcane plantations developed, namely: Is it possible to go beyond the total reported by Friar Vicente do Salvador and Antonil and effectively demonstrate how the growth of the *engenhos* in seventeenth century Rio de Janeiro took place, identifying rhythms and trends? What was the size of those estates, plantations and mills? Who were their owners and what social relationships did they sustain within the captaincy? What was the importance of sugarcane planters with no mills in the total of sugar production? What about free and freed workers employed in the production process? What about sugar production? Is it true, as stated by some authors, that Rio de Janeiro's mills specialized more in the production of spirits, used in the slave trade with Angola, than in the production of sugar? Was there, indeed, a great economic crisis in the second half of the seventeenth century as some authors maintain, or would it have been much less severe, as do others?

Other questions that remain unanswered regard the spatial dimension of these processes: Where were Rio de Janeiro's *engenhos* located? Did they form clearly identified production areas? What was the participation rate of these producing areas in the regional economy? How did the construction of the agrarian landscape of Rio de Janeiro in the first days of colonization take place? What environmental impacts did the plantations cause and what social relationships maintained them? To which extent were both city and countryside affected by the demands and rhythm of the sugar economy?

Due to the lack of documentary sources, many of the issues raised above will never be satisfactorily answered. Even so, it is imperative to discover a little more of what the first two centuries of colonization were like, because only then will we be able to evaluate what the sugar complex in Rio de Janeiro meant in the colonial context as a whole and what role it played in the structuring of the economy of the captaincy and in the daily lives of its inhabitants. We must, in short, dive deeper into that distant past and draw from it the information we need to understand the city's history and geography.

To fulfil this difficult task we can rely not only on the documentary sources already processed by other authors – which need to be confronted with others so as to produce multiplying effects – but also on the huge database that we have been able to build on Rio de Janeiro of the sixteenth and seventeenth centuries, the result itself of extensive research conducted over twelve years of work in the archives of Brazil, Portugal and the Vatican.[5] Such research allowed us to find sparse pieces of information, fragments of which we were able to put together; we must concede, however, that we were not totally able to complete the puzzle.

In this chapter, we discuss only two issues raised by the database: (1) is it possible to go beyond the total reported by Friar Vicente do Salvador and Antonil and demonstrate, effectively, how the growth of the *engenhos* in seventeenth-century Rio de Janeiro occurred, identifying rhythms and trends? and (2) Where were Rio de Janeiro's *engenhos* located?

The construction of Rio de Janeiro's database of *engenhos*

Academic production on colonial Brazil has grown considerably lately. Regarding Rio de Janeiro, this research effort has resulted in works of high quality, which have expanded considerably our knowledge about the city and the captaincy.[6] It should be acknowledged, however, that all this effort has privileged especially the eighteenth century. The seventeenth century is still hidden by mists that urgently need to be eliminated – or at least partially dissipated – if we are to get a complete picture of the historical and territorial formation of Rio de Janeiro.

Fragoso has been an exception to that rule, as he has contributed much to the understanding of the constitution of Rio de Janeiro's society and economy in the seventeenth century (see Fragoso 2000, 2001). However, despite the wealth of proposals offered by this author – and also by Sampaio, as already noted – many questions remain unanswered or need further clarification.

Rio de Janeiro's database of *engenhos* consists of several information-aggregating individual records. Each of these records in its final form corresponds to an *engenho* that we managed to identify and was the result of a long and painstaking process of grouping data scattered in time and in different documents and archives.

The work began with the decision that information found in the documents would be the basis for each single record, each record corresponding to a singular *engenho*. Let us consider the example of the purchase of a sugarcane plot (*partido de cana*) located in Irajá in 1664, which was part of a non-identified *engenho*. Initially the information about this *engenho* generated a record under the key

---

5   In Brazil, the survey was conducted in the following archives: Arquivo Nacional, Biblioteca Nacional, Arquivo Geral da Cidade do Rio de Janeiro, Instituto Histórico e Geográfico Brasileiro. Arquivo da Cúria Metropolitana do Rio de Janeiro, Santa Casa de Misericórdia of Rio de Janeiro, Arquivo Público do Estado do Rio de Janeiro, and Arquivo Publico do Estado de São Paulo. In Portugal, the survey was conducted at the Arquivo Histórico Ultramarino, and at the Arquivo Nacional da Torre do Tombo. Consultations in the Vatican Archive were limited to the Archivum Romanum Societatis Iesu (Society of Jesus).

6   See, among other contributions, Bicalho 1997; Soares 1997; Cavalcanti 1997; Almeida 2000.

'*Engenho* in Irajá, 1664'. On the other hand, records from a public auction which took place in 1683 indicated the purchase of an *engenho* with no declared location, denominated Nossa Senhora do Rosário. An additional piece of loose information originated a third record '*Engenho* Nossa Senhora do Rosário, 1683'. Gradually, however, it was possible to detect through several indicators that many of these loose pieces of information referred to the same *engenho* and we could put together all information relating to this *engenho* in the same record. This happened, for instance, with the above-mentioned *engenho* in Irajá. The reassembling of loose information allowed us to discover that the *engenho* of Nossa Senhora do Rosário located in Irajá, was already standing in 1664 when it belonged to so-and-so, and continued to exist in 1683, the year it was purchased by such-and-such person in a public auction. This grouping of information into a single record allowed us, moreover, to introduce diachronic data into the analysis and to recover the trajectory of the *engenho* throughout time.

The information presented in the records was obtained mostly from deeds recorded in the city's notaries, especially regarding the sale, donation and mortgaging of *engenhos* and sugarcane plots. These documents are part of a set of 45 volumes of notes from the seventeenth century, the majority of which are deposited in the Brazilian National Archives. We also aggregated information obtained from ecclesiastical and private documents, many of which mentioned the existence of *engenhos* and sugarcane plots. With every new piece of information filling a row of records, it was then possible to gradually recover the temporal trajectory of each identified *engenho*. In some cases, we were even able to determine who had built them and the approximate date they stopped their operations. Given that many *engenhos* changed hands through inheritance, we also relied on the rich genealogical information database provided by Carlos G. Rheingantz (cf. Rheingantz 1965, 1967, 1993-1995).

The information contained in the records gave rise to three types of sequential changes in the *engenhos* along time. The first one refers to the case of inheritance and succession of a given *engenho* which is clearly documented in the data. The second type of sequential changes included gaps of information that could be completed with great chances of accuracy; as it is the case, for instance, of an *engenho* associated with the name of an owner at a given moment, and years later, with the name of a son or son-in-law, indicating succession through inheritance. Finally, there were situations in which we managed to recover the changes only for certain period of time and could not extend this any further either forward or backward.

It was rarely possible to identify the exact moment when a *engenho* was built or discontinued, and abandoned. Therefore, we adopted the artifice of considering every *engenho* for which the first (or last) piece of information obtained referred to up to three years from the beginning (or end) of a given decade as having appeared (or disappeared) in the previous (or subsequent) decade. Thus, an *engenho* whose first piece of information concerned, for instance, its sale in 1653, was considered as having appeared from 1641 to 1650; likewise, a *engenho* that was sold in 1668 on which we never obtained subsequent news was considered as having existed as a productive unit at least until the decade of 1671-1680. It is quite possible that with this decision we have shortened the 'useful life' of some *engenhos*, which may have appeared (disappeared) well before (after) the first (last) piece of information on them indicate. However, given the lack of data, we could not have acted otherwise.

The process of grouping information into individual records for each *engenho* allowed us likewise to locate each of them with relative accuracy, either because the location was mentioned in a deed incorporated into the chart, or the neighbouring owners could be identified, or in some document we found references to chapels, roads and/or geographical landmarks still recognizable or possible to recover. We had, furthermore, great concern not to mistake existing toponomy with their former meanings.

Finally, to better take account of the spatial dimension of the analysis, we segmented the territory of the captaincy of Rio de Janeiro into eight producing areas,[7] which were identified based on geomorphologic configuration criteria and, particularly, on the constancy of their toponymic attributions. Next, we allocated to each of these eight areas the *engenhos* that belonged to them. The sum of the binary results provided by the records of each single *engenho* of each producing area then provided the total number of *engenhos* that were in operation in that area in each decade; the sum of the totals of each producing area allowed us, in turn, to reach the total number of producing *engenho* in the captaincy in each decade.

The producing areas, presented here with the help of current place names, were the following:

1. The outskirts of the city of Rio de Janeiro;
2. The Jacarepaguá lowlands;
3. Inhaúma and Governor's Island;
4. Irajá and Meriti;
5. Campo Grande and Guaratiba;
6. The Western Bank of Guanabara Bay (presently Niteroi and beyond);
7. Tapacurá and Cacerebu;
8. Guaguaçu and Guapimirim.

The *engenhos* of Rio during the sixteenth and seventeenth centuries

The location of the 160 identified active *engenhos* in the Captaincy of Rio de Janeiro in the period 1571-1700 is shown in Figure 11.1.

---

7   We used as a cartographic basis for analysis the territory of the captaincy that effectively existed in the seventeenth century, excluding that incorporated into the captaincy of Cabo Frio after its creation.

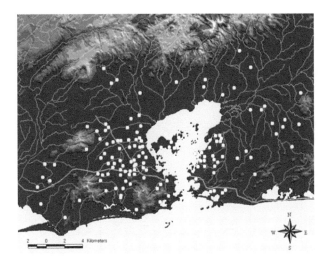

**Figure 11.2 Spatial distribution of engenhos in operation in the
Captaincy of Rio de Janeiro in the period 1571-1700**
Source: Database of the Núcleo de Pesquisas de Geografia Histórica, UFRJ.

As shown in Table 11.1, as well as in Figure 11.2, there was a continuous increase in the number of *engenhos* throughout the 13 decades studied, especially in the 1640s.

It is worth noting that the growth in the number of *engenhos* also occurred in the decades that historiography argues about decline (post-1640) or near collapse (1660 and 1670), which leads us to conclude, reinforcing what other authors have already stated, that the theses which relate the performance of colonial economies exclusively to external prices of sugar and put forward the idea of a widespread decline of sugar cane production in the second half of the seventeenth century have, indeed, to be better discussed (Figure 11.3).

The participation of the eight producing areas in the whole of the Captaincy of Rio de Janeiro was differentiated in terms of continued growth in the number of the *engenhos*. However, as the growth in some producing areas always exceeded the decline in participation of the others, the end result was always positive.

Among the producing areas, Irajá/Meriti was undoubtedly the largest sugar-producing area of the captaincy in the seventeenth century, closely followed by the Western Bank of Guanabara Bay. Indeed, the concentration of *engenhos* in these two producing areas was remarkable since the 1630s. It should be noted, also, that the growth of *engenhos* there was steady. At the end of the seventeenth century, more than half of Rio de Janeiro's *engenhos* were located in Irajá/Meriti and the Western Bank, the former concentrating 38 of the 131 *engenhos* in operation (29 per cent of the total) the Western Bank congregated another 30 (22.9 per cent). It should be noted, moreover, that two important producing areas lost importance

Table 11.1 Active *engenhos* according to producing area of the Captaincy of Rio de Janeiro, by decades, in the period 1571-1700

Producing areas	1571-1580	1581-1590	1591-1600	1601-1610	1611-1620	1621-1630	1631-1640	1641-1650	1651-1660	1661-1670	1671-1680	1681-1690	1691-1700
Outskirts of the City	1	1	1	4	7	8	7	7	6	6	4	4	5
Jacarepaguá Lowlands			1	1	1	2	4	8	10	10	10	11	11
Inhaúma and Governor's Island	1	1	1	1	4	4	6	9	12	12	16	17	16
Irajá and Meriti				2	2	5	11	28	29	32	35	37	38
Campo Grande and Guaratiba			1	1	3	4	3	6	7	9	10	8	10
Banda d' além				3	4	9	12	22	22	22	26	29	30
Tapacurá and Cacerebu								4	4	8	9	11	10
Guaguaçu and Guapimirim	1	1	1	1	2	3	6	14	16	15	11	12	11
Total	3	3	5	13	23	35	49	98	106	114	121	129	131

Source: Database of the Núcleo de Pesquisas de Geografia Histórica, UFRJ.

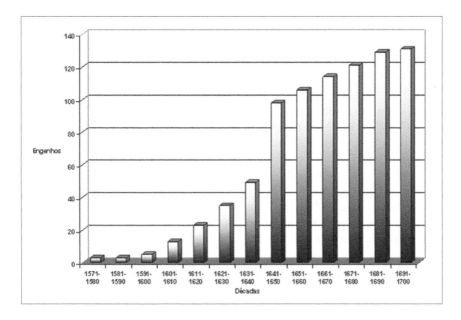

Figure 11.3 *Engenhos* in operation, by decades
Source: Database of the Núcleo de Pesquisas de Geografia Histórica, UFRJ.

during the period studied. The first one was the outskirts of the city, which concentrated many of the *engenhos* in the early days of the captaincy but whose participation rate declined systematically from the third decade of the seventeenth century. The other one was the producing area of Guaguaçu/Guapimirim, which showed an upward trend until the 1660s, declining thereafter, all indicating that if any area was seriously affected by the economic turmoil of the second half of the seventeenth century, it was certainly this one.

Concluding remarks

Although the available documentation is scarce in comparison to that of the eighteenth century, it certainly has not yet been sufficiently explored and may reveal many secrets about the historical process of Rio de Janeiro's social and territorial formation. This requires, however, that investment in basic research, i.e., research dealing directly with primary sources, be intensified. It is a considerable effort, not only regarding the time allocated to data collection, but also in terms of organizing that information which is now scattered through several documentary sources and whose parts need to be connected with each other, so that they can provide the answers we seek.

We hope that we were able to fill some of the gaps that still exist regarding Rio de Janeiro during the seventeenth century. Supported by a rich database that required more than a decade to be complete, we have managed to go beyond the mere quantitative citation of *engenhos* supplied by Friar Vicente do Salvador and Antonil, and reconstruct the remarkable process of proliferation of *engenhos* that took place in the captaincy during the seventeenth century. We have also been able to identify where and when this process was materialized in space. However, many other issues regarding the ancient world of the *engenhos* remain to be discussed and require additional research efforts.

By presenting in detail the methodology used for the construction of the database we have produced, which will be published soon as a research tool, we believe we have also offered the elements that will allow its future improvement. This will depend, however, on a continuing effort to gather data from primary sources and on their correct interpretation.

Archival manuscripts

AHU-RJ – Arquivo Histórico Ultramarino, Lisbon. Rio de Janeiro, Avulsos. Box 2, Nº 42; Nº 57.
AN, 1ON – Arquivo Nacional, Rio de Janeiro. 1º Ofício de Notas, Rio de Janeiro, Books 62, Nº 62, Nº 81.

References

Almeida, M.R. 2000. *Os índios aldeados no Rio de Janeiro colonial – Novos súditos cristãos do Império Português*. Campinas, Unicamp [PhD Thesis]. Published in 2003 as *Metamorfoses indígenas: Identidades e cultura nas aldeias coloniais do Rio de Janeiro*. Rio de Janeiro: Arquivo Nacional.
Antonil, A. (João Antônio Andreoni, S.J.). 1982 [1711]. *Cultura e opulência do Brasil*. 3rd ed. Belo Horizonte and São Paulo: Editora Itatiaia and Editora da Universidade de São Paulo.
Bicalho, M.F. 1997. *A cidade e o Império: O Rio de Janeiro na dinâmica colonial portuguesa – séculos XVII e XVIII*. São Paulo: USP [PhD Thesis]. Published in 2003 as *A cidade e o Império: O Rio de Janeiro no século XVIII*. Rio de Janeiro: Civilização Brasileira.
Boxer, C.R. 1973. *Salvador de Sá e a luta pelo Brasil e Angola, 1602-1686*. São Paulo: Editora Nacional, Edusp.
Cavalcanti, N.O. 1997. *A cidade de São Sebastião do Rio de Janeiro: As muralhas, sua gente, os construtores (1710-1810)*. Rio de Janeiro: UFRJ [PhD dissertation]. Published in 2003 as *O Rio de Janeiro setecentista: A vida e a construção da cidade da invasão francesa até a chegada da corte*. Rio de Janeiro: Jorge Zahar Editor.

Fazenda, J.V. 1924. Antiqualhas e memórias do Rio de Janeiro. *Revista do Instituto Histórico e Geográfico Brasileiro*, Tomo 95 (149).

Ferlini, V. 2003. *Terra, trabalho e poder: o mundo dos engenhos no Nordeste colonial*. Bauru: Edusc.

Florentino, M.G. 1991. *Em costas negras: Um estudo sobre o tráfico atlântico de escravos para o porto do Rio de Janeiro, c.1790 – c.1830*. Niterói: UFF, Instituto de Ciências Humanas e Filosofia, Curso de Pós-Graduação em História [PhD Thesis]. Published in 1997 under the same title, São Paulo: Companhia das Letras.

Fragoso, J. 1992. *Homens de grossa aventura: Acumulação e hierarquia na praça mercantil do Rio de Janeiro (1790-1830)*. Rio de Janeiro: Arquivo Nacional.

Fragoso, J. 2000. A nobreza da República: Notas sobre a formação da primeira elite senhorial do Rio de Janeiro (sécs. XVI e XVII), in *Topoi. Revista de História*, 1. Rio de Janeiro: UFRJ, 2000, 45-122.

Fragoso, J. 2001. A formação da economia colonial no Rio de Janeiro e de sua primeira elite senhorial (séculos XVI e XVII), in *O Antigo Regime nos trópicos: A dinâmica imperial portuguesa (séculos XVI-XVIII)*, edited by João Fragoso et al. Rio de Janeiro: Civilização Brasileira, 29-71.

Fragoso, J. and Florentino, M.G. 1993. *O arcaísmo como projeto. Mercado atlântico, sociedade agrária e elite mercantil no Rio de Janeiro, c.1790- c.1840*. Rio de Janeiro: Diadorim.

Godinho, V.M. 1990 [1953]. Portugal, as frotas do açúcar e do ouro, 1670-1770. *Revista de História*, São Paulo, 15. Reproduzido em Vitorino M. Godinho, *Mito e mercadoria, utopia e prática de navegar – séculos XIII-XVIII*. Lisboa: Difel, 477-495.

Mauro, F. 1989. *Portugal, o Brasil e o Atlântico, 1570-1670*. Lisboa: Imprensa Universitária, Editorial Estampa, 2 v.

Rheingantz, C.G. 1965-1995. *Primeiras famílias do Rio de Janeiro, séculos XVI e XVII*. Rio de Janeiro: Livraria Brasiliana Editora, Vol. 1 (1965); Vol. 2 (1967); Colégio Brasileiro de Genealogia, Vol. 3 (1993-1995).

Salvador, V. do (Frei). 1982. *História do Brasil, 1500-1627*. 7th ed. Belo Horizonte and São Paulo: Ed. Itatiaia and Ed. da Universidade de São Paulo.

Sampaio, A. de. 2000. *Na curva do tempo, na encruzilhada do Império: Hierarquização social e estratégias de classe na produção da exclusão (Rio de Janeiro, c. 1650-c.1750)*. Niterói: UFF, Programa de Pós-Graduação em História [PhD Thesis]. Published in 2003 as *Na encruzilhada do Império: Hierarquias sociais e conjunturas econômicas no Rio de Janeiro (c. 1650-c. 1750)*. Rio de Janeiro: Arquivo Nacional.

Schwartz, S.B. 1988. *Segredos internos: Engenhos e escravos na sociedade colonial*. São Paulo: Companhia das Letras/CNP.

Soares, M. de Carvalho. 1997. Identidade étnica, religiosidade e escravidão. Os 'pretos minas' no Rio de Janeiro (século XVIII). Niterói: UFF. Published in 2000 as *Devotos da cor: Identidade étnica, religiosidade e escravidão no Rio de Janeiro – século XVIII*. Rio de Janeiro: Civilização Brasileira.

Chapter 12

Striking Roots in Soil Unknown: Post-War Transformations of Cultural Landscape of Former German Towns in Poland

Barbara Borkowska

Introduction: Cultural landscape of former German towns

In the anthropological sense, cultural landscape can be understood more as an ongoing cultural process than a stable spatial structure; as an interrelation between human and space which is never finished and never complete (Bender 2002). Spatial layouts of paths, streets and squares, as well as the appearance of houses, cemeteries, monuments and churches can thus be regarded as material manifestations of the way people understand and engage with the world surrounding them, and as expression of their values, norms and needs. What happens, however, if a town is absorbed into a different country, is abandoned by its former dwellers, and resettled with people who arrive from culturally different regions?

That was indeed the case of western and northern Polish territories that had been part of Germany before the war.[1] The result of the after-war policy for creating homogenous nation states was mass displacement of the pre-war inhabitants of those territories (predominantly Germans, but also people of undetermined nationality), while their place was taken by Polish settlers.[2]

The major point of concern which arises here is the process of cultural landscape transformation, including not only the changes in the appearance and functions of specific parts of the town, but also the gradual transformation of the inhabitants' perception and valuation of the urban landscape.

---

1   The term 'Western and Northern Territories' means here former German territories located east of the river Oder. They were assigned to Poland pursuant to the decisions taken at the Yalta and Potsdam conferences, as part of compensation for eastern Polish territories that were annexed by the Soviet Union.

2   It is estimated that approximately five million of pre-war inhabitants left the Western and Northern Territories between 1945 and 1989 (Jankowiak 2006).

Surveyed area: Neidenburg/Nidzica

Some aspects of the above changes will be analysed here in the example of Nidzica, a small town with a population estimated now at 15,000 inhabitants, located in the northern part of Poland, in the Mazury Lake District.

The town, named 'Neidenburg' before 1945, was founded by the Order of Teutonic Knights in the late fourteenth century, at the foot of the monumental Teutons' castle. Since the border between the State of the Teutonic Order (later Duchy of Prussia and German province of East Prussia) and Poland was just 7 km to the south of the town, the area has always been a typical cultural border zone where influences of Polish and German culture interspersed for ages. During its history, Neidenburg experienced a number of catastrophic events, including fires, plagues and invasions, which tore at the town with various degrees of destructiveness. However, the largest transformations of the cultural landscape were due to both world wars of the twentieth century, as the town was almost entirely destroyed by Russian troops in 1914 and 1945. After 1945 the larger part of East Prussia, including Neidenburg, whose name was changed to Nidzica, was incorporated into Poland.

As with other former German towns, most pre-war dwellers left Nidzica, and soon after the destroyed town became a destination point for waves of new inhabitants.

1945-2009: Factors contributing to cultural landscape transformations

Factors determining the formation process of the post-war landscape of Nidzica may be divided into two categories. The first comprises external factors, including mainly Poland's political, social, and economic situation. The second category encompasses internal factors, i.e. local social and cultural processes that were connected directly or indirectly with needs, aspirations and decisions of the community of Nidzica.

Political and social processes in Poland

Territories that were incorporated into Poland in 1945 differed from old Polish territories in infrastructure development, spatial layout of towns and villages and type of architecture, among others. For new inhabitants, their novel environment was an obvious reflection of different cultural patterns, associated then with hostile Germans – aggressors and invaders, whose tradition and memory were not worth preserving. Consequently, the cultural landscape of those regions could not have become a natural element of the individual and group identity of the Polish dwellers, for it did not contain any positive cultural content that could be identified and recognized as their own.

Thus, one of major aims of state and local authorities was to unify western and northern regions with other territories of Poland, and to convince new inhabitants of their unalienable right to live there. According to post-war communist propaganda, it was claimed that the newly incorporated territories had belonged to Poland centuries before; therefore after 1945 they were not 'gained', but 'recovered'.[3] Every effort was made to uncover (or create) traces allegedly evidencing a former Polish status of those territories. At the same time, almost everything that was left by the Germans who had lived there for ages, was removed or ignored. Such actions were referred to as 'de-germanization' and 're-polonisation'. De-germanization was first and foremost the removal of pre-war inhabitants (Germans) and the eradication of any traces of their presence in those territories. Re-polonisation ('return to Polish character') primarily meant settling a Polish population in those territories and transforming the cultural landscape to make it the greatest possible expression of Polish culture and tradition (Rutowska 2000).

That long-term process was visible both in the policies (i.e. creation of a certain vision of the territories' history and after-war social processes) and in specific actions that aimed at transforming the cultural landscape. With time, the rhetoric of 'recovered territories', 'de-germanization' and 're-polonisation' was gradually abandoned. The change in political situation in 1989 and the abolishment of censorship in studies on Western and Northern Territories permitted a creative discussion of their pre-war history and cultural heritage. It also became possible to carry out a diligent analysis of post-war spatial transformations made on Western and Northern Territories, including both positive and negative aspects of those changes.

Economic situation

Also, the economic situation and vast destruction caused by the war throughout the country had no small impact on the shape of the former German territories' cultural landscape. Nidzica, restored with great care after the First World War, was thrown into ruins by the Soviet troops again in 1945. More than 75 per cent of its urban fabric was damaged, including almost the entire town centre. A large part of machines and equipment from local enterprises was taken away, while buildings were burn down and seriously destroyed. Unfortunately, then, the restoration was neither that fast nor that advantageous for the town's appearance as it was after the First World War. On the one hand, pursuant to government's instructions, priority was given to the restoration of towns and cities that were strategic for Polish economy and politics. For this reason, almost all bricks, either retrieved from the ruins or collected during the demolition of damaged houses, were transported to Warsaw to assist the capital's reconstruction. On the other hand, for many years

---

3 Indeed, some parts of these territories belonged to Poland or were under its political or cultural influence for some periods in their history. However, the argument about their 'original Polish status' was in most cases largely inappropriate.

Nidzica did not receive sufficient state funding to carry out works necessary to improve its infrastructure. In numerous cases this led the residents themselves to renovate houses on a makeshift basis, without taking care – understandably, in that situation – to retain their pre-war form. Up until the 1970s the town was speckled with empty spaces where destroyed houses had stood, while the dominant feature in the centre was the unsightly, mutilated remains of bourgeois tenements, often deprived of the original ornaments (and often even without the top floor, destroyed during the war), covered by makeshift roofs made from boards and tar paper.

The disastrous economic situation of the town, frequently combined with the local authorities' ineptitude at that time, was the reason why the most necessary investment projects remained uncompleted for years.

The town did not start transforming its appearance rapidly until the end of the 1960s and the beginning of the '70s. At that time, a number of industrial enterprises of great importance for the region, and also a local housing cooperative, were set up in Nidzica. Unfortunately, in designing new multi-family buildings, the priority was output rather than aesthetics. Moreover, the said practice overlapped with general trends in the housing construction industry to erect large-panel prefabricated buildings of a uniform, monotonous appearance, arranged in rows and clustered in groups, forming small housing estates. Consequently, buildings of the 1970s and 1980s, despite having met the housing needs of that time, left a permanent blot on the urban landscape. In the late 1980s newly built structures started to be more diversified to somewhat better match the local architectural tradition.

Analogous processes can be also observed in single-family housing. In the town's current panorama, one can easily distinguish buildings that predate 1990. Before that date, owing to the limited access to building materials of good quality and restricted freedom in the selection of architectural designs, single family buildings usually took the form of a cuboid with a flat roof.

Together with the change in the political system, the year 1989 brought, not only freedom of speech, but also economic liberation: command economy was replaced with market economy. Access to higher quality building materials, larger freedom in the selection of architectural designs, elimination of the state monopoly in the building industry, as well as the development of local self-government, strongly influenced the town's transformations. Regrettably, some of those changes were more degrading than enriching for Nidzica's cultural landscape. That process is particularly visible in single-family housing, which started to reflect architectural trends dominating the whole country of Poland quite fast, thus excluding the possibility of incorporating any local style or tradition. Therefore, as everywhere in Poland, there are 'manor-house' style buildings in Nidzica,[4] as well as plastic

---

4   Erecting houses resembling gentry's manor houses has recently become popular everywhere in Poland. Typical features of such a house is its neo-classicist porch with a triangular pediment resting on columns, and a gable roof sometimes covered with thatch instead of tiles (Murzyn 2008).

windows, facades covered with siding, and yellow or blue roofs made of metal tiles. Fortunately, the efforts aimed at investing the town's cultural landscape with the appearance adequate to the town's historical characteristics have been recently more and more visible. The said efforts are not only the result of the inhabitants and local government's growing historical awareness, but have also somewhat more mercantile roots. Indeed, Nidzica is located in a very attractive touristic region, Mazury Lake District. However, the town itself cannot offer many touristic attractions to visitors, for most of its historical monuments were destroyed during the war. For several years after the war, the basis for the local economy was industrial enterprises, most of which went bankrupt in the early 1990s. As a result, there is the need for a more intensive use of tourism as a source of local income, which has directly influenced the actions to enhance the town's touristic and aesthetic qualities.

Local social and cultural processes

In addition to the external factors described above, the after-war transformations of Nidzica's cultural landscape were also influenced by specific internal social and cultural processes.

After the war, as a result of the migration processes, the population of Nidzica was largely diversified. Individual needs and patterns of spatial practices of the new dwellers had a significant impact on how they were shaping Nidzica's cultural landscape. Therefore, they had to either adapt to the new environment (e.g. by acquiring new skills) or adjust some elements of the spatial structure to their own practical and aesthetical needs. Thus, immediately after the war, it was a common practice to convert tiled stoves into bread stoves, widen and reshape windows, re-arrange the layout of rooms, and paint, or cover with plaster, half-timbered or red brick facades, typical of the Mazury region. Frequently, ornamented elevations of pre-war buildings were not perceived as vintage pieces or local cultural heritage, but rather as elements onerous to maintain. Therefore, it sometimes happened that decorations were hacked off from the facades, the parapets or turrets that crowned buildings were removed, or even entire buildings were intentionally brought to ruin to make place for new structures, visually and functionally more attractive for their occupants.

The attitude towards the town's landscape was also influenced by the circumstances accompanying the migration, these being different for each region of origin. Most of the pre-war inhabitants were settlers from poor and traditional villages on the Polish side of the former Polish-Prussian border. For most of them, the reason for migrating was the desire to improve their economic situation (Beba and Pijanowska 1998). Despite the disastrous condition of the town, for those people a new living place meant the advancement of their social and sometimes economic position. However, since the new surroundings, either close (flat, house) or more distant (public space), were frequently treated as a 'gained' and not hard-earned property, they were usually used in an exploitative way, for maximum individual benefit.

Another considerable group of settlers were people displaced from the former eastern territories of Poland (the so-called Eastern Borderlands), lost after 1945 to the Soviet Union and now part of Ukraine, Belorussia and Lithuania. Mostly, they were farmers and workers from small villages and towns. Their migration was forced by circumstances of either a direct nature (escape from physical danger) or an indirect one (loss of possessions as result of military conflict and the unwillingness to adopt Soviet citizenship after the war). In the post-war propaganda, they were called 'repatriates' to suggest that they were 'returning to their homeland'. In fact, they perceived themselves rather as expellees driven out of their forefathers' land; therefore they often referred to themselves as 'depatriates', 'deportees', i.e. 'those banished from the homeland'. Such a situation contributed to their deep feeling of injustice, as well as aversion to the new environment that shared no similar feature with their previous place of living. Additionally, some of them firmly believed that their new situation was short-term and uncertain and that the Eastern Borderlands would be soon returned to Poland, possibly as a result of a third world war. The fundamental reason for such specific perception was distrust in the definitiveness of the Yalta and Potsdam settlements, combined with concerns about a permanent Polish status of the Western and Northern Territories. These fears remained with some people until 1970, when the governments of the Polish People's Republic and the Federal Republic of Germany signed a treaty providing the basis for the normalization of their mutual relations, and for the first time the German government officially accepted the inviolability of the existing Polish borders.

Despite the completely different circumstances that accompanied their migration, both groups developed quite similar attitudes towards the town's cultural landscape. For many years, the urban space was perceived as an area for exploitation and realization of one's individual needs rather than as a space for creating a local identity or common achievements and heritage that should be taken care of. Consequently, the dynamics of cultural adaptation and integration processes, as well as the relevant transformation of the individual or group identity, were adversely affected.

Changes in Nidzica's cultural landscape – examples

All these circumstances had a significant impact on specific actions that were carried out in the post-war period and were targeted at the town's cultural landscape.

Interference with the cultural landscape has been performed by transforming forms, functions or meaning of particular architectural structures or entire urban developments. In specific cases those kinds of transformations varied in intensity, ranging from total destruction to minor alterations, and occurred in various configurations. Such processes will be presented here in three examples: the market square, the castle and the local cemeteries.

The Market Square

Looking at a view of Neidenburg on a pre-war postcard, one can see a centrally located market square with a town hall, surrounded by a regular row of two-storey townhouses, all situated at the foot of a monumental castle. Since the very beginnings of the town, the Market Square has always been the town's centre, not only due to its particular location but also to its functionality. It has thus been a centre of local politics (seat of municipal authorities in the town hall), economy (stores, craftsmen' workshops, restaurants and hotels in houses around the square), entertainment (place of mass events), and *sacrum* (location of the church in the immediate vicinity of the square).

Before the Second World War, townhouses clustered around the Market Square were of rather uniform design. In 1945, most were seriously damaged. In post-war times, some were provisionally rebuilt and designed for residential and customer service purposes. At the same time, their elevations were deprived from the original shape and appearance. In the early 1970s, dilapidated remnants of pre-war buildings on the western side of the market square were demolished, although they could have been renovated in their pre-war arts. Unfortunately, because of urgent housing needs combined with the ideological reasons described above, two four-storey residential blocks were erected instead. That investment project, though beneficial in the short run, contributed to considerably destroy the town's cultural landscape by depriving the town's centre of its historical flavour.

Figure 12.1 Neidenburg in the 1930s

Note: In the top right-hand corner, the Teutonic Castle, overlooking the town.

Source: Postcard from the author's collection; photograph by Bruno Weist, publisher unknown.

Figure 12.2 The Market Square in 2006
Note: At the bottom, the western side of the Market Square, with two blocks of flats built in the 1970s. Opposite, houses resembling the pre-war architecture, built after 1994.
Source: Reprinted with permission from the Municipal Office of Nidzica, www.nidzica.pl.

Figure 12.3 Eastern side of the Market Square in 2009

After 1990, construction works on the eastern frontage of the Market Square started. That project caused a very positive change in the town's landscape, for the new buildings draw on pre-war architectural style. Although individual townhouses were built by private investors, they were in line with the local zoning plan and instructions by the District Conservation Officer. The town hall also recently regained its original shape, when a turret was added on its roof in 2007, which looked exactly as it did in pre-war times.

Nidzica's Market Square, despite considerable changes in its appearance, continues to be the most important place in town and fulfils its original political, economic and integrative functions. Its largest part is usually occupied as municipal car park, but for special occasions it serves for fairs, festivities, religious processions and other mass events, just as it did ages ago.

The Castle

The Castle was built by the Teutonic Order in the late fourteenth century with the town alongside it. Despite several serious destructions and successive renovations and reconstructions, the building has preserved its original shape. From its foundation to 1945, the Castle served predominantly for administrative purposes. Originally, it was a seat of the Teutonic prosecutor and then, after the Order's secularization in 1525, it was the head office of the prince's governor. In the nineteenth century the Castle was earmarked as a seat of the local administration and until the end of the Second World War housed the court, prison and private apartments.

In 1945 the Castle was seriously damaged. For ideological reasons, no decision was immediately taken to have it rebuilt. Being the manifestation of the Teutonic State's power and German culture, it was to suffer the same fate as many other monuments of the former German territories, destroyed by the new Polish possessors. Fortunately, in the early 1960s the first renovation works began and were completed in 1965. Despite its almost unaltered form, the Castle totally changed its functions. After the renovation, the Castle became a seat of cultural and tourism institutions: the municipal centre of culture, library, museum, art gallery, sculpture atelier, hotel and restaurant. It thus lost its significance as governmental structure and became a popular place of entertainment and social life. The Castle's symbolic meaning was also transformed. In post-war Polish historiography and journalism, the Teutonic Order was equated solely with military and religious expansion, which had been targeted at Poland and started in the thirteenth century. The Teutonic Castle, being direct evidence of the Teutonic state's power (perceived as direct ancestors of the German Reich) could not have been included in the symbolic resources of Nidzica's new inhabitants. The fact that the Castle was deprived of the significance of a governmental centre (through the change in its functions) had to be strengthened by the appropriate modification of its history. Thus an episode was used regarding Wladyslaw (Ladislaus) Jagiello, king of Poland, who occupied the Castle of Nidzica and established a Polish garrison there, right before the victorious battle of Grunwald in 1410.[5] That event allowed the Castle to be connect with Poland's history and, even more importantly,

---

5   On 14 July 1410, in the battle of Grunwald (known in the German literature as the first battle of Tannenberg), one of the largest battles of medieval Europe in terms of the

with its glorious moment. Therefore, at the foot of the Caste hill, a memorial to King Wladyslaw Jagiello was unveiled and the Polish episode in the history of the Castle became one of the most important events from Nidzica's past.

For the last twenty years, there have been more and more intensive references to the Castle's Teutonic history. Unfortunately, as that tendency is the result more of the town's tourism promotion than appreciation for the Teutonic Order's role in the development of civilization in that region, it is mostly presented in simplified and spectacular forms, like staging knights' fights and organizing 'medieval feasts' in the Castle's chambers and restaurant.

The Castle constitutes the most distinguishing and recognizable element in Nidzica's cultural landscape. Its outline, visible both from the road that leads to Nidzica and from almost every place in town, has become the town's symbol present in many graphic representations such as postcards, brochures or guidebooks, and even in the newly created coat of arms of the Nidzica District.

The cemeteries

There were five cemeteries in Nidzica just before 1945, all of which were the charge of local religious communities.

The largest cemetery, belonging to the Evangelical Church of Augsburg Confession, was founded in 1830 in the southern peripheries of the town. Most graves took the form of stone tombs surrounded by wrought iron fencing. Within that cemetery (in the immediate vicinity of former burial places), a municipal cemetery was opened after 1945. In the early 1970s almost all pre-war graves were levelled, as part of the 'de-germanization of the Recovered Territories', and only few of them have survived until today. Nowadays, the place left by the destroyed Protestant cemetery is not used as a burial ground and plays rather the role of a park. Although Nidzica's current inhabitants are aware of the existence of the pre-war cemetery there, its destruction does not seems to be subject to any public debate.

The second pre-war cemetery (the Roman-Catholic one) is located close to the former Protestant cemetery and up to the early 1950s occasionally served as burial place. Its specific location in an out-of-the-way place made it avoid destruction after 1945 and survive in quite good shape until today. To its survival might also have contributed the fact that the Catholic cemetery was perceived as being a Polish graveyard and thus, contrary to the Protestant cemetery, it was not deemed an element that represented a different culture or ideology.[6]

---

number of participants, the Teutonic Knights were defeated by the allied armies of Poland and Lithuania.

6   In the Mazury Lake District, Catholicism was associated with Polish national identity, while Protestantism was linked to German national identity. Such simplified equation of religion and nationality did not always correspond with other criteria of ethnical identity, these including language or national self-identification.

There was also a small Jewish community in Neidenburg before the Second World War (counting 66 people in 1939) that possessed two cemeteries, both of which established in the nineteenth century. Unfortunately, the cemeteries are at present almost entirely destroyed, as a result of intentional devastation and natural corrosion. Only singular *matzevas* (tombstones) may be found there now. The place of the older cemetery can be easily located and distinguished, for a historic cemetery gate has been preserved and the cemetery itself remains under the care of the Regional Office for Conservation of Monuments. By contrast, the more recent cemetery, situated on the hill surrounded by newly built houses, can hardly be located for it is almost entirely overgrown with bushes and trees.

The fifth of Nidzica's pre-war cemeteries was the cemetery of Christian-Protestants (Baptists). Our present knowledge about this graveyard is scanty and by large limited to information in possession either of local history enthusiasts or of very few families that have been living in Nidzica for generations. It is difficult, if not impossible, to find any materials documenting the cemetery's history, save for the fact that its location is marked on the pre-war map of the town.

There is awareness among Nidzica's inhabitants of all cemeteries referred to above, except for the Baptist one. These cemeteries, however, play a minor role in rituals connected with veneration of the dead. That is the result of the fact that the majority of the Nidzica population are incomers with no blood or sentimental ties with the pre-war inhabitants buried in the local cemeteries. Still, in recent years several campaigns have been organized to clean up pre-war necropolises, initiated predominantly by schools, cultural associations and scouts. In 2005 the local community also founded a monument commemorating pre-war citizens of Neidenburg who were buried between 1830 and 1945 in the former Protestant cemetery, with inscriptions both in Polish and German. This symbolic gesture can be regarded as a very positive sign of changes in perception and valuation of the town's very complicated history.

Final remarks

The feeling of territorial belonging to and identification with cultural landscape is one of the most significant factors determining personal or group identity. The answer to the question 'who am I?' usually correlates with the answer to another question 'where am I from?' and 'my history' is inseparably interwoven with 'the history of my place of living'. In former German towns, including Nidzica, ties that once bound the inhabitants with their place of origin were broken, while attempts to artificially create such ties with a new place of living were for a long time less effective than expected.

With time, the strangeness of the living space was gradually overcome and new environmental ties and social bonds were created. However, it seems that only the generations born and brought up in Nidzica are able to fully realize the need for putting down roots. In contrast to their parents and grandparents, they are able to consider those territories as their natural homeland and have no reason to prove, through propaganda slogans or actions, their rights to those lands. Since 1989, a public dialogue about the town's and the region's pre-war history has also been possible. Consequently, growing interest may be observed in local history and the pre-war appearance of the town. Contributing here are, among other, popularising campaigns by the local press, which raise great interest among Nidzica's community.[7]

Any community inhabiting a given place feels a natural need to make it functional and understandable. That was indeed the case for post-war inhabitants of Nidzica, who were gradually attempting to grow accustomed to, or familiarize themselves with the town's strange landscape to make it truly their own. In a number of cases, their actions were controlled by ideological factors. However, they also reflected economic possibilities and internalized cultural patterns. Where adaptation or adjustment of the new inhabitants' resources to certain existing conditions was prevented, particular components of space had to be modified. Thus, elements that were deemed unaesthetic were either given new appearance or were removed. Useless ones received new functionalities and usually some symbolic meaning, just like the Castle of Nidzica. Those that were incomprehensible or ideologically incoherent with post-war reality were very frequently destroyed or fell into oblivion, like Nidzica's pre-war cemeteries.

The extent to which Nidzica was ruined by war determined two different ways in which relations between humans and space were subsequently shaped. Paradoxically, the almost total destruction of the town's buildings and structures facilitated to some extent the adaptation of the post-war inhabitants to new conditions, for it offered an opportunity to reconstruct the town in line with those inhabitants' functional and symbolic needs. On the other hand, however, the new inhabitants' ignorance of, or intentional disregard for, pre-war development models resulted in chaotic urban planning which was reflected in the maladjustment of post-war buildings to both regional specificity and the remnants of the original spatial plan.

It is understandable that Nidzica's contemporary inhabitants, most of which are descendants of the post-war settlers, are not able either to feel that any historical continuity (which is usually maintained through constant referring to a direct source) exists between them and the pre-war inhabitants of Neidenburg, or to make the latter's history a part of their group or individual identity. However, present inhabitants can surely try to understand and accept the pre-war heritage, with the cultural landscape being its inalienable element. The post-war period showed that relations between humans and space cannot be shaped through either

---

7   From 2002 to 2009, approximately 100 articles have been published in the local weekly magazine by local historians and enthusiasts presenting Nidzica's pre-war and post-war history, the pre-war appearance of the town, and mementos of Nidzica's material culture. As supplement to the weekly magazine, a series of several reprints of pre-war postcards was issued.

manipulation of a given place's history or concealment of certain aspects of the said history that fail to meet valid political requirements. Despite resulting from the need to eliminate the strangeness of the new place of living, and thus being understandable to some degree in the realities of early post-war time, such approach considerably hindered the social integration of Nidzica's post-war inhabitants and their adaptation to the new conditions. For twenty years now, changes have been observed in how the local authorities and the community of Nidzica approach the town's space, these changes being the result of the need to make the town's history and image coherent. Currently, care about the town's space is manifest not only in beautifying it, i.e. keeping squares, streets, and greens tidy, but also in the harmonized integration of new elements to prevent serious dissonance which might affect the perception of the landscape as a whole.

References

Beba, B. and Pijanowska, J. 1998. Demograficzne uwarunkowania rozwoju kultury na Warmii i Mazurach, in *Tożsamość kulturowa społeczeństwa Warmii i Mazur*, edited by B. Domagała and A. Sakson. Olsztyn: Ośrodek Badań Naukowych im. W. Kętrzyńskiego, 23-33.

Bender, B. 2002. Time and Landscape. *Current Anthropology (Supplement)*, 43, 103-112.

Jankowiak, S. 2006. Bilans wyjazdów Niemców z Polski w latach 1945-1989. in *Ziemie Odzyskane/Ziemie Zachodnie i Północne 1945-2005*, edited by A. Sakson. Poznań: Instytut Zachodni, 143-154.

Murzyn, M. 2008. Heritage Transformation in Central and Eastern Europe, in *The Ashgate Research Companion to Heritage and Identity*, edited by B. Graham and P. Howard. Farnham: Ashgate, 315-346.

Rutowska, M. 2000. Polityka wobec niemieckiej spuścizny kulturowej, in *Wspólne dziedzictwo? Ze studiów nad stosunkiem do spuścizny kulturowej na Ziemiach Zachodnich i Północnych*, edited by Z. Mazur. Poznań: Instytut Zachodni, 167-200.

Chapter 13

The Cultivated Mire Landscape as a Mirror of Finnish Society

Minna Tanskanen

Introduction

Finland is a northern country, but its climate is milder and more maritime and its climatic conditions are more favourable than in most mainland areas located equally far north, between the 60th and 70th latitudes (Tikkanen 2005). Due to the considerable excess of precipitation over evaporation, Finland has an abundance of peat-based soils. However, the climate is favourable for forest growth over most of the country. Many crops can also be grown farther north in Finland than anywhere else in the world. In many references, Finland is mentioned as the northernmost agricultural country in the world.

Different languages have many words describing the various types of peat-based soils. In the Finnish language, the general word for mire or peatland is *suo*, but there are also plenty of words for the different *suo* sites. In English, there are also words (such as bog, swamp, fen, marsh, and so on) which more precisely define the nature of a mire area. In this chapter, I use the word *mire*, which best describes the general nature of the Finnish peat-based soils.

In the Finnish context, mires are usually defined botanically, as sites supporting a peat-producing plant community. According to the ecological definition, mire can be defined as an ecosystem, sustained by humid climate and a high water table, due to which partially decomposed organic matter accumulates as peat (Laine and Vasander 1996).

The Finnish physical landscape is a mixture of forests, lakes and mires. According to many researchers, Finland is the richest country in mire in Western Europe. Before the extensive drainage of mires, the original biological mire area was 10.4 million hectares, which means that almost one third of the Finnish land area was comprised of mires. Nowadays, the total coverage of geological peatlands (with peat layer over 30 cm) is 7.7 million hectares.

Finnish mires are post-glacial formations and many natural processes have transformed them into the physical shape which has in many ways been utilized by human beings. When people first came to the north, mire nature was in its pristine condition, untouched by humans, and the use of mires adapted to the conditions of nature. Throughout the ages, mires have had many uses but the most important ways of using mires have been forestry, agriculture and peat harvesting. In the twentieth century, approximately six million hectares of Finnish mire were drained for forestry (cf. Tanskanen 2000) and 0.7-1 million hectares for agriculture. In 2007, about 66,000 mire hectares were in peat production and approximately 10,000 hectares were prepared for production (Pohjois-Pohjanmaan ympäristökeskus 2009). According to estimates, over 600,000 hectares of mires will be suitable for peat harvesting in the future. Mires have also been utilized under road construction, as water reservoirs, and as landfills. Nowadays, 1.13 million hectares of Finnish mires have also been protected (Ympäristöministeriö 2009).

The goals and framework of the text

The aims of this chapter are to introduce the history of the agricultural use of Finnish mires and to outline the social and the political contexts of the Finnish mire landscape cleared for agriculture. My aim is not to represent the physical or visual changes in mire landscape but to outline the frames for the changes. I am not interested in describing landscapes in and of themselves, but in showing and analysing how mire landscape has reflected the Finnish society and its political and legal contexts.

The concept of landscape is defined here as a venue for social, political and historical events. All those events have left their mark on the physical landscape, and it is quite justified to claim that landscape is always a stratified entirety. As in a papyrus, something is obviously present but, at the same time, something is almost lost – strata and single signs can be read only if the reader knows what to look for. In the recognition and outlining of layers, one needs to know the history of a society, the aims and the principles of it, and also the values which have directed the legislation and the way of using the land.

By concentrating on rural policy, legislation and also values I introduce the main periods of the agricultural use of mires. My interpretation is based on historical literature of the Finnish agriculture and rural areas, statistics, texts of laws, and also fiction. I outline the turning points in the agricultural history of Finland and especially the way of taking mires as a part of it. I leave the physical landscape in a role of an abstract venue of history. I analyse change processes by writing vertical themes, in which the idea is to explain the change in landscape by paying attention to the context in which the change has happened.

Awakening of expectations

Originally, mires were used as meadows where natural grass vegetation was harvested for fodder. The first occasional references to Finnish mire cultivation date from the fourteenth century (Myllys 1996). However, mire cultivation can be said to have started in the seventeenth century in western Finland.

At that time, the general atmosphere for mire cultivation was very positive. Up to the seventeenth century, mires were cleared for agriculture on a small scale only, but in a relatively short time a large expansion of mire cultivation began to take place. Academic studies spread among ordinary people and the message was that mires should be put to use (Kivinen 1959). Because of night frost, mires were seen as an enemy and it was necessary to fight them. Drainage of mires was clearly seen as a way to a better life. This view was also greatly supported by fictional literature. For example, the Finnish national author Aleksis Kivi (1870/1993) in his novel *Seitsemän veljestä* (Seven brothers) created characters who cultivated mires, fought against frost, and with furious labour finally succeeded and became wealthy.

The state was also behind the reform. Already in 1740 a law was passed for minimizing taxes on new mire cultivations (Huokuna 1994). In 1816, the Finnish Government published a plan to drain all Finnish mires at the state's expense (Lukkala 1938). The plan was based on three points: 1) increasing arable land; 2) fighting frost; and also 3) helping social life by providing jobs for rural inhabitants.

Suomen Suoviljelysyhdistys (Finnish Mire Cultivation Society) was founded in 1894 and, with good reason, one can view that as one of the pivotal moments in the agricultural use of mires. The main aim of the new society was to develop mire cultivation in all parts of the country (Huokuna 1994). From the outset, the society was very active in the fields of research and experimental work and it also spread information about new methods and tools. As a result of the educational work, more rural people cleared Finnish mires.

Many small farmers, tenants and other rural inhabitants actively worked on mires. The results were not always good, but the atmosphere was confident and great hopes were set on the action. At the same time, in the north and east, farmers began to specialize in cattle breeding and butter production, which caused a growing need for arable land. Naturally, the process increased an interest in more active uses of mires.

Birth of the Finnish small-farm system

At the turn of the twentieth century, Finland was still an autonomous part of Russia. At the same time, approximately 90 per cent of the population lived in the countryside. The percentage remained almost the same during the first decades of the twentieth century, during which Finland struggled in the middle of the worst social problem of the autonomous period (Rasila 1982). The fast increase in population had created a huge number of landless people. At the beginning of the twentieth century, only 23 per cent of farmers owned their land, 34 per cent were tenant farmers, and 43 per cent of the rural population was landless (Virtanen 2003). Population increase and unemployment caused hardship and confusion in the rural areas of the country. Normally, the oldest child inherited the property and the younger children lost their rights to the land. Without land property, a person's social path pointed only downhill.

The social problem increased over a period of decades but all the time it was underestimated and the settlement of the question was put off to a later time (Haatanen 1968). The core of the problem, the huge number of landless people, became emphasized in eastern Finland where the geomorphologic circumstances favoured small cultivations, and the lack of arable land was obvious.

The problem needed to be solved quickly, and a new structure of property holding based on private ownership was seen as a solution. The problem of the landless people gave a push to the creation of the Finnish smallholder system. At the beginning of the twentieth century, Finland started to get more and more small farms and that period is called the time of voluntary transferences of property and land partitions. In the middle of that, Finland became independent in 1917. In 1918, the process was speeded up by a new land reform, a law for redemption of tenant lands (VNL 15.10.1918/135), and tenant farmers had the right to redeem their land. That period was also a time of general transformation in agribusiness.

The main result of the process was a large number of new small farms with only a few hectares of land and some cows. In the 1920s, a period of strong attachment to land began. The rural population worked hard; they cleared new meadows and fields, drained mires, and channelled and dammed natural water systems. The enthusiasm for land clearances was strengthened by a great transformation in rural policy due to which the focus of the basic agricultural production shifted from cultivation of bread cereals to favouring fodder plants and hay, which served livestock farming better. The transformation was based on the years of famine (the worst were 1866-67) when it was realized that in a northern country such as Finland it was safer to base agriculture on livestock than on vulnerable cereal production (Kupiainen and Laitinen 1995). Other reasons that supported land clearances were the remarkable reforms in cultivation techniques and the emerging interest in forestry.

In the 1930s, Finland sank into a severe economic depression. Difficult unemployment and social problems, especially in rural areas, caused a change also in the evaluation of mires. Old values, beliefs and myths attached to the mire were pushed aside and mires started to be seen as waste lands. The environment and nature were examined from the viewpoint of economy and in that context the environment began to be classified as productive areas, potential areas, and unproductive areas. Mires were mostly seen as an unproductive part of nature but their economic potential was recognized. This caused activity among private landowners who started to exploit their mires toward becoming a part of Finnish forestry and agriculture.

More and more mires were cleared as productive fields and, at the same time, the prevailing forest policy directed the generation of more productive mire forests (see Tanskanen 2000). Small farmers cleared new arable land and also ditched mires for forestry use. Huge mire areas were brought into production. It was not only a matter of clearing land but also of striving for a better future and an affluent society. Agriculture and forestry together were recognized as a foundation of Finnish rural life.

One can easily get the impression that the pioneering spirit was born among the rural population but, in fact, it was an express endeavour of the state. Both the famine years at the end of the nineteenth century and the starvation caused by the First World War had emphasized the importance of self-sufficiency in food, and thus the guarantee of it was placed at the centre of Finnish agricultural policy (Jutikkala 1982; Kananen 1986). Increase in arable land was at the core of the definition of the policy. From 1919 onwards, the state began to pay rewards to farmers who cleared new fields (Ahvenainen and Vartiainen 1982), and from 1928 land clearances were promoted by both rewards and cheap loans (Jutikkala 1982). In many cases, money was sufficient incentive to clear mires for agricultural use.

On most small farms, forestry and agriculture constituted an entirety. Income from forestry was invested in the development of dairy cattle agriculture. Soininen (1982) argues that the transition from grain to dairy cattle production would not have been possible without forest income. From a small farmer's viewpoint, agriculture and forestry formed an entirety which was complemented by secondary occupations depending on needs and seasons. Kupiainen (1995) uses the term *humble folks* to describe people who earned their bread from a combination of living strategies.

Kananen (1986) states that the area under cultivation increased by about 600,000 hectares (more than 30 per cent) from the First to the Second World War, especially in eastern and northern Finland. Statistics include all land clearances so it is impossible to know the exact area of mire clearances during that period. Naturally, mineral soils were the most desired but, in many places, a small farmer's livelihood depended on mire clearances.

Hunger for land

The Second World War caused total confusion in rural life. The Winter War (1939-40) and the Continuation War (1941-44) between Finland and Russia caused shortage of labour, horses and fertilizers. In addition to that, in the Moscow Peace Treaty in March 1940, Finland was forced to cede some parts of the country. Finland lost about 11 per cent of its fields and therefore became dependent on importation of food cereals, which strengthened the view that it was necessary to be self-sufficient in food production. State policy-makers made it clear that Finland needed large areas of new arable land.

Essentially, self-sufficiency in food was equated with securing the existence of the Finnish nation. Both the nation and its citizens desired security. Lack of food and money, starvation, inflation, and prices directed by black markets irritated people, who then started to turn to the land for security and means of support. In an uncertain situation, a piece of land meant continuity and certainty for everyday life. A small parcel of land and one or two cows were not only a dream but also the reality of an individual's life (Laitinen 1995). *Land hunger* started to grow.

The great hunger for land was not driven only by a private but also by a national will to rise to one's feet after a painful wartime. This hunger for land was strengthened by the resettlement of evacuees; approximately 430,000 individuals were resettled in different parts of the country. The evacuees, comprising more than 11 per cent of the Finnish population, were relocated through a resettlement program that was commended by the international community (Karjalan Liitto 2009).

All evacuee families were partly compensated for their losses by being given a small farm, or a plot for a detached house, or an apartment. The land used for these grants was confiscated by the state from municipalities and private owners. Tens of thousands of Finnish people had to give land to the evacuees. In order to implement the resettlement of the evacuees, guidelines had to be established and necessary laws enacted. One of the most important was the Land Acquisition Act (VNL 5.5.1945/369), which, among other things, prescribed rewards for clearances of new fields and dwelling places.

The biblical term *wandering in the wilderness* accurately reflects the progression of the action because most of the new holdings formed through the settlement policy were cleared in the deep forests beyond existing settlements. Many authors, such as Eino Säisä (1972), described how evacuees and also war veterans cleared new small farms, how the landscape changed and, especially, what reasons, meanings and feelings lay behind the work they had to carry out. In a few years, many mires changed dramatically; the evacuees did not bemoan their hard fate but started to clear mires and forests in order to build a new life for their families.

Between 1941 and 1959 about 150,000 hectares of land were cleared for fields based on the Land Acquisition Act, as well as further 200,000 hectares based on independent decisions by farmers (Laitinen 1995). In the statistics, clearances are not categorized but the land areas include clearances both on mires and mineral soils. Relatively speaking, the field area increased most in eastern and northern Finland.

For example, in 1959, the number of farms with over one hectare of field was over 330,000. On average, a farm had only 5-10 hectares of field (Laitinen 1995). Mostly, the owners of small holdings continued the tradition of the humble folks' lifestyle – dairy farming was at the core but they also needed secondary occupations. At the same time, the rural settlement spread to unsettled areas, and by favouring the small-farm system Finland secured food production and strengthened people's relationship to the land.

In post-war Finland, forestry was strongly led by an ideology called *with forest to better days* (see Tanskanen 2000). At the same time, the agricultural sector developed alongside forestry and the powerful will to build an affluent society engendered hunger for land, and people were more attached to land than ever. More and more mires and underproductive forests were cleared and brought into cultivation.

Depopulation of the countryside

The resettlement and the clearance of arable land delayed the structural change of the Finnish countryside. In the 1950s, Finland was still a rural country and its national economy leant almost completely on basic production. In the early 1960s, the tendency changed and urbanization became stronger, also in Finland. Machines, modern cultivation techniques, better fertilizers and more nutritious cattle feeds decreased the need for labour on farms and, in addition to that, overproduction started to strain the Finnish agriculture (Laitinen 1995).

The land clearance activity continued but the culmination of action came in 1968 when the total field area had reached 2.75 million hectares. Approximately one third of that area had been cleared from mires. After that, everything changed. Already in 1969, restrictions on field uses were regulated by law (VNL 11.4.1969/216). As prescribed by law, the state compensated the farmers if they kept their fields uncultivated for an agreed number of years. At the same time, all the rewards and subsidies for clearances were abolished. The situation was peculiar because first the laws had justified the establishment of a new farm and clearance of new fields (VNL 5.5.1945/396; VNA 2.6.1945/506) but after some years the farmer could claim compensation for abandoning the use of the same fields.

Suddenly, the hunger for land was replaced by rural depopulation, and the age structure of the rural population became distorted. In 1950, about 45 per cent of the population made their living from agriculture and forestry, but in 1970 the percentage was no more than 20, and in 1975 it was only 15 (Wiman 1982). Depopulation of the Finnish countryside, the small size of fields, lack of continuators for the farms and a decrease in secondary occupations were among the reasons that led many farmers to give up agriculture.

Regarding mire landscapes, the situation was a new one. After a few decades of hard work, mire clearances came to a sudden end. At first, farmers abandoned the small, oligotrophic mire fields susceptible to flooding that were not located close to their main cultivations. The need for new arable land also dissipated. There was no need for small and, in many places, low-producing mire fields and therefore they began to be replanted into forests. The Finnish forestry was led by the Forestry Improvement Programme (*Metsätalouden Rahoitusohjelma* MERA, 1965-75) and both it and the general Finnish forestry policy encouraged private land owners to plant straight-line birch and spruce forests on the abandoned mire fields (Tanskanen 2008). Many mires had only a relatively short life cycle as fields.

Modern EU agriculture

In 1991, the parliament passed a law (VNL 5.12.1991/1385) forbidding new field clearances. It has been estimated that a total of 0.7-1 million hectares of peat-based soils had been cleared before that. Four years later, in 1995, Finland joined the European Union, and the era of the present European agriculture commenced also in this country. According to Myllys (1996), about 200,000 hectares of mire fields still remained in agricultural use, which meant that approximately 10 per cent of EU Finland's arable field area was peat-based.

With the European Union, effectiveness became the keyword in Finnish agricultural production. This requires broad and productive fields and an increase in farm and field sizes; heavy machinery and specialization have become common even in Finland. With the EU's rural policy aims, small-sized and, in many cases, low-producing and distant mire fields have been abandoned as agricultural lands and turned into forest lands.

Siiskonen (2003) calculates that at the turn of the millennium 53 per cent of the population lived in the capital city region and other urban centres in southern Finland. She also estimates that in 2030 the corresponding share will be 60 per cent. Only 16 per cent of the population will consistently live in the Finnish countryside, which tells of the still-continuing, powerful depopulation trend begun in the 1960s. Nowadays, the scions of the rural *humble folks* live in towns and the former small farms are their summer houses.

It is easy to find two clear trends in using the former mire fields. The first is their use for forestry; some of the fields have naturally been forested but most have been afforested as *wood fields* by the landowners. Statistics do not reveal the exact land area of the afforested mire fields. One of the reasons is that they were done while afforesting fields in mineral soils. In this text, I have emphasized the abandonment of mire fields but one has to remember that the same also happened on mineral soils – small-sized and distant fields have found new lives as forest lands, especially in northern and eastern Finland.

The second trend is keeping suitable mire fields as part of present agriculture. Between 1990 and 2008, the number of Finnish farms decreased from 130,000 to fewer than 66,000 (Luonnontila 2008; Tike 2009). During the same period, the average field area per farm increased from 18 to over 34 hectares: there are fewer farms but they are using and cultivating bigger land areas. Because of large investments and mechanization, the size of farms continues to grow. This means that the existing farms not only lease all available fields but also clear new arable land. Mire fields cleared decades ago have been seen as a potential for present development; if they remain productive enough, it is realistic that they can still have a future as part of modern agriculture.

According to the latest inventory of cultivated organic soils (including both peat-based and loam fields), their area was about 300,000 hectares, which is almost 14 per cent of the arable land area in Finland (Myllys and Sinkkonen 2004). During the last decade or so, the common agricultural policy of the EU and its financial support system have affected Finnish agriculture and the way land is used. Old agreements not to cultivate fields have been forgotten and farmers are free to use the present financial subsidies regulated by the EU and the state. The subsidies have encouraged farmers both to return abandoned fields under cultivation and to clear new mire and mineral fields. According to calculations, during Finland's EU membership (1995 to the present), about 10,000 mire hectares have been cleared for agricultural use. On both mineral and peat-based soils, the annual rate of clearances has been approximately 10,000 hectares.

Conclusions

The history of the use of mires for agriculture has five main periods (Figure 13.1). From the research material I could outline five clear turning points – the early eighteenth century, the beginning of the twentieth century, war time (1939-44), the mid-1960s, and the late 1990s. The vertical themes between the turning points can be named as: awakening of expectations, birth of the Finnish small-farm system, hunger for land, depopulation of the countryside, and modern EU agriculture.

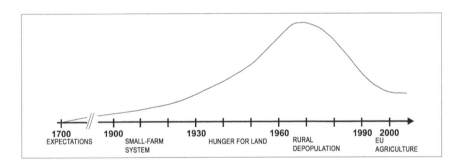

Figure 13.1 The main periods of the agricultural use of Finnish mires
Note: More relevant are the time scale and the periods, not the land area of the clearances. The line expressing the clearances is only suggestive.

By the early twentieth century, changes in Finnish society began to be strongly reflected in the landscape. As with other landscape elements, the mire landscape, too, was experiencing a great change. Small farmers cleared mires for agricultural use and brought them into production. At first, a powerful attachment to land was born and, after that, a huge hunger for land, strengthened by the hope of a new affluent society, came into existence.

At their peak, a third of Finland's cultivated areas were peat-based lands. Most of them were used for a relatively short period only. From the end of the 1960s, urbanization increased and the former land hunger transformed into depopulation of the countryside, while the need for new arable lands decreased. The use of Finnish mires for agriculture seemed to come to a sudden end. Many areas cleared years or decades before were naturally forested or afforested by landowners. The 'wood fields' occupied the rural landscape. After much hard work and individual effort, the period of small farming and the heyday of rural life had ended and the landscape was closing again.

Nowadays, the European Union's rural policy aims have set the framework also for the use of mires. Under the control of the rural policy of the state and the EU, small-sized and in many places under- or low-productive mire fields have been set aside and the areas are in use by forestry. Many of them are already growing young birch or spruce forests. From the viewpoint of productivity, the best mire fields have been kept or returned to cultivation. As preferred by the EU, the average size of the Finnish farms is growing and the existing farms are enlarging their field areas not only by leasing and buying fields but also by clearing new arable land.

Mires are an essential element in Finnish rural landscape. Although the present landscape reflects today's effectiveness, it also displays marks and elements from the previous periods of mire use. The most essential socioeconomic and political endeavours and conventions of certain periods are creating layers on the Finnish mire landscape even if they are not necessarily dominating it. The Finnish mire landscape is carrying on the story of the history of the Finnish mire clearances.

In the end, one has to remember that this narrative of the socioeconomic and political context of the agricultural use of mires is a result of my personal choices and emphasis. My subjectivity is obviously present and, if written by someone else, the narrative could be different. Given that, what is its value and what can it offer for general landscape research? With this exemplary excursion I wished to emphasize both the layer structure and the cultural nature of every landscape. One of the bases for understanding and interpreting landscape is the recognition of a social context. Analysing landscape structures or visual changes is not enough but one has to understand in what context changes have happened or are happening. Over the decades, political aims have changed with the prevailing social situations. The turning points in social life have always caused changes in land usage practices. The present landscape consists of layers that reflect the socioeconomic and political history. In my excursion, the cultivated mire landscape exists as a mirror of the Finnish – and nowadays also the European – society.

References

Ahvenainen, J. and Vartiainen, H.J. 1982. Itsenäisen Suomen talouspolitiikka, in *Suomen taloushistoria 2*, edited by J. Ahvenainen, E. Pihkala and V. Rasila. Helsinki, Finland: Kustannusosakeyhtiö Tammi, 175-191.
Haatanen, P. 1968. *Suomen maalaisköyhälistö tutkimusten ja kaunokirjallisuuden valossa*. Helsinki, Finland: Wsoy.

Huokuna, E. 1994. *Sata vuotta suotutkimusta ja viljelyneuvontaa: Suoviljelysyhdistys 1894-1994*. Mikkeli, Finland: Länsi-Savo Oy.

Jutikkala, E. 1982. Omavaraiseen maatalouteen, in *Suomen taloushistoria 2*, edited by J. Ahvenainen, E. Pihkala and V. Rasila. Helsinki, Finland: Kustannusosakeyhtiö Tammi, 204-221.

Kananen, I. 1986. *MTK ja Suomen maatalouspolitiikka: Elintarvikepulasta omavaraisuuteen 1917-1949*. Helsinki, Finland: Kirjayhtymä.

Karjalan Liitto ry 2009. *Karjalan Liitto – Briefly in English*. [Online] Available at http://www.karjalanliitto.fi/english [accessed: 1 June 2009].

Kivi, A. 1870/1993. *Seitsemän veljestä*. Helsinki, Finland: Otava.

Kivinen, E. 1959. Tulevaisuuden näkymiä soiden käytön alalta. *Suo*, 10(2), 24-30.

Kupiainen, H. 1995. Ihmisten ja yhtiöiden maannälkä: Egyptinkorven ja Nurmijärven maanomistusolot isojaosta vuoteen 1939, in *Rintamalta raivioille: Sodanjälkeinen asutustoiminta 50 vuotta*, edited by E. Laitinen. Jyväskylä, Finland: Atena Kustannus Oy, 161-183.

Kupiainen, H. and Laitinen, E. 1995. Suomalaisen asutustoiminnan juuret, in *Rintamalta raivioille: Sodanjälkeinen asutustoiminta 50 vuotta*, edited by E. Laitinen. Jyväskylä, Finland: Atena Kustannus Oy, 29-51.

Laine, J. and Vasander, H. 1996. Ecology and Vegetation Gradients of Peatlands, in *Peatlands in Finland*, edited by H. Vasander. Finnish Peatland Society, Helsinki, 10-19.

Laitinen, E. 1995. Vuoden 1945 maanhankintalain synty, sisältö ja toteutus, in *Rintamalta raivioille: Sodanjälkeinen asutustoiminta 50 vuotta*, edited by E. Laitinen. Jyväskylä, Finland: Atena Kustannus Oy, 52-138.

Lukkala, O.J. 1938. Nälkävuosien suonkuivausten tuloksia. *Metsätietoa II*, 4, 145-161.

Luonnontila. 2008. MA1 Maatilojen määrä ja peltoala. [Online] Available at http://www.luonnontila.fi/beta/[accessed: 16 June 2009].

Myllys, M. 1996. Agriculture on peatlands, in *Peatlands in Finland*, edited by H. Vasander. Helsinki, Finland: Finnish Peatland Society, 64-71.

Myllys, M. and Sinkkonen, M. 2004. Viljeltyjen turve- ja multamaiden pinta-ala ja alueellinen jakauma Suomessa (Summary: The area and distribution of cultivated organic soils in Finland). *Suo*, 55(3-4), 53-60.

Pohjois-Pohjanmaan ympäristökeskus. 2009. *Turvetuotanto ja ympäristö*. [Online] Available at http://www.ymparisto.fi/default.asp?contentid=103463 [accessed: 25 May 2009].

Rasila, V. 1982. Väestönkehitys ja sosiaaliset ongelmat, in *Suomen taloushistoria 2*, edited by J. Ahvenainen, E. Pihkala and V. Rasila. Helsinki, Finland: Kustannusosakeyhtiö Tammi, 132-153.

Siiskonen, P. 2003. Miten maaseudulle oikein käy? Avoimia kysymyksiä vuosikokouksen alla. *Maaseudun uusia aika* 1, 64-65.

Soininen, A. M. 1982. Maa- ja metsätalous, in *Suomen taloushistoria 2*, edited by J. Ahvenainen, E. Pihkala and V. Rasila. Helsinki, Finland: Kustannusosakeyhtiö Tammi, 27-51.

Säisä, E. 1972. *Kukkivat roudan maat 2*. Helsinki, Finland: Tammi.

Tanskanen, M. 2000. Näkyvän takana: Tutkimus metsäojitetun suomaiseman kulttuurisuudesta. (Abstract: Behind the visible: A research on the culture in the landscape of peatland drained for forestry.) Joensuu, Finland: University of Joensuu, Department of Geography, Publications 8.

Tanskanen, M. 2008. Suoviljelymaisema yhteiskunnallisena kertomuksena, in *Maaseutumaiseman muutos, arvottaminen ja eurooppalainen maisemay-leissopimus* (Abstract: Rural landscape change, valuation and the European Landscape Convention), edited by K. Soini, E. Pouta, T. Kivinen and M. Uusitalo. Maa- ja elintarviketalous 135. Helsinki, Finland: MTT, 96-112. [Online] Available at www.mtt.fi/met/pdf/met135.pdf [accessed: 16 June 2009].

Tike, Information Centre of the Ministry of Agriculture and Forestry. 2009. Number of farms by agricultural support area on 2000-2008. [Online] Available at http://www.mmmtike.fi/fi/index/tiedotteet/090128_tiedotteet2009_23/090213_maatilarekisteri.html [accessed: 16 June 2009].

Tikkanen, M. 2005. Climate, in *Physical Geography of Fennoscandia*, edited by M. Seppälä. Oxford: Oxford University Press, 97-112.

Virtanen, P.V. 2003. Maareformit itsenäisessä Suomessa. *Maankäyttö* 3, pp. 9-14. [Online] Available at www.maankaytto.fi/arkisto/mk303/mk303_134_virtanen.pdf [accessed: 1 June 2009].

VNA 2.6.1945/506. Asetus maanhankintalain toimeenpanosta. Prescribed in Helsinki 2.6.1945. Statutes of Finland 506/1945: Pp. 921-964.

VNL 15.10.1918/135. Laki vuokra-alueiden lunastamisesta. Passed in Helsinki 15.10.1918. Statutes of Finland 135/1918: 1-25.

VNL 5.5.1945/369. Maanhankintalaki (Land Acquisition Act). Passed in Helsinki 5.5.1945. Statutes of Finland 369/1945: 669-698.

VNL 11.4.1969/216. Laki pellon käytön rajoittamisesta. Passed in Helsinki 11.4.1969. Statutes of Finland 216/1969. [Online] Available at http://www.finlex.fi/fi/laki/alkup/1969/19690216 [accessed: 12 June 2009].

VNL 5.12.1991/1385. Laki pellonraivauksen määräaikaisesta rajoittamisesta. Passed in Helsinki 5.12.1991. Statutes of Finland 1385/1991. [Online] Available at http://www.finlex.fi/fi/laki/alkup/1991/19911385 [accessed: 12 June 2009].

Wiman, R. 1982. Maalta kaupunkiin, in *Suomen taloushistoria 2*, edited by J. Ahvenainen, E. Pihkala and V. Rasila. Helsinki, Finland: Kustannusosakeyhtiö Tammi, 494-505.

Ympäristöministeriö. 2009. *Soiden suojelu*. [Online] Available at http://www.ymparisto.fi/default.asp?node=747&lan=fi [accessed: 25 May 2009].

Chapter 14

Barroso Revisited: Long-Term Consequences of Emigration on Mountain Landscape in Northern Portugal

Bodo Freund

A unique case study for Portugal

Many rural areas in the EU result from an accumulation of adverse factors such as peripheral situation in relation to economic centres, meagre natural resources, demographic and socio-economic conditions unfavourable for agriculture. Mountain areas of the Mediterranean countries where people have moved to urban areas of their country or emigrated are widespread, with many individual cases.

In Portuguese publications, changes in landscape due to population decline are often mentioned, unfortunately in a rather general way due to lack of exact documentation. There are two reasons for this: First, for extensive areas of the country and nearly the whole northern part there are no land registry plans (*cadastro geométrico*), so that the basis for mapping land use is scarce. Second, this type of fieldwork has nearly never been practised by Portuguese geographers. That is why land-use maps detailed on the level of parcels are extremely rare (e.g. Carvalho 1999).

The Barroso mountain region on Portugal's northern border is an interesting case because it is the only example where changes in the rural landscape originating from rural exodus can be exactly documented and analysed. In 1967 the author surveyed the land-use of seven village districts by registering the utilization of every plot (Freund 1970). At that time it was very difficult to have access to American air photos of 1954 (!) in black, grey and white at an approximate scale of 1:16000 with slight distortions. Amplifications made by the author on copies were naturally of extremely moderate quality but had to serve as basis for surveys on the ground. Meanwhile, the maps thus elaborated after much fieldwork have gained value of documentation for land-use before mass exodus caused visible effects.

In 2007 three of the villages were revisited for new surveys. Now, new satellite photos transformed into chromatic photo-maps at a scale of 1:5000 were used, evidently a much better basis for subsequent mapping of the landscape. Nevertheless, fieldwork remained difficult and time-consuming. Inevitably, results are not perfect, especially because plans of land registry that might be useful for checking and correcting are still not available.

Location, ecology and former agriculture

At about 75 km from the Atlantic coast the Barroso massif rises from less than 200 to more than 1,500 meters. The high block is the most northern part of a line of mountain areas that receive extraordinarily high amount of precipitation (Alvão, Marão, Montemuro, Caramulo) and where agrarian structures and current problems are quite similar (Portela 1992; Black 1993, Syrett 1996). The *Alto Barroso*, often misleadingly named *planalto*, is the most extensive of these mountain regions and the only one with about a hundred villages at an altitude between 800 and 1,100 meters. Traditional agriculture is strikingly similar in adjacent south Galicia (Bouhier 1979).

As for agricultural ecology, climate is the most limiting factor. The meteorological station of the municipal centre of Montalegre at 1,005 meters registered a mean of 1,041 mm of precipitation per year from 1931 to 1960, mainly falling from October to March, and a short Mediterranean-type summer period of water deficit (Figure 14.1).

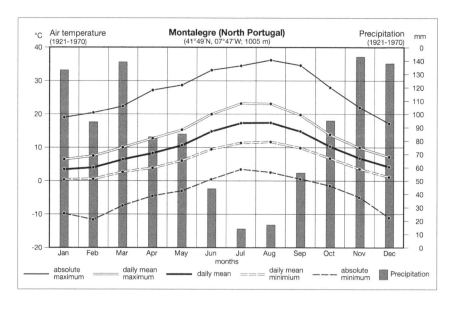

Figure 14.1 Montalegre, temperature and precipitation

Most important, however, is the extraordinary large eight-month period, from October to May, when frost can occur (Daveau 1978). The white frost and ice needles near the bottom of valleys are definitely more than frosts registered at the (former) meteorological station on a slope above Montalegre. Therefore, neither olives nor vines are grown on Alto Barroso and fruit trees are rare because frosts too often destroy blossom. Even the range of possible field crops is extremely limited.

On the poor soils that have developed on granite and schist, rye is the only bread-cereal traditionally grown since for centuries it has proven to be the most reliable staple food. Corn cultivation yields more per area. Above 800 meters, however, results are volatile because of frosts, and in the eastern parts of the region there is lack of water for irrigation. Potato cultivation, first documented in very few villages in mid eighteenth century, remained something of a garden crop for self-consumption till the early twentieth century.

Since the region had no connections by rail or road it remained very isolated and crop production was almost exclusively for local (self)sufficiency. Only animals driven over considerable distances could be sold. This is why regional mountain farms specialized in the production of the famous *Barrosão* cattle, used in the neighbouring Minho province as draught animals and finally fattened for slaughtering.

Already a hundred years ago, the Barroso became known even abroad for various communal practices (Peixoto 1908a, b). Most villages had a small baking house (*forno da aldeia*) roofed with big flagstones, a stable for the communal breeding bull (*boi do povo*), a flock of sheep and goats (*vezeira*), in some cases also a herd of cattle. Especially in the eastern and central parts of Barroso, with relatively flat terrains, collective rotation of rye and fallow (*ano sim, ano não*) was widely practiced up to the thirties. Then potatoes and corn entered the alternation with rye, but the local community continued to follow the two-field system. Village councils (*coutos*) decided on the spatially differentiated use of the commons (*baldio*), on common activities organized in the fields and on scheduling public works, like path repair.

Only when a state road across the region (N 103) was finished in 1932 was there a better connection to the rest of the country so that potatoes could be sold as cash crop. The decisive impulse for expansion came in the Second World War, when imports of seed potatoes stopped and self-supply was the national priority. It turned out that the high, relatively humid and windy region had the best conditions for the production of certified seed material, uncontaminated by plant diseases. Under these extraordinary historical circumstances, natural conditions generally considered adverse suddenly became favourable. But when Portugal entered the EC/EU in 1986 the potato cycle was definitely over.

The authoritarian regime rhetorically disseminated a rural ideology but in reality rural areas were neglected until 1974. Even worse, technocrats did not understand the traditional farming systems and were unsympathetic towards the apparently 'backward' population. In the 1940s and 1950s they contributed to its impoverishment by three interventions.

First, there was the *de facto* dispossession of the villagers by compulsory forestation of wide parts of the commons traditionally used as rough pasture, source of fire wood and of bedding for animals, and thus necessary for dunging the tilled land (Santos 1992). Second, from 1951 to 1964 much productive land was flooded by five reservoirs for hydroelectricity production. The expropriated owners received low financial compensation and the then state-owned *Electricidade de Portugal* (*EDP*) did not electrify the villages at once. At the end of the sixties many villages were still completely dark at night and households could not use electrical appliances.

And finally parts of the communal land suitable for tilling were attributed by the *Junta de Colonização Interna* (*JCI*) to seven new settlements.

Profoundly rural society, extremely weak agrarian structure

Today, the location is much less disadvantageous for the transport of commodities. But the time-distance to non-agricultural jobs remains extremely unfavourable and has very negative consequences for rural society and agrarian structure. The next town, with about 20,000 inhabitants, is Chaves, 45 km east of Montalegre. To the west the link to Braga, with 155,000 inhabitants, is made by the nearly 100 km long winding national road. To Porto (330,000 inhabitants), one can choose 150 or 175 km routes with different lengths of motorways, 50 or 120 km respectively. As a consequence, people face unattractive alternatives – weekly commuting, resigning to live as poor farmers or leaving the region.

From 1930 to 1960 the region's population increased as a consequence of extremely reduced emigration (Figure 14.2). Traditional destinations – Brazil and North America – were nearly closed since the world economic crisis and continued closed during the Second World War. On the other hand, the 'potato boom' brought some relief. The increase in rural population by 50 per cent in one generation implied terrible lack of land for the still largely dominating subsistence farming. As a result, the agrarian structure became greatly dominated by very small holdings.

For the early 1960s, Santos (1992) made a classification of local socio-economic farm types. Contrary to former authors, who preferred to describe communal practices as typical of the remote parts of Portugal and Spain (Dias 1948, 1949; Cabo Alonso 1956) and thus indirectly drew a picture of social equality and harmony, he followed O'Neill (1984) and stressed the very clear hierarchy of the agrarian society of that time.

Since the 1960 census, however, a complete structural change has occurred. On the one hand, industrialized countries in Western Europe were eager to welcome millions of guest-workers, even if many were unskilled. On the other hand, colonial wars in Africa (1961-1974) increasingly motivated young men to

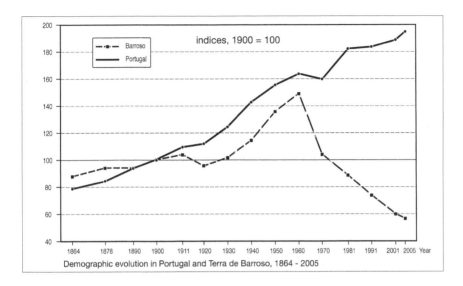

Demographic evolution in Portugal and Terra de Barroso, 1864 - 2005

Figure 14.2 Population in Portugal and Barroso region

flee abroad. As the Barroso region directly neighbours Spain with a long, difficult to control border-line, undocumented emigration to France was rampant. After more than 100 years of rural overpopulation and labour excess, there was suddenly a shortage. Not only small peasants emigrated in masses, but also agricultural workers. Farmers who traditionally employed hired labour faced serious problems and the small upper class of '*casas grandes*' could no longer exist.

The exodus of young adults in the 1960s induced a reversal in age structure because few of the emigrants returned to their villages for (early) 'retirement', thus nevertheless contributing to the increase in the elderly people (Figure 14.3). Quite a few of them had often changed employers and jobs, and after a complicated life did not have the documentation required to receive a complete pension. Nearly all the pensioners get a larger or smaller part of their total income from agriculture in money or produce for self-consumption.

At the end of 2008, 17,140 inhabitants lived in the Barroso region, which is constituted by two administrative units (*concelhos*), Montalegre and Boticas. The mean population density is only 15 inhabitants per square kilometre. The two tiny municipal towns, Montalegre and Boticas, have about 1,800 and 1,100 inhabitants, respectively. The remaining 14,240 people live in 49 communes (*freguesias*) with 170 villages or hamlets. It follows that the average number of inhabitants per settlement is hardly 85, actually much less than spontaneously estimated in view of the number of buildings. Only five villages have over 200 inhabitants. Consequently, the opportunities for local and regional commodity supply and service provision are very limited, obviously a negative factor for the quality of life.

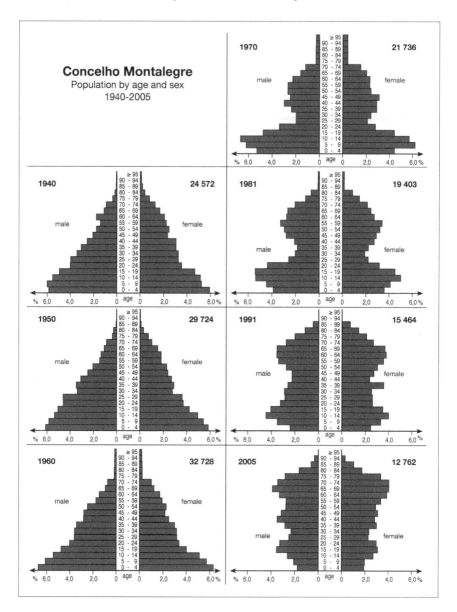

Figure 14.3 *Concelho Montalegre*: **Age structure**

According to the results of the 2001 census, the population is rather aged. It is true that 62 per cent belonged to the 15-65 age group and thus could be considered potentially active; but at least 27 per cent were 65 years old or over. Actually, many people are early pensioners, surviving relatives or to a certain degree handicapped, often due to labour accidents in the locality or abroad, so that 42 per cent of the total population were pensioners, in one way or another.

Population is declining by a little over 1 per cent yearly, predominantly a consequence of high death rates (16.6) and extremely low birth rates (6.1), at least partly due to the loss of people in procreative age (INE 2005).

In 2001 about 85 per cent of the total population lived in villages and hamlets in a profoundly rural milieu. Over 60 per cent of the households and nearly 70 per cent of the population were linked to agriculture. Only 15 per cent of the potentially active population had (also) non-agricultural work, mainly in the two small towns. One must conclude that there are extremely few commuters and the majority of people lead a relatively isolated life.

As the results of the 2009/2010 agricultural census are not yet available, the agricultural structure can only be analysed on the 1999 data base. From the households then linked to farming (3,758), agriculture was the main source of income for about 72 per cent, although the exclusive basis for only 24.5 per cent. As 75.5 per cent of households had additional income it must seem astonishing that only 13.2 per cent of peasants and 13.8 per cent of their family members were economically active outside the farm.

Agricultural households that receive some form of pension must be estimated at 48.5 per cent because so-called 'other incomes' (like rents from real estate) are of minimal importance. The high frequency of pensions, already found in the population as a whole, is somewhat confirmed for the agricultural sector by two other indicators. The first is the farmers' age structure: 34 per cent were 65 years or older and 26 per cent in the 55-64 age group. In the latter group there were probably many who had left paid employment early, especially re-migrants. The second confirming indicator is the time dedicated to agriculture: whereas 87 per cent said that this was their main economic activity, only 19 per cent declared themselves full-time farmers, with at least 40 hours per week or 240 days per year. The majority (57.7 per cent), however, do more than half the workload of a full-time farmer. The physical capacity of the elderly land-owners and the generally low pension, of only €250 per month on average (2001), motivate or force them to do relatively much work. This is also why 47.3 per cent of households get their dominant income by the value of the agricultural produce, much of which is directly consumed.

From traditional to current agriculture: The statistical approach

On the basis of the data of both counties that form Barroso, thus including lower and warmer parts of the region, one may construct a hypothetical farm representing the statistical means. It is a property of 8.1 ha, of which only 2.5 ha are arable and garden land. Half of this area is cultivated with rye, half a hectare is occupied by irrigated corn or potatoes. It is evident from the diminutive area of rye, potatoes and corn that no commercial goal is involved; in fact, produce is for consumption by the family and for feeding its animals. Irrigated natural meadows would be the most important part of the land (3.0 ha). Statistics present another 2.2 ha of 'non-irrigated pasture' which in reality is private heath (*bouça*). The remaining 0.4 ha is wood, brooms, chestnut trees and – only in the lower parts of the region – vines, olive and orchard trees.

Roughly speaking, farmers' income is calculated in the EU on the potential value generated by animals and areas under cultivation minus the necessary inputs. Not astonishingly, Barroso is thus a region of extreme smallholdings in economic terms. In 1999, by this form of assessment more than half the peasants (51.5 per cent) earned under €2,400 in cash or equivalent produce yearly, less than €200 monthly. Only 23.1 per cent and 16.6 per cent of farms came in the next classes of up to €400 or €800 a month, respectively. Not even one in ten peasants earned more from agriculture.

Considering the rapid decrease in farm units to 45.8 per cent in the three decades from 1968 to 1999, it might be surprising that holdings do not have more land. But as initially there was an extreme smallholder structure there was not much land to be mobilized and even this was not totally absorbed. Given the methodological reservation that 1968 and 1999 data can only compared with restrictions (INE 2001: 143), one can calculate that the land occupied by fields and meadows (as a whole) has been reduced by over a quarter (26.7 per cent). This is almost exclusively due to the reduction in tilled land by 41.4 per cent; about 5 per cent of former arable land has been converted to meadows and pastures, which have gained in value and have thus increased by 10 per cent.

The 1999 agricultural census does not show the extension of land fallen into disuse because none of absentee owners' abandoned land is registered. The remaining farmers declared only 3.4 per cent of their own land idle. Thus most derelict land in the landscape seems to belong to people living abroad or in other parts of Portugal.

As cropland has been significantly reduced and marginal parts have been giving above-average, yields have increased and are not really bad compared to Portuguese averages. But in a unified European market this is not the decisive reference. The very unfavourable situation becomes clear by comparing these with results in other countries. Regional means in kg/ha, officially estimated by local experts, are 1,300 for rye, 2,200 for corn and 18,000 for potatoes. These are poor results compared to recent (2004-2008) averages calculated on EU data for neighbouring Spain (roughly 2,000/9.900/27,500) or for Germany (4,800/9,000/41,000). And it should be kept in mind that these poor results are attained by much more labour input.

Farms with animal husbandry have decreased more than farms as a whole. From 1968 to 1999 peasants with cattle declined only by some percentage points to 51.5 per cent of all farmers, but those with pigs dropped from 82.2 to 40.6 per cent. Only one in six or seven farmers still kept sheep or goats. Conversely, stocks and consequently animals per holder have increased. On average, there are ten cattle per holder, still extremely few by international standards.

In 1968 most villagers had small ruminants since tending did no absorb much time. In nearly every village there were common flocks (*vezeiras*) which pastured nearly the whole year round on the commons and the stubble fields. In the course of the year every holder had to tend the flock for a few days proportionally to the animals he had put in (*à vez*) but often owners assigned this task to their children. Today, school is attended rigorously. Keeping sheep or goats individually is only worthwhile with many animals to make cash (including EU subsidies) or as a hobby.

Some farms specialized to one or another grazing species. Regarding pigs, however, the average number per farm remained low (3.9) and in fact there are very few peasants with 30 or 40 animals. This is surprising because the region is famous for home-made smoked sausages and ham and this might have good chances of expanding sales.

Central and eastern Barroso: Two-field system in dissolution

Statistics permit conclusions concerning agriculture and the socio-economic structure of households, but they hardly give an idea of the rural landscape and not at all of its change. This is only possible by comparing coloured maps with 15 classes of land-use in tiny plots.

Even when revisiting a village district with much attention for a second survey it is not possible to recognize what exactly has changed during the last 40 years. Only as soon as the new map is compared with the old one do the differences become striking. And additionally, even the new land-use map does not fully reflect some minor but nevertheless significant changes that have been realized during fieldwork.

Peireses, a village located in mildly hilly relief south of Montalegre, may serve as an example for the central part of Alto Barroso and to a certain extent for the east as well. In 1967 all plots had been used without exception. The division of arable land into two tracts or 'leaves' (*folhas*) of equal dimension was very clear: the one side was totally cultivated with rye, the other dominantly with potatoes and somewhat less with corn as 'fallow crops'. Arable land on fairly level terrain had the aspect of a typical open field structured in the furlong system with narrow parallel strips indicating that tilling with ploughs and land division through inheritance has been practiced for generations. Peireses was therefore an excellent example for collective biannual rotation.

Directly adjacent to the village lie humid tracts with tiny open plots that are not subject to collective rotation. They are still taken up by vegetable gardens (cabbages, beans) or by potatoes and grass. Gentle or nearly flat valleys are filled with meadows always surrounded, but not always divided, by walls. Generally, irregular blocks proved to be typical for land that is not tilled, e.g. meadows and plots with somewhat woody vegetation. Especially at a relative distance bigger blocks enclosed by walls of irregular thick stones were occupied by oaks (*carvalhos*), shrub (*mato*), broom (*giesta*) or heath, which are destined for firewood or animal bedding.

The 2007 survey leads to a rather different pattern of land-use: meadows were expanded on former cropland. The two-field system once practised by all owners in the common is still somehow discernable in the map but no longer visible in the landscape because homogeneity has dwindled by strips being cultivated in deviate ways or fallen fallow. This happens more often on the side intended for rye. The area for fallow crops is clearly better used, now dominantly by corn and not by potatoes. This indicates that cultivation of these two plants is considered more worthwhile. Some marginal parts of formerly tilled land have been abandoned.

Interest in cultivating rye as bread-cereal for self-consumption has dwindled because households have become smaller and consumer habits have changed during the life in other countries. Nowadays even in small villages households are provided with bread and cakes by itinerant sellers who come in special cars. Fields where combined harvesters cannot work are generally no longer used for rye cultivation. Besides, in many places the cost for paid work with machinery is not justified by the value of the harvest; this means that land with weak capability is no longer cultivated.

Cultivation of potatoes has significantly decreased as well, although yields per area are above the national average and the region is famous for this crop. But labour inputs are high because it is impossible to mechanize the harvest due to mostly stony fields. In general, even the area under cultivation for self-supply has been reduced because Barroso is no longer an overpopulated region with a majority of very poor families in desperate need of food. The area for corn has been expanded despite very low yields in grain. Animal husbandry is again the most profitable form of business for those who produce for the market.

As only a few fields in Peireses are located more than 1 km away from the village, distance cannot be the main factor for land abandonment; week land capacity is more influential, but most important is absenteeism. This is reflected in the rather irregular scattering of waste fields as was noticed in studies of social fallow in Germany 50 years ago. In those cases, however, social fallow was rightly considered an indicator of transition among rural land owners; here it is more likely an indicator of definitive abandonment and decay. Most of the remaining people are not interested in cultivating emigrants' plots because crop production for more than self-consumption is not profitable. The lack of interest explains that informal parcel concentration and growth of holdings have remained rather insignificant.

During the first survey, land-use classification was somehow easier, especially concerning woody vegetation. Types of land-use could be ascertained as heath, broom, shrub or oaks because it was productive land really managed. General enclosures and dividing walls were visible. Now there are gliding transitions from idle land to heath and then to brush and woodland with young oak-trees. Looking at plots with broom it is often difficult to decide whether it is an abandoned field invaded by this plant or a former cultivation of broom abandoned for years.

Additionally, vegetation is often so dense and high that local orientation is extremely difficult.

On the other hand, it is astonishing that individual chestnut trees (*castanheiros*) along paths and in a few groves (*soutos*) continue to be clearly visible as dark spots in satellite images, as in former air photos. Some of these trees are large, old and impressive elements in the landscape and should be protected as landmarks.

It is evidently (nearly) impossible to represent on a map that many parcels are not completely used and spontaneous vegetation is progressing inside the enclosure, beginning at the edges and continuing along the walls. In the meadows rush expands as weeds when drainage stops being managed. Neglect in ploughing at first glance seems to be a sign of incompetence and carelessness but it is more likely due to elderly users' and disabled owners' physical weakness.

After 40 years it seems strange that some of the paths framed by walls are now completely filled with dense spontaneous vegetation and can no longer be used. Other tracks are in poor condition because they are not maintained as in former times when the local community did this work together after the village council's decision (*couto*). Old villagers remember and spontaneously mention that previously the agrarian landscape was cleaner.

In the sixties, Peireses was a sprawling but nevertheless rather compact village with a cemetery at the southern end and a tiny school at the northern end. From this point it is one kilometre of local road to the main road leading to the municipal *vila* of Montalegre. At the junction there was a group of perhaps three houses. Now a little row of houses has been built along the main road and eight houses have been scattered along the local road leading to the junction. Besides, some houses have been added on the northern and the southern ends of the village. Conversely, several houses in the old part are no longer inhabited. Thus signs of dissolution and decay can be detected in the settlement and in the agricultural landscape as well.

Expanded villages, socially hollowed out

For the structural change in villages, evidence can be given by the case study of Mourilhe, located 4 km west of Montalegre, about 1,000 m altitude.

In 1967, at first glance the compact irregular village had a rather homogeneous aspect because nearly all the buildings were constructed with grey granite and half of them were still roofed with straw. This produced a primitive, severe look which was gloomy when the sky was covered.

On a closer look, a wide social differentiation could be perceived after the first misleading impression. Most farmhouses had stables on the ground floor and living space on the first floor which normally was reached by outside stairs. This type of building dominated the densely built-up core area of the village. Besides, there were 18 buildings without stables. These tiny houses were occupied by the very poor people called *cabaneiros*, those who lived in a hut and could not be considered peasants. Farmers that were better off (*lavradores abastados*) lived in the village too. A gateway gave access to a – not always regular – *patio* surrounded by buildings for dwelling, animals, produce, ox-carts and firewood. At the periphery of the village and even somewhat detached, two large buildings of well-pared granites were remarkable. These estates, '*casas grandes*', belonged to the regional upper-class families. In Mourilhe one can see the still inhabited Casa dos Chóios with an enormous threshing floor made by large granite flags and a small grain barn dating to 1868. More striking than this building of the formerly richest farmer is the rural hotel which was originally the Casa do Outão, with a private chapel built in 1761.

In 1967 there were about 70 barns, nearly all in the thinly built-up half of the village closer to the mountain pastures and never in the dense core. The separation was probably intended to prevent that hey and straw from catching fire. There were only 36 separate stables, mostly at the lower fringe of the village and more close to the fields to be dunged. The number of minute grain barns, normally unobtrusive annexes to other buildings, has not been exactly documented; there were probably 13 to 21 of them. For corn cobs there were at least 10 *espigueiros*, little storehouses on pillars, generally well ventilated in locations outside the densely built-up parts of the village.

In 2007 Mourilhe was revisited for a new survey and an additional register of the owners' residences outside the village. Now numerous changes surfaced. Many new buildings, discernable by the light and sometimes colourful outward appearance, had been constructed at the periphery or on plots where old houses were demolished. Besides, many ruins are striking, especially former hey barns, but also houses of all dimension in the core area. Inside the ruins, trees often reach the place where straw-roofs disappeared long ago, and rubbish was thrown inside. In some cases ruins and new houses stand side by side and make a strong contrast.

It is not at all immediately visible that over half the dwellings are not occupied for most of the year or even for years. This applies to both old and new houses. Most of the absentee owners live in France (68) and in the United States (about 34). Other countries like Switzerland (3), Belgium (2), Brazil (1) and Germany (1) are of little importance, as are other parts of Portugal (12).

Emigration from the commune of Mourilhe, to which a second village (Sabuzedo) belongs, dates at least to the nineteenth century. After a first population apex in 1878 the number of inhabitants decreased in spite of the excess of births over deaths and then stagnated until 1930. Afterwards, emigration was nearly impossible so the population peak was reached in 1950 with 613 inhabitants. In the fifties, drift from the land resumed and during the sixties increased enormously. Although emigration waned from 1973 to 1983 as a consequence of restrictions in receiving European countries, local population continued to decrease, probably by families reuniting abroad. In 1981 the villagers were already less than half the 1950 number. At the last census, in 2001, in both villages only 144 inhabitants remained in 61 households, 44 of which were constituted by elderly couples and a few families.

Especially from about 1974 to 1995 many houses were built, but others were demolished or just collapsed. As a balance the number of dwellings has increased by 15 per cent since 1950 whereas population of the parish has diminished by 77 per cent. Such opposite developments occurred in the Barroso region as a whole. Data of the 1999 agricultural census show that population had decreased by 70 per cent, but 40 per cent of the houses had been built since 1950.

All forms of supply business are no longer profitable due to the extremely reduced and still decreasing local population. The small shop and the small pub registered in 1967 do not exist anymore. The house for bread-baking and the stable for the bull, both belonging to the village community, are out of use and empty. The little school, probably built only in the 1950s, was already given up in 1991. Even the mayor is not really living in 'his' commune anymore. He has his first residence in Braga and comes merely at weekends. Normally only some elderly people can be seen in the village because no more than two peasant families with children have remained, just one of them in full-time farming. Intensive social activity is restricted to August when emigrants come for their annual visit.

As most dwellings are not occupied, the buildings for agriculture are generally not used either or used for other purposes. Only by exact registration does one realize that there are about 60 garages. They are integrated into new houses or are rebuilt former stables. What is the purpose? Most of the elderly people have no car. But cars are the emigrants' means of transport and still also have significance as status symbol. They cannot be left in the narrow lanes of the village, so garages are absolutely necessary.

Belated and controversial public investment,
doubtful socio-economic prospects

Since 1975, and especially since Portugal's entry in the EC/EU in 1986, the infrastructure has quickly been improved. Even the last villages have road access and when the 1999 agricultural census was made nearly 99 per cent of farmhouses had electricity and 97 per cent water connection, though only 82 per cent had sanitation. Certainly, one must add that not all farmhouses are continuously inhabited. But what has been welcomed by local populations as enormous progress had already been found by emigrants abroad as traditional basic equipment.

There remains a grievous lack of public transport. School buses that take children to Montalegre arrive at this central place long before shops and services open and often do not transport other passengers at all. People who do not have their own car are dependent on scarce neighbours or taxi drivers. Commuting without driving is impossible, but to buy a car a level of income is necessary which is difficult to reach in the regional labour market. In 2002, the average monthly income in Montalegre was €653 in the secondary and €625 in the tertiary sector, in Boticas even less. Understandably, people with high income hardly have the motivation to live in a half-deserted village.

The two tiny municipal towns of Montalegre and Boticas have been rather conspicuously equipped by special investments. There are new buildings for the market places and the bus terminals, and new parks financed by the Polis-Program embellish them. Boticas has a big new Town Hall (*Câmara Municipal*), Montalegre has a huge multi-purpose complex and an eco-museum (2009). All roads leading to that *vila* run into five roundabouts that are decorated with monuments representing regional topics (potato-growers, herdsmen, fighting bulls, dolmen, arches); at least two further monuments (celebrating emigrants and soldiers of the carnation coup that ended the colonial wars) can be found in the tiny municipal centre of 1,800 inhabitants. An attentive visitor may get the impression that there is a desperate attempt to make micro-towns attractive by landscaping and by public art aiming at strengthening the residents' identification and attracting visitors. Nevertheless, both *vilas* have lost inhabitants since 1960, undoubtedly relatively less than their hinterland. Critical minds may suppose that ties to the region cannot be intensified this way and that politicians have made monuments for themselves with public money instead of improving factors fostering economic activity.

What has been lacking so far are jobs. Therefore, emigrants of the region are forever hindered from coming back to their *terra* before they receive pensions. Regional youths find hardly any economic basis in acceptable distance. Industrial investment in such peripheral regions of Portugal is extremely unlikely ever since countries of Eastern Europe have become EU members, and Asian producers are fierce competitors of Portuguese ones too. Besides, regional labour force is not a match in quantity, age and skills.

There are some promising efforts to promote tourism. Prehistoric, geological and biological features, elements of popular architecture and regional cuisine are presented both in folders and on the Internet in Portuguese, English and sometimes French as well. Along the roads wooden signs indicate sites to visit, but unfortunately they are often absent from the field paths. In some parishes old water mills and common baking houses have been restored. Probably unique are some *fojos,* built traps for former battues on wolves.

Certainly, more attention is stimulated by the increase in events deliberately spread over the year. For decades, the region is known in Portugal for the tradition of letting bulls of different villages fight on some open space. Meanwhile a ring has been built in a fenced area near Montalegre. On Sunday afternoons from early June to mid-August bullfights are organized now by private *Barrosão* breeders. About 200 spectators come by car and even in coaches to about ten afternoons of *chegas de bois.*

In June there are competitions in the national hang-gliding championship from the near-by Larouco (1,515 m) and motor-cross races at the foot of this mountain. Many more people come even from afar to the various annual markets (*feiras*). The most important is the *feira do fumeiro* or market for ham and smoked sausages that are still home made.

For an assessment of tourist potential it is necessary to make comparisons. Mountain areas are not at all rare in the northern half of Portugal. The Serra da Estrela is the highest, the best known and most visited range because snow remains around the Torre (1,993 m) for weeks. The Estrela is situated nearly in the centre of the country, relatively close to the metropolitan region of Lisbon and is traditionally the object of a certain admiration.

West of Barroso and also bordering on Spain is the Serra do Gerês. This impressive mountain range is also well known as worth seeing because it has very steep slopes, rocky pinnacles and no settlements in its inner parts. The National Park Peneda-Gerês was founded in 1971 and is still the only one in Portugal. Some parishes in north-western Barroso are part of it, but in fact a very deep valley makes a strict divide from the Gerês chain proper. On the Barroso side the landscape is less impressive because the relief is mighty but smooth.

The five reservoirs have clearly contributed to intensify scenic amenity but they are not used intensively by tourists. Water levels vary and of course there are no sandy beaches. Regional people come to take a bath and there are usually a few foreign campers, especially from the Netherlands. Certainly, it is possible to go for a sail, especially on the most extensive (10 km, 2,212 ha) water surface of the 'Lake of Pisões' (Barragem do Alto Rabagão). The Portuguese, however, definitely prefer the seaside.

There is indeed some tourism in the region. A few mansions have been remodelled for *turismo rural* with good technical equipment and traditional furniture. Some little land-hotels and pension are scattered, not very visible either and poorly frequented. They can serve as starting points for wandering, sightseeing tours or relaxing in a quiet and rather isolated environment.

Away from the national road and the two municipal *vilas*, however, it is difficult to find any restaurant. The very few establishments are rather simple and normally offer meals only at weekends. Daily guests are men of the village who meet for a drink and conversation whilst television is uninterruptedly offering loud entertainment. That is why the atmosphere is not really comfortable for tourists in search of rural silence. Those coming from abroad most probably can make themselves understood in French because quite a few of the inn-keepers are returnees.

In general, the Portuguese and foreigners prefer different aspects of the region. For foreigners only a stay in summer is acceptable. They may appreciate wandering in a 'natural' environment with seemingly 'prehistoric' elements of popular architecture and like to have fairly good individual traditional-style accommodation. Brochures are available for seven pedestrian trails (*trilhos pedestres*), which are marked and kept free indeed. This is surprising because wanderers are extremely rare. Recently guided tours have been offered, for example, along former smugglers' paths (*rota do contrabando*). There is probably an unused potential for horseback-tours through the calm agrarian landscapes and the extensive heath-lands of the mountains. One can also imagine mountain bike tours. The most appropriate season is from the end of May to mid-September. June is the month of blossoms, especially yellow brooms. In August crowds of emigrants bring intensive social life into the region, especially to the small centres.

The Portuguese, in turn, prefer the winter season. They hope to see some snow even if only on the hilltops. For many hunting and fishing are favourite activities. The region has undoubtedly a potential for hunting-tourism because wildlife has considerably increased due to the abandonment of formerly tilled land and the reduced use of the commons as rough pasture.

Most plentiful are rabbits and partridges. Wild boars have become numerous and are already considered a nuisance especially to corn fields. Roes, which had not been seen for generations, are coming back but are still strictly protected. The rivers and the reservoirs are attractive for fishing. Barroso men consider fishing and hunting important for their quality of life. So there may be a conflict of interests, keeping people in the region and attracting tourists. It is not certain that local hunting and fishing societies will give licenses to other people.

Even if tourism expands considerably it cannot become the basis for many jobs and most of them will be part-time. Limiting factors can easily be seen. Barroso is not (yet) endowed with enough traits to attain an attracting singularity among the competing mountain regions in the vast northeast area of 'high Portugal'. Most villages are no more picturesque under architectural and social aspects; this can hardly be compensated by staying in the *patio* of a well-kept noble farmhouse.

Barroso does not have good prospects for secondary homes as required by people from other regions. In Portugal there is no lack at all because construction has boomed for decades at the long coastline whereas in the interior an enormous stock of houses has been left unoccupied as a consequence of migration to national centres or abroad. That is why demand concentrates on regions and premises with high attraction elements.

Barroso is far from the Lisbon metropolitan region, which is the dominant source of recreation demand, and people of that region prefer to go south and to the seaside. Even the Porto metropolitan region is at considerable distance. Houses in the Barroso that are not occupied during the many months of low night temperatures and high humidity deteriorate quickly; by the way this is certainly an additional factor to depreciation of premises built by emigrants.

The conclusion is that agriculture will remain the most important regionally bound economic activity. Inhabitants of the Barroso will continue to get much of their income by transfers in various forms, as subsidies to agricultural activities, pensions and payments to public servants. Consequently, conditions for agriculture and living must be improved, but with caution to prevent undue expenditure.

References

Black, R. 1993. *Crisis and Change in Rural Europe. Agricultural development in the Portuguese mountains*. Aldershot: Avebury.

Bouhier, A. 1979. *La Galice. Essai géographique d'analyse et d'interprétation d'un vieux complexe agraire.* La Roche-sur-Yon/Vendée: Imprimerie Yonnaise.

Cabo Alonso, A. 1956. El colectivismos agrario en la Tierra de Sayago. *Estudios Geográficos,* 17, 593-658.

Carvalho, E.M. de. 1999. *Basto (Sta Tecla) Uma Leitura Geográfica (do século XVI à contemporaneidade).* Guimarães: Universidade do Minho, Instituto de Ciências Sociais.

Daveau, S. 1978. Le périglaciaire d'altitude au Portugal. *Colloque sur le périglaciaire d'altitude du domaine méditerranéen et abords,* 1977. Strasbourg: Association Géographique d'Alsace.

Dias, J. 1948. *Vilarinho da Furna. Uma aldeia comunitária.* Porto

Dias, J. 1949. Minho, Trás-os-Montes, Haut-Douro. *Congrès International de Géographie. Livret-guide de l'excursion.* Lisbonne

Freund, B. 1970. *Siedlungs- und agrargeographische Studien in der Terra de Barroso/Nordportugal (=Frankfurter Geographische Hefte, 48).* Frankfurt am Main: Waldemar Kramer.

INE/Instituto Nacional de Estatística 2005. *Anuário Estatístico da Região Norte.* Lisboa: INE.

INE/Instituto Nacional de Estatística 2002. *XIV Recenseamento da População, IV Recenseamento da Habitação 2001.* Lisboa: INE.

INE/Instituto Nacional de Estatística 2001. *Recensamento Geral da Agricultura 1999, Trás-os-Montes.* Lisboa: INE.

INE/Instituto Nacional de Estatística 1971-1972. *Inquérito às Explorações Agrícolas do Continente 1968. Explorações de menos de 20 hectares. Estimativa a 20 % (*1971*)/Explorações de mais de 20 hectares* (1972). Lisboa: INE.

INE/Instituto Nacional de Estatística 1962-1969. X *Recenseamento Geral da População 1960.* (Especially Tomo I, vol.1°, dados retrospectivos) Lisboa: INE.

O'Neill, B.J. 1984. *Proprietários, Lavradores e Jornaleiros – Desigualdade Social numa Aldeia transmontana. 1870-1978.* Lisboa: Edições D. Quixote.

Peixoto, A. 1908a. Formas da vida comunalista em Portugal. *Notas sôbre Portugal,* Vol. I, 73-84.

Peixoto, A. 1908b. Survivances du régime communautaire au Portugal. *Anais Científicos da Academia Politécnica do Porto,* III, 4, 205-221.

Portela, J. 1992. The Terra Fria Farming System: Elements, Practices and Neglected Research Domains, in *Endogenous Regional Development in Europe. Theory, Method and Practice,* edited by H. De Haan and J. Douwe van der Ploeg. Brussels: European Commission DG VI, 263-286.

Santos, J.M. 1992. *Mercado, economia e ecossistemas no Alto Barroso. Um estudo de sistemas de aproveitamento de recursos naturais.* Montalegre: Edição da Câmara Municipal de Montalegre.

Syrett, S. 1996. *Local Development. Restructuring, locality and economic initiative in Portugal.* Aldershot: Avebury.

PART III
Landscape Assets, Resources and Services

Chapter 15

From Landscape to Tourism and Back: The Emergence of a Greek Landscape Conscience

Theano S. Terkenli

Introduction

Greek landscapes have been plagued by much neglect, misuse or even irreparable destruction, throughout the country's history (Beriatos 2009; Manolidis 2008; Kizos and Terkenli 2006; Terkenli 2004; Doukellis 1998) and especially since Greece's era of rapid urbanization (1950s and 1960s) (Kizos et al. 2007; Terkenli 2004; Simaioforides 1989). Moreover, the country currently finds itself in the uneasy position of being at the receiving end of any and all international postmodern globalizing trends in urban and regional planning and spatial organization. The latter processes of external, top-down impacts of the workings of a new global *cultural economy of space* (Terkenli and d'Hauteserre 2006; Terkenli 2002) are coupled with long-standing tensions between the 'traditional' and the 'novel', in all aspects of contemporary Greek life. This pseudo-dilemma, namely 'traditional' vs. 'novel', takes on many forms in the case of the Greek landscape: between the continuation of age-old productive uses of the rural landscape versus its museumization as an object of cultural value and aesthetic pleasure, or between its trading as a commodity in various consumption industries versus its reverence as a place of family or ethnic roots, an object of contemplation through the arts, etc. In the meantime, for all practical purposes, the landscape has been absent from most expressions of everyday private or public life in Greece, at the same time as, in most European countries, it has repeatedly been attributed properties of an essential context and product of high quality in life. Indicatively, although it has signed the ELC, Greece has been stalling its ratification ever since.

The first objective of this study is to sketch an outline of the intertwined relationship of landscape with tourism. The second is to argue that the lack of a well-developed lay landscape conscience in Greece is slowly and steadily being overturned at present, through the growth of domestic tourism. It is proposed that, in place of a full-fledged industrial revolution, tourism has been the main venue of the development of countryside awareness and the generation of a landscape conscience among Greeks. As is well-known, Greece relies on tourism for much of its revenue, with tourism ranking first among its sources of foreign income – the country's first industry.

From landscape to tourism and from tourism to landscape

It has been widely admitted that tourism has greatly contributed to the worldwide diffusion of landscape form, function and symbolism, through time and place (Towner 1996). Roberts, for instance, argues that ideas about the aesthetics of landscape in mid-eighteenth century Britain prompted the development of tourism in Wales, and still continue to condition public policy toward the protection of such areas, internationally. 'Particular landscapes were chosen, modified, promoted and marketed for tourism purposes, lending support to the claim that the organizers of these "picturesque" tours were among the first to develop what we call today a "package holiday"' (2009). A number of European artists visiting Greece in the mid-1800s 'showed less concern with the pleasurable, scenic side of Greece, and instead attempted to describe their own personal vision or experience of the country' (Tsigakou 1991: 20), a definition often provided for the activity of tourism itself. Today, since statistically most people travel for purposes of 3S (sea, sand, sun) tourism, landscape also becomes the primary international tourism motive.

Characteristics of modern European cultural landscapes were already established by the seventeenth century, imbuing the definition of landscape with notions of vistas, prospects, or views of scenery of the land (Terkenli 2002). Likewise, conventions of sightseeing performance were first developed in Europe between 1600 and 1800, through the shift from touring as an opportunity for discourse to enthusiasm for travel as public 'eyewitness' observation. During the initial period of sightseeing, landscapes were described by travellers then 'in the manner of a painting' (Adler 1989: 14). By the eighteenth century, through an aesthetic transformation, 'a new canon of 'picturesque travel' added natural landscape to the other 'things' which an aesthetically trained eye might hope to grasp' (Adler 1989: 22): a democratization and demystification of a human-landscape relationship starting to open up for all.

Such processes and notions have accompanied the development of landscapes until our times and set the stage for twentieth-century organized contemporary mass tourism. The evolution of particular trends in tourist demand inevitably led to and reciprocally stemmed from appropriate interventions in the visited landscape, through very specific principles and strategies of landscape design and planning that grew out of the art or science of the perspective. In all these cases, the landscape has been staged through tourism planning and development initiatives for purposes of tourist consumption (Terkenli 2004; Lofgren 1996). The transformation of landscapes for purposes of leisure and tourism has been central to contemporary socio-spatial change stemming from the explosion of transportation possibilities and the representation of the past within the present in leisure and tourism industry (Aitchison et al. 2000). The enduring intensity of pleasure sought and found in landscape since the Renaissance, in the context of an emerging European bourgeoisie, expresses something profound and constant about the human condition (Rose 1996), linking landscape, aesthetics and pleasure inextricably together, highlighting the significance of the human emotional component in the relationship visitor-landscape. *Sightseeing* has always been an integral part of the enjoyment of touring.

Among poles of tourism appeal, the landscape has always retained a primary position, either in terms of landscape forms, functions and values as destination points, or in terms of whole landscapes, constituting the tourism scene itself. As the image or representation of a place, landscape is the first and most enduring medium of contact between tourist and prospective or consumed place of travel. Through acquired photographs and other mementos it becomes a traveler's lasting memoir (Terkenli 2002). The tourist landscape thus emerges as the product of tourism activities which tend to dominate an area's appearance and organizational structures and functions. It becomes a social interface where local and global perspectives, the sides of supply and demand, production and consumption etc. come together in the ready construction and consumption of place identity (Terkenli 2000: 185-6) (Figure 15.1).

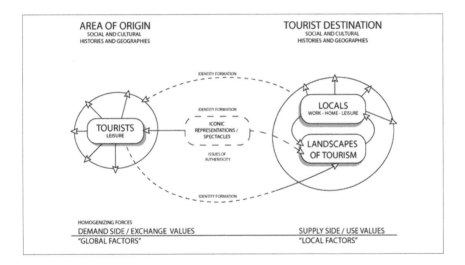

Figure 15.1 The formation of tourist landscape
Source: Terkenli 2002.

Tourist landscapes are both natural and human-made, as they are specifically designed to serve all needs of tourism development (Wall apud Jafari 1982; Gunn 1979). Moreover, tourist landscapes are currently advertised, replicated, signified and communicated around the world electronically or through travel literature, through their images, prompting tourists to become amateur semioticians and collectors of signs (MacCannell 1976). Tourism marketing reproduces images and discourses about landscapes, through representations of cultural signs, on the basis of which the tourist – through processes of experiential re-interpretation of the sign – may assess the sight and validate the meanings of the visited landscape, within the predominant discourse (Norton 1996). Such processes serve to produce and reproduce 'distinctive' characteristics of the tourist destination and to shape tourist landscape identities.

What ensues through these processes, however, is a tendency towards landscape uniformity, commercialization, banality, inauthenticity, placelessness, disposability, and/or deconstruction: in short, all processes of *a new cultural economy of space* (Terkenli 2006), often leading to a spatial re-organization, with uncertain, unproblematic and unstable impacts on the landscape. Such processes, supported and supplemented by other visual means of contemporary mass media – such as TV and video – blur geographical differentiation and the distinction 'authentic'-staged and familiar-exotic in the landscape images conveyed. They obviously have an adverse impact on the landscape itself, it terms of loss of its cultural character, its physical/ecological constitution and other geographical particularities. Worse, they indoctrinate the viewer/visitor in stereotypical, sterile landscape expectations (white sand beaches and palm-tree-lined boulevards, international-style shopping malls and casino lands, etc.), thus perpetuating the spread of cultural globalization, to the detriment of local landscape distinctiveness.

Greeks and their landscapes: Dissection of an unfulfilled relationship

In most circumstances of local or public life, regarding cultural trends, economic activities, political initiatives, social issues, urban and regional development, planning and management, the Greek landscape almost appears to be a non-entity. As a rule, local interests, input and decision-making concerning the landscape are normally ill-informed, marginalized, or – more commonly – non-existent. Under these conditions, landscape matters tend to remain overwhelmingly dependent on public or private economic or political interests.

In Greece, systematic physical planning interventions on the landscape have generally tended to be restricted to metropolitan and urbanized areas and have predominantly been a long-standing tradition of the design sciences (Figure 15.2). Mobilization in matters pertaining to the agricultural landscape has only been very recently instigated through European Union legislation and subsidized interventions (through C.A.P.) that enforce rural landscape protection and preservation (Kizos and Terkenli 2006; Louloudis et al. 2005). The lack of institutional support, in terms of landscape planning, policy and management, is evident in the absence until recently (October 2009) of a ministry for the environment. Meanwhile, Greek urban centres suffer much from a lack of green: indicatively, Athens has the lowest percentage of per capita green space in the EU (2.3-3 m^2).

Figure 15.2 Athens, view of Lykabetus Hill, 2007

In the 1990s landscape science underwent a shift from the fragmentary, peripheral and haphazard preoccupation of the so-called design sciences (architecture, landscape architecture, urban and regional planning) with practical landscape issues, as they developed out of related design and planning initiatives and spatial interventions, to a more concerted, focused and systematic landscape approach by several more disciplines and practitioners (Terkenli 2004). This was mainly a qualitative shift, characterized by its very limited extent and impact on actual landscape problems, as well as on effective application in landscape policy. With regard to Greek landscape education and science (Terkenli 2004), despite serious, but rather sporadic and fragmented, efforts in tertiary education institutions around the country, landscape education is still lacking, at all levels of the educational system. Processes of establishing landscape science, research and practice as such have gradually been gaining ground in very recent years.

Normally, negative impacts on the landscape tend to be recognized as such only when they become catastrophic or lead to calamities (i.e. 2007 and 2009 fires), eliciting some form or other of top-down or bottom-up reaction. However, the gravest impacts of human activity on the Greek landscape continue to scar or deform it unabatedly. Perhaps most significant among these are the ongoing rural-urban migration, with consequent abandonment of agriculture and livestock breeding, and the widespread proliferation of second-home construction. The unabated exodus of rural population from the Greek countryside is robbing it of its guardians and stewards, while the burgeoning – often illegal – secondary residence growth by an increasingly affluent middle class was 'exacerbated in recent years by the country's adhesion to the European Union and the influx of North Europeans in search of holiday homes' (Stathatos 1996: 18).

Turning to the underlying causes of Greece's problematic relationship with its landscape, these may be traced to the lack of a defined and well-developed landscape conscience in the country, with deep roots in its history as a modern nation-state (Terkenli 2008). *Conscience* here is defined as a mixture of perceptions, thoughts and emotions, which presupposes the existence of an external world (IDP 1990), whereas *landscape conscience* refers to the distinctive conscious or unconscious bonds that characterize a person's or a people's relationship with their landscapes. As argued elsewhere (Terkenli 2008), the causes of such a conscience deficiency in Greece may be many; of these, only a number will be addressed here, those that seem to have played the most crucial role in the problematic relationship of Greeks with their landscapes. What follows then is a brief tracing of the historical, aesthetic and socio-cultural trajectory of the relationship of modern Greece with its landscape, during the past 150 years, in search of the – essentially – *urban* origins of a landscape conscience.

Greece never went through a Renaissance, an urban rebirth, a baroque phase or even an optical revolution, as did most other European nations. It rather adopted aspects of modernity in certain realms of life *a posteriori*, by implanting and overlaying them on pre-existing cultural particularities and local ways of life. From an aesthetic perspective, the landscape ideal and most characteristic form of representation of this cultural realm remained the two-dimensional, apparently flat, but actually inverted, perspective of the Greek Orthodox art: ecclesiastical iconography, powerfully evocative and compelling to all Greeks to-date. Amidst the deeply religious populace, under the four-century-long Islamic domination, Greek Orthodox art seems to have imprinted its highly influential worldview on the Greek mind and psyche: a bottom-up landscape ideal and way of relating to the world still pervasive in Greek life and art. At the same time, mainly during the nineteenth century, the Greek landscape was being re-constructed by Western painters, in accordance to top-down romantic ideals, at the basis of the then emergent Greek cultural identity: a) archaism and b) orientalism (Terkenli et al. 2001).

Greece also never went through a full-fledged industrial revolution. At the time when the new nation-state of Greece was being created (1820s-30s), there existed no middle class to speak of, in order to re-invent the landscape concept through an urban-rural contradistinction (the urbanites' nostalgia for the loss of the countryside), as was the case in Western/Northern Europe. Elsewhere in the First World, such stretches of urban growth succeeded an industrial revolution and led to the development of a sense and an ideal (conscience) of landscape (Bunce 1994). In Greece, however, the period of high urbanization rates (1960s and 1970s) did not give rise to or stem from a wholesale industrialization of the country's economic system, but rather the rapid adjustment of the Greek society and economy to a tertiary-sector-led nexus of activities. The incentive for the development of a Greek landscape conscience, as we shall see below, rather emerged through the rise of internal tourism and a more general turn among urbanites back to 'nature' and the countryside.

Nonetheless, perhaps the most significant underlying cause of such a landscape conscience deficiency among Greeks has been the lack of a lay sense of the *landscape as a common good*. This may be defined as the integrated set of material and non-material dimensions and features of the landscape, at the disposal of a particular social group, where its use by one user diminishes its amount available to all others, but for which the exclusion of additional users is difficult or impossible (Bromley 1991). In accordance with this tendency, contemporary Greeks seem to fail to think of landscape as part of their common home (Terkenli 1995), a trend indicative of a disintegration of their previously existing 'traditional' environmental conscience vis-à-vis outdoor resources, including the landscape. Instead of a cooperative, ethically active and vigilant strategy of generous mutuality (*community*) (Ostrom 1990), a 'marketplace principle' has persisted among cultural characteristics of Greek social life, throughout its history, up to the present (McNeill 1978). Instead of being acknowledged as a *common* economic, cultural and environmental resource, the Greek landscape had been taken for granted till the end of the 1970s. It first started to be acknowledged at that time, through interconnections then emerging between agricultural modernization and the rural landscape (nature vs society), through urbanization and through tourism.

The role of tourism in the development of a Greek landscape conscience

As elsewhere in the First World, and as the preceding discussion indicates, urbanization has been the main triggering force towards the development of a Greek landscape conscience – in terms of landscape awareness, perception, attachment, valuation, and engagement with the landscape as such. In Greece, however, the re-discovery of the Greek landscape was instigated mainly via agricultural modernization, via environmental destruction or deterioration, via the arts, but mostly – as the last section of this essay will argue – via tourism and recreation. This trend has been materializing largely as an attempt to overcome urbanites' growing physical and psychological distance from and nostalgia for the countryside. This landscape re-discovery has been resting on, triggering, but also materializing through various forms of re-visiting the landscape: through actual re-discovery (domestic tourism), through imaginary re-discovery (a nation's and its diaspora's symbolic constructions of landscape ideals) and through iconic re-discovery (communicated images/texts via advertisement, the press and other media). The last part of this chapter will attempt to substantiate this argument, through a part-hermeneutical, part-phenomenological investigation of this recently emerging trend. The objective will be to trace indications of such a shift in lay landscape conscience, through an analysis of indications that the Greek public is beginning to become cognizant and concerned about the country's landscapes.

This analysis ought to rest on an obvious disclaimer, namely that there is no single 'Greek landscape', but rather a number of real or symbolic Greek landscapes, depending on the position and situation of the viewer/analyst. For example, the

Greek island landscape may be conceived as a cultural image of tourist consumption for its visitors, as a home ridden with problems for the local populations, or as a cultural or family hearth. In the latter case, the island landscape as a national symbol and as a cultural and family hearth, is constructed in collective Greek imagination with an orientation towards a historical past: a homeland, laden with vestiges of an ancestral land and a rich cultural (historical, archaeological or sacred/religious) heritage. According to this myth, it is perceived as an essentially uninhabited landscape for most of the year, while, during holidays and especially summer, it becomes 'vacationland', a playground for both domestic and international tourism (Terkenli 2001, Tsartas 1989). Such perceived and/or constructed qualities of the Greek island landscape mainly derive from its visual characteristics. They have also tended to be expropriated and exploited for various 'development' purposes, often with negative impacts on the appearance and undermining the very essence of the landscape that attracted tourism there in the first place.

Some first signs of mobilization in rural landscape matters in Greece have only very recently been arising, through EU legislation and subsidized interventions (C.A.P.) that enforce the protection and preservation of the rural landscape (Vlachos and Louloudis 2008; Louloudis 1992). In rural studies, rural geography, rural sociology and local/rural development, the landscape has been conceptualized as a productive resource – either in primary economic production or in tourism – and as a vital living space for its rural inhabitants. There has been little concerted effort, however, by Greek academics, authorities or professionals, towards practical intervention in rural space, such that would help raise public landscape conscience (Vlachos and Louloudis 2008; Louloudis 2006; Louloudis et al. 2005). Both at a research level and in practice, any rural landscape measures or interventions have stemmed from the EU (i.e. implementation of C.A.P. regulations in the context of agro-environmental measures) or from state environmental projects, such as Regional Development Plans. C.A.P. has been the main driving force behind most change in rural space – change that has proven to be detrimental to the Greek landscape (Louloudis et al. 2005) – and has, moreover, not instigated any substantial shift in human-landscape relationships in rural Greece.

No grand-scale transformation in lay landscape conscience has been effectuated through the arts, either. According to art historian Haris Kambouridis, as concerns 'high culture', top-down landscape renderings by Greek painters, 'the European gaze on the Greek landscape has never ceased to hold the leading role in our mentality' (2009: 6). Greece always held the imagination of those western artists that have instilled a landscape model in Greek landscape painting, in terms of a romantic ideal. Long periods of war and foreign domination had resulted in the impoverishment of both its people and the landscape, leading nineteenth century visiting artists to attempt simply 'to evoke 'the spirit of the place [...] as a means to recreate a sense of nostalgia for a bygone age and to satisfy the literary imagination' (Tsigakou 1991: 16-17). Instead, the highly anthropocentric tradition of Greek Orthodox Ecclesiastical art, as presented above, still remains the dominant paradigm in lay landscape perception and depiction.

Similarly, Greeks' mentality towards the environment, generally speaking, is slow to change (Pettifer 1993). In the Greek Constitution, the landscape has, legally and institutionally, been allocated to environmental law (Issues of Government Gazette 160/16-10-1986, 3257 and 3258), with only other references to it appearing in master plans and regulatory statutes concerning the protection of archaeological spaces, as well as in legislation on traditional settlements, aesthetic forests and national parks. Contemporary Greeks seem to be developing an environmental conscience only very recently, principally on the basis of concern for climatic change, air and water pollution and forest fires. Indicatively, a recent nation-wide longitudinal survey on Greeks' environmental conscience elucidates not only broader emerging trends in their relationship with the environment, but also skirts around or touches on Greeks' relationship with the landscape, in multifold ways. One of the survey's primary findings was that, since 2001, environmental issues (concern, interest, prioritization, education, responsibility and action) come in first or second place among predominant social issues (Greeks' Environmental Conscience 2008). These signs of a growing environmental conscience were markedly higher among inhabitants of Greece's two largest urban centres (Athens and Thessaloniki) and of tourism destinations (also featuring higher and rising standards of living), such as the Aegean Islands and Crete. Most significantly, when it came down to the specific environmental aspects of concern, the landscape was totally absent from lay people's conceptualization of their environment. Only when questions of aesthetics were included in matters of environmental deterioration did the landscape appear to be of a prominent concern to respondents, together with ecological concerns. In these cases, as in all cases of heightened public concern for the environment, it was the more well-informed urban young-to-middle-aged and middle-aged Greeks that showed most pronounced signs of an environmental/ landscape conscience.

The preceding discussion inadvertently leaves tourism as the principle source and means of a slow, but clear-cut and steady, re-establishment of a relationship between – mostly urban – Greeks and their landscapes, and to the first signs of an emergence of a contemporary Greek landscape conscience. From the early to mid-1990s, the balance of tourism demand in the country between international and domestic tourism has been changing. Whereas, until then, international tourism had been the most significant and well developed form of tourism in the country (over 75 per cent of total tourism arrivals in 1989), the trend has since been partly reversed – although domestic tourism still remains very difficult to measure. In the past 10-15 years, Greeks have been rediscovering their landscapes *en masse*. The underlying causes of this trend must be found in a) a higher standard of living and dissatisfaction with conditions of city life, b) aggressive advertisement and promotion of Greek destinations by the state and the tourism supply side and c) the combination of emerging alternative forms of tourism and other tourism growth opportunities, provided by Greeks' discovery of 'long-weekend' tourism (Figure 15.3).

Figure 15.3 The village of Oia, Santorini, view over the Caldera, 2007

Thus, contemporary Greeks seem to have started noticing landscapes toured or viewed in the travel press and in various tourism memorabilia. A detailed study of domestic tourism shows that the 'environment'- generally speaking – is among the three most significant motives of domestic tourism in Greece (Tsartas et al. 2001). Since there is no explicit 'landscape' category, the 'environment' comes closest to it and third in people's preferences, after 'holidays and entertainment' and 'rest and relaxation'. This comes as no surprise, since it is 3S tourism that most Greeks pursue during their established vacation period in July/August. Through domestic tourism, Greek urbanites are slowly beginning to develop a landscape conscience, to rediscover landscape, and, through it, to rediscover family roots, local histories, 'authentic' Greece and to 'return to nature'. For instance, 'traditional' settlements, archaeological sites and monasteries currently constitute powerful poles of tourist attraction, thus creating new functions and values for pre-existing cultural landscape forms and features.

Where the trend towards a reconnection with the landscape through tourism becomes most apparent, however, is in alternative forms of tourism, and especially agro- and eco-tourism. In the context of domestic tourism in Greece, these two alternative forms of tourism tend to be highly interconnected (i.e. through sale of local forest products, hiking, olive picking, etc) and highly beneficial to landscape preservation and management. Agrotourism, for instance, is relatively new to the Cyclades – one of the top tourist destinations in the country: it is not yet as developed as other more familiar types, but it is growing fast, for both foreigners and Greeks (Kizos et al. 2007). According to Kizos et al., on many Cycladic islands, there is a growing demand for traffic-free tracks and footpaths, permitting the enjoyment of rewarding, interesting, stimulating and/or special-interest activities (e.g. bird-watching). Besides sightseeing and the enjoyment of landscape, such activities have also contributed to landscape-related actions in the context of local policy initiatives, i.e. to conserve, signpost, clean and repair old paved footpaths (European-funded LEADER projects), etc. Although few and sporadic, these initiatives lead to a more widespread landscape awareness and promotion among both inhabitants and visitors and directly contribute to the preservation of 'traditional' landscape forms, functions and meanings/values for both locals' and tourists' purposes. 'Policy initiatives are only recently beginning to emerge, in response to growing concern about the future of deteriorating landscapes [...] Once old landscape elements acquire new functions (due to new values regarding the landscape or activities in it), they have better chances of preservation' (Kizos et al. 2007).

Concurrently, these – mostly rural – tourism and recreation landscapes have been related to a series of significant transformations, in connection not only with conventional agricultural production, but also with landscape symbolism and representation. Specifically, they have been turning into urban or semi-urban consumption spaces, through recreation activities and second-home construction (Hadjimichalis 2008). Thus, rural landscapes – whether coastal, island, mountain or other – have been adjusting to their new uses, through a) preservation mostly of their formal attributes, for purposes of visual appeal (Terkenli 2000), and through b) protection of selected features, for purposes of tourism consumption (i.e. promotion of local products, provision of aesthetically seductive views and sites, etc.).

The loss of spatial specificity, an inevitable outcome of changes imposed by capitalism on social structures, seems to be currently exacerbated, through new processes of landscape transformation: processes of the *new cultural economy of space*. As a result of the industrial revolution, 'in order to respond to the now undifferentiated landscape, the urban middle classes invented the concept of landscape, a concept soon associated with the picturesque [...] It was precisely this urbanization, and the increasing distancing from nature to which were subjected the population of societies in the process of industrialization, which almost simultaneously created the need for contact with some substitute, however false' (Stathatos 1996). For Greeks, this need and rediscovery was instigated by domestic tourism. Stathatos describes how post-war Greek governments were quick to realize the vast profits that could be made out of a modernized tourism industry; the result was an intensive promotional campaign at home and abroad, still active today, through publicity photographs, posters, postcards and other representations of Greek space. These representations promote an imaginary country upon which the sun always shines brightly, where the sea is always blue and placid, the houses – of a uniformly Cycladic style – are invariably freshly whitewashed, and all of whose inhabitants are permanently cheerful, welcoming and colourful (Stathatos 1996). The problem, however, is that Greece has been exporting, but also consuming this distorted image of itself (Figure 15.4) for four consecutive decades, and there is reason to believe that it may be becoming innate (Stathatos 2008).

Figure 15.4 Cape Sounion, Attica, an imaginary landscape depiction from the campaign of the National Tourism Organization of Greece, 2006
Source: Reproduced with permission.

Conclusions

The primary objective of this study has been to examine the ways in which tourism has been the driving force behind the development of a landscape conscience in contemporary Greek society. For this purpose, the study has argued that this shift in lay attitudes towards landscape has been materializing mainly through domestic tourism, and especially eco- and agro-tourism. As contemporary rural and natural spaces urbanize at a fast pace, landscape analysis and management are becoming increasingly pressing in Greece through general concern for the environment (agro-ecosystem degradation, housing construction and urban expansion). Landscape issues are also becoming increasingly imperative in the context of the growth of eco-tourism and agro-tourism, signalling a re-discovery of the Greek natural and cultural heritage. Finally, they are also surfacing through tendencies emerging among urbanites of a *return to nature* and to rural and environmental values and cultural roots.

There are, however, additional challenges, if Greece is to catch up with the rest of First World countries in sustainable landscape management and concerted landscape planning and policy implementation. Much more so because, in the face of current forces of globalizing transformation, instilled by a new cultural economy of space, tourist landscapes increasingly tend to resemble one another at an alarming rate: Utopian paradises with tropical characteristics, widely and uniformly (re)produced, through standardized, homogenizing processes of landscape replication, exclusively for purposes of recreational consumption (Trova 2008, Terkenli and d'Hauteserre 2006, Terkenli 2002). Obviously, the solution lies in re-defining and developing lay landscape conscience. This task requires concerted nation-wide mobilization in all sectors and strata of society towards increased landscape knowledge and education, active public participation in decision-making and, most of all, urgent action in re-configuring our landscape geographies – a task long overdue.

References

Adler, J. 1989. Origins of sightseeing, *Annals of Tourism Research* 16, 7-29.

Aitchison, C., MacLeod, N.E. and Shaw, S.J. 2000. *Leisure and Tourism Landscapes: Social and Cultural Geographies*. London: Routledge.

Beriatos, E. (ed.) (forthcoming). *Proceedings of the Scientific Symposium on the Protection and Promotion of the Greek and Mediterranean Landscape* (in Greek), Cephalonia, Oct. 26-28, 2007.

Bromley, D. 1991. *Environment and Economy*. Cambridge: Blackwell Publishers.

Bunce, M. 1994. *The Countryside Ideal: Anglo-American Images of Landscape*. London: Routledge.

Doukellis, P. 1998. Mentalities, ideologies and modern Greek landscape management, in *A European Geography*, edited by T. Unwin. New York: Longman, 78-79.

Gunn, C. 1979. Landscape assessment for tourism, in *Proceedings of Our National Landscape: A Conference on Applied Techniques for Analysis and Management of the Visual Resource*. Incline Village, Nevada Vol. 23-25/04, 409-414.

Greeks' Environmental Conscience 2008. *Report of the Hellenic Society for the Environment and for Culture*. Athens.

Hadjimichalis, K. 2008. Geographical representations/fantasies of countryside landscapes (in Greek), in *Proceedings of the Scientific Meeting on 'The Claiming of the Countryside'*, edited by K. Manolidis. Volos: University of Thessaly, 27.

IDP (The International Dictionary of Psychology) 1989, edited by N.S. Sutherland. New York: Continuum.

Issue of Government Gazette (in Greek) 160/16-10-1986, 3257. Athens: National Publishing House.

Issue of Government Gazette (in Greek) 160/16-10-1986, 3258. Athens: National Publishing House.

Jafari, J. (ed.) 2000. *Encyclopedia of Tourism*. London: Routledge.

Kambouridis, H. 2009 *Modern Greek Landscape Painting from the 18th to the 21st Century: Vision, Experience and Reconstitution of Space* (in Greek). Athens: B. & M. Theocharakis Foundation of Plastic Arts and Music.

Kizos, T., Spilanis, I. and Koulouri, M. 2007. The Aegean islands: A paradise lost? Tourism as a driver for changing landscapes, in *Europe's Living Landscapes*, edited by B. Pedroli, A. van Doorn, G. de Blust, M.L. Paracchini, D. Wascher and F. Bunce. Wageningen: KNNV Publishing, Zeist.

Lofgren, O. 1999. *On Holiday: A History of Vacationing*. Berkeley: University of California Press.

Louloudis, L. 2006. Agricultural landscape reconstructions: Land and farmers in the community of Anthili, in the Spercheios Delta (in Greek), in *Proceedings of the 9th Panhellenic Conference of Agricultural Economy, 'Economy and Society in the Face of Challenges to the Global Agro-Nutritional System'*, 2-4 November 2006. Athens: Agricultural University of Athens (in digital form).

Louloudis, L. 1992. Agricultural Modernization and Rural Landscape Transformation (in Greek). Topo.s 4, 135-56.

Louloudis, L., Beopoulos, N. and Troumbis, A. (eds) 2005. *The Rural Landscape: The Palimpsest of Centuries of Agricultural Labor* (in Greek). Athens: Ktima Merkouri, Korakohori Hleias.

Manolidis, K. 2008. The countryside and the traumas of post-war Greece (in Greek), in *Proceedings of the Scientific Meeting on 'The Claiming of the Countryside'*, edited by Manolidis, K. Volos: University of Thessaly, 54.

McNeill, W.H. 1978. *The Metamorphosis of Greece since World War II*. Chicago: The University of Chicago Press.

Norton, A. 1996. Experiencing nature: The reproduction of environmental discourse through safari tourism in East Africa. *Geoforum*, 27(3), 355-73.

Ostrom, E. 1990. *Governing the Commons: The Evolution of Institutions of Collective Action*. Cambridge: Cambridge University Press.

Pettifer, J. 1993. *The Greeks: The Land and People Since the War*. Harmondsworth: Penguin.

Roberts, G.C. 2009. Landscape aesthetics and the tourist mind. Paper presented at the 4th International Scientific Conference *Planning for the Future – Learning from the Past: Contemporary Developments in Tourism, Travel & Hospitality*, Rhodes, 3-5 April 2009.

Rose, G. 1996. Geography and the Science of Observation: The Landscape, the Gaze and Masculinity, in *Human Geography: An Essential Anthology*, edited by J. Agnew, D.N. Livingstone and A. Rogers. Oxford: Blackwell, 341-350.

Semaioforidis, G. 2005. *Passages: Texts for Architecture and Metapolis* (in Greek). Athens: Metapolis Press.

Stathatos, J. 1996. *The Invention of Landscape: Greek Landscape and Greek Photography 1870-1995*. Thessaloniki: Camera Obscura.

Stathatos, J. 2008. Humans and landscape in contemporary Greek photography (in Greek), in *Proceedings of the Scientific Meeting on 'The Claiming of the Countryside'*, edited by Manolidis, K. Volos: University of Thessaly, 30.

Terkenli, T.S. 1995. Home as a region. *The Geographical Review*, 85(3), 324-334.

Terkenli, T.S. 2000. Landscapes of tourism: A cultural geographical perspective, in *Tourism and the Environment: Regional, Economic, Cultural and Policy*

Issues, edited by Briassoulis, H. and van der Straaten, J. Revised Second Edition. Dordrecht: Kluwer Academic Publishers, 179-202.

Terkenli, T.S. 2001. Towards a theory of the landscape: The Aegean landscape as a cultural image. *Landscape and Urban Planning*. 57, 197-208.

Terkenli, T.S. 2002. Landscapes of tourism: Towards a global cultural economy of space? *Tourism Geographies*, 4(3), 227-54.

Terkenli, T.S. 2004. Landscape research in Greece: An overview, *BELGEO*, (2-3), 277-288.

Terkenli, T.S. 2006. Landscapes of a new cultural economy of space: An introduction, in *Landscapes of a New Cultural Economy of Space*, edited by T.S. Terkenli and A.-M. d'Hauteserre. Dordrecht: Springer.

Terkenli, T.S. 2008. Greece and landscape: An unfulfilled relationship, in *Landscape Research Group Newsletter, 'Sheffield Conference on the European Landscape Convention 19-20 Nov. 2007'*, edited by B. Young and P. Howard, P. Manator, East Dartmoor: Moor Print, 11-13.

Terkenli, T.S. and d'Hauteserre, A.-M. (eds) 2006. *Landscapes of a New Cultural Economy of Space*. Dordrecht: Springer.

Terkenli, T.S., Tsigakou, F.-M. and Tsalikidis, I. 2001. The physical landscape of Greece in nineteenth-century painting: An exploration of cultural images. in *Art and Landscape: Proceedings of the Symposium of IFLA*, edited by G.L. Anagnostopoulos. Athens: Panagiotis and Effie Micheli Foundation, 618-634.

Towner, J. 1996. *An Historical Geography of Recreation and Tourism in the Western World 1540-1940*. New York: John Wiley and Sons.

Trova, V. 2008. From the anguish of greekness to the happiness of the tropics (in Greek), in *Proceedings of the Scientific Meeting on 'The Claiming of the Countryside'*, edited by K. Manolidis. Volos: University of Thessaly, 39.

Tsartas, P. 1989. *Social and Economic Impacts of Tourism Development on the Prefecture of Cyclades and Especially on the Islands Ios and Serifos, During the Period 1950-1980* (in Greek). Athens: EKKE.

Tsartas, P., Christou, E., Sigala M. and Chalkiti, K. 2001. *Electronic Services and Applications in Tourism: Current Situation and Perspectives. Report for Information Society EU Project*, www.ebusinessforum.gr. Athens: E-Business Forum.

Tsigakou, F.-M. 1991. *A Romantic Vision of Greece. Through Romantic Eyes: European Images of Nineteenth-Century Greece from the Benaki Museum of Athens*. Alexandria, Virginia: Art Services International.

Vlachos, G. and Louloudis, L. 2008. From the field to the space: The emergence of the agricultural landscape in the post-productive countryside (in Greek), in *Proceedings of the Scientific Meeting on 'The Claiming of the Countryside'*, edited by K. Manolidis, Volos: University of Thessaly, 24.

Chapter 16

Biodiversity and Land Abandonment: Connecting Agriculture, Place and Nature in the Landscape

Ruth Beilin, Regina Lindborg and Cibele Queiroz

Introduction

In this chapter we present three study sites – Hållnäs, Sweden; Sistelo, Portugal; and Poowong, Australia – to provide insight into three different temporal and spatial scales of change (James et al. 2000); and we present an argument for an integrated approach to how ecology and sociology frame ideas and practices around agricultural landscape change affecting biodiversity values. As Hobbs and Cramer (2007: 2) note, 'land abandonment has been a feature of humanity's relationship with the world's ecosystems for as long as history has been recorded'. The concept of 'abandonment' masks landscape changes directly related to local management practices. These management practices are made up of complex interactions between society and place. It is this complex social and ecological landscape that we explore.

As Naveh (1998) argues, from the moment that humans begin to change the landscapes they encounter, the consequences are incremental. In a system, the flow-on effects of small changes cannot always be understood or even experienced at the time they occur. This complexity encourages us to re-examine the idea of nature and its relationship to agriculture. In Sweden, we see thousands of years of agriculture in one place. The nature preserved here is associated with meadow flowers and grasses that have evolved at the edges of fields and roadsides. Abandoning these fields – meaning no agriculture – results in woody shrub succession and meadow species are overtaken. This is a very particular representation of nature associated with the history of Sweden's development. Social and ecological perspectives are active in constructing Swedish practices resulting in the agricultural landscape and its maintenance as a cultural landscape. Each of our study sites evolves from this cultural understanding of a relationship between nature and production that manifests in expectations of how a landscape is best managed for acceptable nature and agriculture outcomes.

Biodiversity has become a synonym in science and especially in ecology of a particular idea of 'nature' and of its physical manifestation (Deverre et al. 2007). There is wild nature or 'wilderness', and there is biodiversity as 'isolated' in protected areas and as 'nature' in agricultural landscapes. In Australia, biodiversity is associated almost exclusively with indigenous flora and fauna, separated in practice from agricultural production values (Beilin et al. 1997). In Europe, biodiversity is widely understood as the result of agricultural practices as reinforced by the EU subsidies to protect 'biodiversity hotspots' in agricultural landscapes. Each of our sites has issues associated with the separation of conservation and production values; however, each also demonstrates the potential to reunite values within the same landscape. In Australia, for example, the indigenous flora of pre-European settlement is still in evidence despite the reality of 200 years of massive forest habitat clearing to create European-style agricultural landscapes. Programs like Landcare acknowledge biodiversity values in planting out riparian zones and re-establishing billabongs with indigenous material without impacting production systems (Beilin 1997, 1999). Lefroy and Smith (2004) attribute flow-on benefits between revegetation zones and production, and suggest an integrated approach is important for conservation and for production.

Just as 'abandonment' is associated with the loss of agricultural production, and not necessarily with a loss of ownership, a sense of place evolves with historical management practices. In Sistelo, local memories of place represent a past aligned with agricultural production and animal husbandry rather than one that promotes a park landscape. The drivers of change, therefore, are both internal and external to place. As Harvey (1996) suggests, place provides an opportunity to disaggregate the historical process of its construction and evolve a radically different landscape. It is this task that we ultimately foresee as emerging as both a possibility and a necessity across these sites.

Section I

Biodiversity or 'nature' science is historically separated from the practices associated with production sciences – horticulture, agriculture and resource management. Disciplines construct ideas about nature – its representations and symbolic meanings in society – and then examine these ideas as existing in a place. Existing biophysical systems assist in supporting such visions, but as the story of Australia shows, early white settlers imagined the landscape was highly fertile and well watered and a perfect setting for European farming systems. They did not see signs of previous land management by Aboriginal people or the predominantly thin and delicate soil profiles that did not tolerate hoofed animals or the extreme weather conditions that included long seasons of drought (Bolton 1981). Science knows how to create the conditions that are required for the agricultural production system to prosper and we manage it as 'command and control' (Holling and Meffe 1996), often with little reference to the ecosystem.

We discuss the three sites not in a 'compare and contrast' manner but by exploring the co-dependent qualities of nature and culture that arise as part of agricultural practices (Table 16.1). The design of this study provides insights for management of abandoned land, and explores 'abandonment' in the context of each locale. The landscape is a continuum (Crowe 1958) and, as Eaton (1990) states, 'a signpost for what a culture values'. In considering place as a way of delineating our sites in a global landscape, the continuum is an important metaphor, indicating the global and local at both social and ecological levels. Each place – Sistelo, Hållnäs and Poowong – has experienced markedly different landscape transitions; and while we focus on the agricultural aspects of these changes, the continuum demands we remain aware of how we boundary each site for discussion.

Each of these sites has some characteristics that define it as isolated or 'remote', while at the same time being subject to global penetration of politics and markets, and the vagaries of demographic shifts. They demonstrate the complexity of place. As Hay (2008) notes, the diversity of place ensures multiple interests that are both regressive and progressive in terms of change; and importantly, there are multiple opportunities for land use that are not associated with continuing traditional agricultural production systems though contributing to increased biodiversity and alternate production values. Fischer et al. (2008) provide evidence of multiple social expectations around productive yield and the interaction of farmland with biodiversity that suggest the need for more complex sets of criteria in decision-making by land managers; and these we argue manifest as characteristics of place.

Landscape ecologists focus on floral and faunal changes, emphasizing the importance of scale. In recognizing an ecosystem or describing it in relation to land associated with an oak forest, for example, the forest can define the place. However, practices such as cattle grazing, fire management and plantation establishment profoundly connect social intentions to this landscape. Ecosystem services provide a way of measuring the contributions of nature to human welfare (Boyd and Banzhaf 2006) and making the connection between direct and indirect use and benefit – for example, pollinating species that ensure ongoing production or clean air or water. The landscape sociologists' methods include illuminating these links between biophysical science and social interaction with nature and biodiversity.

Section II: Social and ecological discourse of land use and abandonment

This study involves three agricultural ecosystems with different environmental, socio-ecological and cultural characteristics. In each country we selected one study site because: at a regional scale they represent issues associated with land use, change and biodiversity; we have extensive existing databases in each area; and there is good literature available on history, landscape change and place attachment associated with these sites. Although the sites have different management and production regimes, they share some salient characteristics. For example, the sites are all affected by the difficult physical terrain and experience ideas about isolation and 'remoteness' affecting management.

Hållnäs, in the County of Uppland, Sweden, is a forested area (c. 40 per cent) with a small-scale mosaic of habitats and farms. It is about 20 km from nearby towns and there are 3,000 inhabitants (Lindborg et al. forthcoming). Arable fields are c. 20 per cent and semi-natural pastures also approximately 20 per cent of the landscape area. Parts of the research area in Hållnäs are protected for conservation under the European Natura 2000 (Slada SE0210270). There is a long tradition of agriculture and declining farming since the 1950s; nonetheless, the landscape still contains fragments of managed semi-natural grasslands. The 'cultural landscape' of historical grave fields, old settlements and rune stones is extensive. Hållnäs is also advertised as an area for recreation (http://www.Hållnäs.se). There are only 11 farms in Hållnäs now. Clark and Stein (2003) note that population size and density have little influence on community attachment. This is confirmed by what Swedish ecologists refer to as the national expectation that landscapes like Hållnäs are to be maintained as traditional agricultural fields. The grassland ecology in this area is nationally significant.

Sistelo, northwest Portugal, is a small parish located in the Peneda Mountain Range (41° 58' N; 8° 22' W). The mean annual rainfall is 2,000 mm and altitude ranges from 180 metres in the river valley to 1,350 metres on the mountain, which, combined with a diversified use of the territory, leads to a great variety of habitats. Agriculture was the main economic activity, with livestock as the most important source of income. The agricultural terraces began hundreds of years ago, with the advent of corn in the region. There is seasonal migration for pasturing livestock from the valleys in the winter to the mountain tops during summer (generally common property areas). These practices shaped the local landscapes and ecosystems leading to significant scenic values.

Drivers of change include the afforestation of common property areas by the Portuguese state in the 1940s, significantly reducing pasture. This, coupled with the attractiveness of labour opportunities in foreign countries, was a trigger for a large out migration in the 1960s (Medeiros 1984). These landscapes were not suitable for mechanization and keeping farming required continual physical labour. Agriculture in Sistelo is now considered uneconomical and there is a lack of alternative sources of income across the region, leading to the gradual abandonment of agricultural practices.

In Australia, the case study site is Poowong in the Strzelecki Ranges, located in southeastern Victoria. The bioregion is 341,862 hectares. Prior to white settlement in the mid-1800s, this area was an extensive forest on duplex soils (Noble 1976), and has been continually cleared since the white settlement began. In the 1920s it was nicknamed 'the heartbreak hills' as it was so difficult to make a living on farms across these steep slopes and with the poor soils that remained. By the 1970s the government bought back land or took over abandoned farms and planted pine and eucalypt forests for pulp logging. Many of the continuing farms are dairy farms. Dry land pasture (rain-fed) is 66 per cent of land use; non-farmland excluding remnant vegetation is 19 per cent; remnant vegetation 9 per cent; horticulture 4 per cent; forestry 3 per cent (Geospatial data DSE 1997 as detailed in http://www.

wgcma.vic.gov.au). Average annual rainfall varies across the region from 800 to 1,400 mm.[1] In the 2006 census there were 587 people residing in the locality of Poowong, with 33 per cent aged 55 or over; 24 per cent of 587 were involved in dairy cattle farming; 10 per cent in sheep, beef or grain farming (http://www. censusdata.abs.gov.au/). The average dairy farm is approximately 150 hectares.

The issues

In Sweden, the process of land abandonment is associated with the aging farming population. This is a European problem, regularly discussed in the EU. Subsidies, while sufficient to maintain formal institutions like schools and health centres are not guaranteed to ensure farming continues; and there is a substantial number of non-farming and amenity migrants (Moss 2006) within these landscapes. The diversity of social change from largely farming communities to diverse landholders with varying expectations of land use has implications for biodiversity. Increased forestry within previously agricultural landscapes, abandonment and intensification can lead to the decline of certain species. Therefore in Sweden abandonment of traditional agricultural practices is seen as negative for biodiversity.

In Poowong dairy deregulation is a significant driver of change. The terrain makes it difficult to switch to other forms of farming because tractoring for cropping, for example, is too dangerous on many of these slopes. The high rainfall relative to other parts of the country is ideal for continued, unspecified agriculture. The terrain also precludes large scale aggregation of holdings and mechanization is relatively limited. There are various scales of forestry activity from farm forestry plots to mountain side plantings. However, powerful consortiums and pension funds have bought vast acres of farmland elsewhere in the state and planted millions of acres of pulp and paper eucalypts on flat land, making Poowong's forest enterprises unviable.

In Sistelo, management options range from a revival of traditional agricultural practices – which in the locals' perspective will not be possible without some intervention from outside the community – to the re-establishment of the native oak forest. The destruction and fragmentation of native oak forests in the past was one of the main causes of the decline of some large species of mammals like the Iberian wolf (*Canis lupus*). Ecologically, the abandoned areas provide the opportunity for forest re-growth and habitat recovery.

In Sistelo and Hållnäs, more obviously than Poowong, traditional agricultural practices are closely associated with a sense of place. Agricultural area preservation is promoted by the EU subsidies both in Sweden and Portugal, and many scientific studies within the EU argue for the importance of biodiversity associated with their fields (Queiroz et al. unpublished ms). In Portugal, regeneration of the previously destroyed native oak forest opens the options on maintaining or improving

---

1 With the continuing drought, rainfall is considerably lower in this region than previously. To date, 470 mm have been received by July 2009. See Ian McNiven 2008.

biodiversity. The identity of the local population is closely associated with the terraced agricultural landscape and it is what they value and want to preserve (Pereira et al. 2005). In Poowong, agricultural innovation and technology have largely dictated the ability of farmers to operate here. Ecologists recognize about 29 remaining temperate rainforest species from the forest that existed at white settlement. Government subsidies focus on assisting non-viable farmers to leave or re-structure. In Australia, land use change is expected to occur in response to market forces.

Table 16.1 describes the main trends in demography, biodiversity and local land use in the last 60 years for the three study sites. The information in the table is a synthesis of findings during previous research in these study sites (Beilin 1999, 2001, 2007; Lindborg et al. 2008; Pereira et al. 2005).

Table 16.1 Perception of locals and other indicators of the past 50 years and future (next 30 years) trends for population, social cohesion, land use and biodiversity in the three study sites

Time period	Site	Population — Local's perceptions	Population — Other indicators	Social cohesion — Local's perceptions	Social cohesion — Other indicators	Land use	Land use — Local's perceptions	Land use — Other indicators	Biodiversity	Biodiversity — Local's perceptions	Biodiversity — Other indicators
Historical	Sistelo, Portugal	↓	↓	↓		Agricultural land	↓	↓	Game animal species (small size)		
						Abandoned land	↑	↑			
						Oak forest	↑	↑	Big mammals		
Present	Hällnäs, Sweden	Farmers ↓	↓	↓		Semi-natural grasslands, meadows and pastures	↓	↓	Species associated with agricultural landscape		
						Coniferous forest	↑	↑			
		Lifestyle immigrants ↑	↑			Number of farms	↓	↓	Species associated with forest		
	Poowong, Australia	Farmers ↓				Eucalyptus forest	↓	↓	From past to recent past (native forest degradation and agricultural pressure)		
				↑	↑	Forest plantations	↑				
		Lifestyle immigrants ↑				Dairy farming	↓	↓	From recent past to present (Land-care program – reforestation, creation of reserves in abandoned land)		
						Abandonment of marginal farmland	↑	↑			

Table 16.1 continued

Future	Place	Actor					Land change			Biodiversity		
	Sistelo, Portugal		↓	↓	↓		Regeneration of native oak forest/re-vegetation of marginal land	↑	↑	Species associated with agricultural landscape		↓
							Agricultural land	↓	↓	Forest species, big mammals	↓	↑
							Abandoned land	↑	↑			
	Hållnäs, Sweden	Farmers	↓	↓			Semi-natural grasslands, meadows and pastures	↓	↓	Species associated with agricultural landscape		↓
							Forest land	↑	↑			
		Lifestyle immigrants	↑	↑	↓	↓	Number of farms	↓	↓	Species associated with forest	↑	↑
	Poowong, Australia	Farmers	↓	↓			Leasing of land from existing farmers	↑		Unmanaged/ Abandoned agricultural areas	Uncertain	↑
							Increase in forest area (for production and carbon sequestration)	↑				
		Lifestyle immigrants	↑	↑	↓		Bigger farms	↑	↑			

Section III: Place, culture and nature

'Place' and community attachment negotiate the meaning of land abandonment and nature-biodiversity. Places are significant in developing and maintaining self and group identity (Davenport et al. 2005), but not necessarily a shared identity or expectations about what the landscape should look like. Instead, it indicates the diversity of landscape use and feelings of an emotional link between themselves and physical places (Tuan 1974). This does not suggest uniform behaviour will ensue in the management of that place. Place theory literature suggests that trust, reciprocity, collective and individual actions and social relationships can create shared meaning or multiple shared meanings with a particular place. In our discussions with local informants we describe these concepts as creating social and landscape coherence (James 2000).

Hay's (2008) study of a forestry cohort in Tasmania found a deep physical and socio-cultural dependence among the bush saw-millers and fellers. It manifests in a close identity with the welfare of the forest and the mountain. This resonates with Horowitz's et al. (2001) findings that the health of the biota can be an indicator of the people's wellfare – i.e. a synergy evolves between those managing the landscape and the landscape itself. Integrating the social and ecological characteristics of place in order to understand these intrinsic values assists in understanding social meaning associated with management practices.

Place dependence (Clark and Stein 2003) requires that an individual uses the resource – as our farmers do for production outcomes; however, it is possible for non-residents to develop a strong sense of attachment without living there or using its resources. In Hållnäs and Poowong, we note the newest migrants are not farming; but they also have strong ideas about landscape values both at an amenity level and as hunters, recreationalists or lifestylers. Portuguese remittance families also value their cultural landscapes. In this context, place dependence and place attachment are linked by social and ecological memory. Horowitz et al. (2001) suggest that any locality has its own biodiversity and endemic features which are part of a person's place attachment and become part of their identity and their community's identity.

Clark and Stein (2003) consider the role of community attachment in resource management and add the physical characteristics of place that communities associate with their agricultural landscapes. These lead to regional institutions such as local breeds or brands, like the '*Barrosã*' meat in Sistelo's Peneda Mountain range. Bonaiuto et al. (2002) indicate in their research linking environmental attitudes, local identity and place attachment in two conservation areas, the relationship between social and place identity and people's understanding of landscape change is not bounded by collective interpretations of good outcomes. In their study, ecotourism businesses had different expectations of land use change from agriculturalists though both were considering the future of abandoned land. This points to the complexity of place identity as it is unlikely to be either exclusionary or reactionary, as Hay (2008: 233) notes; but rather it is able to transform according to perceived advantage or disadvantage at a local scale. In this way communities or networks can mobilize against a global incursion or welcome a previously 'foreign' change.

In Australia, land use change reflects a general shift in government policies from sociospatial equity to emphasize economic efficiency (Tonts 2000). In Poowong and marginal landscapes – marginal because of their small scale rather than their ecological values – farmers are at the mercy of the marketplace, and the value of their land and their commodity depends on its contrariness. Abandonment can result from ecological degradation. Even if there is a landscape culture that is directly transferable to the next or newer inhabitants, it is not clear who that next group is to be in terms of landscape management. In Sistelo and Hållnäs there is almost no in-migration and their populations are aging. The overall demographic decline among farmers is the same at our sites.

The act of abandonment is a physical leaving of humans from the land that in most cases precedes ecological succession and natural forest regeneration. Benjamin et al. (2007) suggest that to understand the change process we need knowledge of history, spatial and ecological dynamics and to understand the views of landowners towards these landscapes. Abandonment can lead to increased fire frequency and intensity as a result of the absence of management; or less desirable regional ecotypes. Hobbs and Cramer write that 'the socioeconomic consequences of land abandonment [...] may be positive in some areas, but negative in others, if the abandonment is accompanied by, or leads to, rural depopulation, loss of traditional industries, and reduced income from tourism' (2007: 310).

In the European context, Benjamin et al. (2007) report that there are two major trends in the perception of abandoned land – rejection by viewers because it is not productive – for example in the Swedish situation, this land is referred to as 'useless', while in Australia, non-productive land is described as 'marginal' but rarely as 'abandoned' by its current owners. For non-farmers, land not being used for production is often interpreted as wild or 'natural' and has romantic connotations associated with a return to nature. We propose that the difference between non-productive use and not being used by people is a cultural transition zone as well as an ecological one. Benjamin's et al. research (2007) indicates that their agricultural sector respondents did not see any biological usefulness in abandoned farmland dominated by shrub, but focused on the loss of economic return. However, eco-centric respondents, pensioners and new arrivals responded with ecological values and did not necessarily associate the abandoned land with profitability in the first place. Landscape change may be triggered by agricultural failure and in the Australian context it is likely to be instigated by changes in agricultural policies which, when combined with a market downturn, create unsustainable debt. In Poowong, land slip and soil erosion contributed to agricultural decline and the government encouraged forestry plantations as an alternative to farming.

Therefore landscape management is dependent on the local inhabitants and their cultural context, regional level policies and scientists with an interest. Residents and non-residents decide what is meaningful and how the landscape is to be managed. These activities are nested in their wider regions. Place, therefore, is also about community networks across the landscape and how these diverse networks of interest and purpose imagine and seek to manage what were formally agricultural landscapes.

Adaptive capacity as understood in the social relationships of daily life necessarily evolves out of local culture and its history. Pelling and High (2005: 309) divide adaptive behaviour into those that 'reinforce existing organizational or system stability and those that modify institutions to add resilience through flexibility'. The former can be compared to the agricultural subsidies to maintain traditional agricultural practices and reinforce the social structures they depend upon; the latter to the management of common property areas in Sistelo which, while maintaining a committee of commons, is now mainly done by the parish authority. However, the issue of the management of the abandoned terraces remains unresolved and is a question for this study. In Poowong, farmers look to farm forestry and conservation zones to assist in managing difficult terrain, so they invoke the memory of the Great Southeastern Forest to provide historical continuity. The conservation zones on farm are not conventionally understood as productive and it has been a twenty-year shift in thinking that has supported locals to see conservation zones as contributing to the welfare of the overall regional landscape. As part of this transformation, 'good farmers' has come to mean those that are not just producing in agricultural terms but are restoring indigenous vegetation. This experience affirms that social capital is adaptive, reflecting changes to community power structures 'as enacted through relational and networked space' (Pelling and High 2005: 314). Similarly, Davenport and Anderson's study (2005) points to the evolutionary character of place meanings and how land use change or insights into ecological function, as in our sites, caused enhanced attachment.

Conclusion

In using place to think about local interpretations of nature we argue for a more sophisticated approach that integrates the complexity of abandonment, culture, and nature into the landscape of agricultural production. As ecology shifts towards socioecological systems construct, the significance of people in the landscape reflects in how they imagine and manage their places. Horowitz et al. (2001: 256) argue that places become part of self-identity so, in the context of our study areas, the challenge is how identities formerly centred on agricultural production can incorporate more biodiverse landscapes. As in Poowong, this transition is nascent through Landcare revegetation projects similar to Deverre et al's (2007) experiences in the Massif of France. With no large scale in-bound farmer migration, these places depend on the local population and interested scientists to begin such transformations.

Landscape coherence has been contested and evolving in each of these sites. In Sistelo, large areas of native oak forests were long ago degraded for agricultural purposes or wood exploitation. Later in the mid-twentieth century, open grazing areas on the mountain tops were nationalized and converted to coniferous forest. In Poowong, the long-gone Great South Forest remains in the present residents' collective memory because of indigenous remnants but also due to using forest images to encourage revegetation on farms. In Hållnäs people remember the 1950s' childhood landscapes of traditional agriculture with open meadows. Horowitz et al. (2001: 257) note that 'feelings of belonging can be invoked by remembering incidents and events' associated with places and these reinforce place identity and the sense of legitimacy associated with being in a place. Therefore, we hypothesize that historical continuity in some residents is valuable for extending the possible options for change; and newcomers and non-residents can experience attachment without having to be there (Brown and Raymond 2007). This suggests that values can be maintained through strong regional differentiation and through other landscape products such as specific foods or plants known to belong to a region because of its location. Therefore, regional legitimacy may be transferred to newcomers.

Pred's extensive review of 'place' literature indicates place has historically been understood as a static thing in the landscape – a tableau of 'frozen scenes' (1984: 279) – but it can no longer be seen in that way either in terms of people or ecology. Clearly, science has diverse expectations of place depending on whether you are a resource manager or an ecologist. Also, those deciding on the 'look' of a place may be only occasional visitors interested in maintaining agrarian or recreational landscapes or combinations of both for their amenity values. However, we argue that resource managers and ecologists can benefit from a more nuanced view of how place meaning resonates from its cultural core to influence decisions. In ecological terms it is possible to think about different temporal and spatial scales in the ideal but it is the need to marry these with resource management, which is still locked in to a 'command and control' paradigm (Holling and Meffe 1996) in landscapes such as Poowong, that challenges us to reflect on the dynamism inherent in both ecological and social systems. As Pred indicates: 'No matter how constrained or 'free' and spontaneous self-expression in the creation and use of landscape may be, it always emerges from biographical and place-specific historical and social contexts at the same time that it contributes to the uninterrupted becoming of biography and place' (1984: 287).

By exposing the constructed, culturally dependent, historically contingent meanings of nature, we have sought to invigorate the importance of place as the dynamic focus for change, innovation and renewal. We seek to differentiate the layers of meaning associated with conservation and agricultural activities. As Liepins argues (2000), communities are not fixed. Social connection, landscapes and place are highly permeable groupings and they coalesce in relation to purpose, serendipity and historical origins. Their response to nature is both opportunistic and spiritual. Clark and Stein's research (2003) found that the physical-natural landscape is critical to how some residents related to their communities. Gupta and Ferguson focus their attention on 'the intertwined processes of place making and people making in the complex cultural politics of the nation-state' (1997: 4). 'A strategy for biodiversity may be built up from various component activities in agriculture, rural farming, entrepreneurship, forestry, leisure, wildlife conservation, as well as social justice [...] Through familiarity and identity, people can then learn to have confidence in the more coupled relationships and issues that full-hearted biodiversity management will inevitably throw up' (O'Riordan and Stoll-Kleemann 2002: 98). Ultimately, then, the challenge is one of engagement. If science wants to promote a certain view of biodiversity, it needs to link that view to a narrative that all the stakeholders can understand.

We hypothesize that the long-term view requires us to step outside the boundaries of protected area management, or the fencing lines of regional farms, to re-imagine a landscape that perceives biodiversity in agriculture and conservation management. In such scenarios we can imagine that the concept of an 'abandoned' landscape would be obsolete and the focus on the quality of the integrated landscape mosaic would be the real challenge.

Acknowledgements

Swedish Research Council for Environment, Agricultural Sciences and Spatial Planning (FORMAS).

References

Beilin, R. 1999/2000. Cultivating the Global Garden. *South Atlantic Quarterly*, December 1999 Issue.

Beilin, R. 2001. Underlying it All: Faceless Landscapes and Commodified Views. *Rural Society*, 11(3), 147-162.

Beilin, R. 2007. Landscape with Voices: Reflecting on Resilience on Farms in the 'Heartbreak Hills', Strzelecki Ranges. *Local Global: Identity, Security, Community*, 4, 141-161.

Beilin, R., Fleming, J. and P. Fleming, 1997. Reservations: Issues of Conservation Practice on Private Agricultural Land, in *Conservation Outside Nature Reserves*, edited by P. Hale and D. Lamb, St. Lucia: University of Queensland Press, 206-212.

Benjamin, K., Bouchard, A. and Domon, G. 2007. Abandoned farmlands as components of rural landscapes: An analysis of perceptions and representations. *Landscape and Urban Planning*, 83, 228-244.

Bolton, G. 1981. *Spoils and Spoilers: Australians make their environment.* London: Allen & Unwin.

Bonaiuto, M., Carrus, G., Martorella, H. and Bonnes, M. 2002. Local identity processes and environmental attitudes in land use changes. *Journal of Economic Psychology*, 23, 631-653.

Boyd, J. and Banzhaf, S. 2006. What are Ecosystem Services? [Online] Available at http://www.rff.org/Documents/RFF-DP-06-02.pdf [accessed: 15 February 2010].

Brown, G. and Raymond, C. 2007. The relationship between place attachment and landscape values. *Applied Geography*, 27, 89-111.

Clark, J. and Stein, T. 2003. Incorporating the Natural Landscape within an Assessment of Community Attachment. *Forest Science*, 49(6), 867-876.

Crowe, S. 1958. *The Landscape of Power*. London: Architectural Press.

Davenport, M. and Anderson, D. 2005. Getting from Sense of Place to Place-Based Management. *Society and Natural Resources*, 18, 625-641.

Deverre, C., Fortier, A., Alphandery, P. and Lefebvre, C. 2007. 'The local scenes' of biodiversity: Building a Natura 2000 network in France. *INRA Sciences Sociales: Research in Economics and Rural Sociology*. No. 4, October.

Eaton, M. 1990. Responding to the Call for New Landscape Metaphors. *Landscape Journal*, 9(1), 22-27.

Fischer, J., Brosi, B., Daily, G. et al. 2008. Should agricultural policies encourage land sparing or wildlife-friendly farming? *Frontiers in Ecology*, 6, 380-385.

Gupta, A. and Ferguson, J. 1997. Culture, Power, Place: Ethnography at the End of an Era, in *Culture, Power Place: Explorations in Critical Anthropology*, edited by A. Gupta and J. Ferguson. Durham: Duke University Press, 1-29.

Harvey, D. 1996. *Justice, Nature and the Geography of Difference*. Cambridge, MA: Blackwell.

Hay, P. 2008. 'Balding Nevis': Place Imperatives of an Invisible Cohort within Tasmania's Forest Communities. *Geographical Research*, 46(2), 224-233.

Hobbs, R. and Cramer, V. 2007. Why Old Fields? Socioeconomic and Ecological Causes and Consequences of Land Abandonment, in *Old Fields: Dynamics and Restoration of Abandoned Farmland*, edited by V. Cramer and R. Hobbs. Washington: Island Press, 1:1-14; 16, 309-318.

Holling, C.S. and Meffe, G.K. 1996. Command and Control and the Pathology of natural resource management. *Conservation Biology*, 10(2), 328-337.

Horowitz, P., Lindsay, M. and O'Connor, M. 2001. Biodiversity, Endemism, Sense of Place, and Public Health: Inter-relationships for Australian Inland Aquatic Systems. *Ecosystem Health*, 7(4), 253-265.

James, P., Ashley, J. and Evans, A. 2000. Ecological Networks: Connecting environmental, economic and social systems? *Landscape Research*, 25(3), 345-353.

Lefroy, E. C. and Smith, F.P. 2004. The biodiversity values of farming systems and agricultural landscapes. *Pacific Conservation Biology*, 10(2), 80-87.

Liepins, R. 2000. New energies for an old idea: Reworking approaches to 'community' in contemporary rural studies. *Journal of Rural Studies*, 16, 23-35.

Lindborg, R., Bengtsson, J., Berg, Å., Cousins, S.A.O., Eriksson, O., Gustafsson, T., Hasund, K-P., Lenoir, L., Pihlgren, A., Sjödin., E. and Stenseke, M. 2008. A landscape perspective on conservation of semi-natural grasslands. *Agriculture, Ecosystems and Environment*, 125, 213-222.

Lindborg, R., Stenseke, M., Cousins, S.A.O., Bengtsson, J., Berg, Å., Eriksson, O., Gustafsson, T. and Sjödin., E. Using scenarios for participatory approach for sustainable management and planning – A case study from Swedish agricultural landscapes (in review).

McNiven, I. 2008. Inclusions, exclusions and transitions: Torres Strait Islanders constructed landscapes over the past 4000 years, northeast Australia. *The Holocene*, 18(3), 449-462.

Medeiros, I. 1984. *Estruturas Pastoris e Povoamento na Serra da Peneda*. Lisboa: Centro de Estudos Geográficos.

Moss, L. 2006. *The Amenity Migrants*. Santa Fe, USA: Oxford University Press.

Naveh, Z. 1998. Ecological and Cultural Landscape Restoration and the Cultural Evolution towards a Post-Industrial Symbiosis between Human Society and Nature. *Restoration Ecology*, 6(2), 135-143.

Noble, B. 1976. The Strzeleckis: A New Future for the Heartbreak Hills. Victoria: Forests Commission.

O'Riordan, T. and Stoll-Kleemann, S. 2002. Deliberative democracy and participatory biodiversity, in *Biodiversity, Sustainability and Human Communities: Protecting beyond the Protected*, edited by T. O'Riordan and S. Stoll-Kleemann. Cambridge: Cambridge University Press, 87-112.

Pereira, E., Queiroz, C., Pereira, H. and Vicente, L. 2005. Ecosystem Services and Human well-being: A Participatory Study in a Mountain Community in Portugal. *Ecology and Society*, 10(2), 14 [Online].

Pelling, M. and High, C. 2005. Understanding adaptation: What can social capital offer assessments of adaptive capacity? *Global Environmental Change*, 15, 308-319.

Pred, A. 1984. Place as a Historically Contingent Process. *Annals of the Association of American Geographers*, 74(2), 279-297.

Queiroz, C., Lindborg, R., Beilin, R. and Folke, C. (nd) *Shadows on the Cave* (ms).

Tonts, M. 2000. The Restructuring of Australia's Rural Communities, in *Land of Discontent: The Dynamics of Change in Rural and Regional Australia*, edited by Bill Pritchard and Phil McManus. Sydney: UNSW Press, 52-72.

Tuan, Y-F. 1974. *Topophilia: A Study of Environmental Perception, Attitudes, and Values*. NY: Columbia University Press.

Chapter 17

Natural Landscape Inside Metropolis: Example of Saint Petersburg

Gregory A. Isachenko and Andrey Y. Reznikov

Introduction

On banks of mosses and wet grass
Black huts were dotted there by chance-
The miserable Finn's abode;
The wood unknown to the rays
Of the dull sun, by clouds stowed,
Hummed all around ...

An age passed, and the young stronghold,
The charm and sight of northern nations,
From the woods dark and marshes cold,
Rose the proud one and precious.
(A. Pushkin, *The Bronze Horseman*[1])

The largest cities (with populations of millions) can be considered territories where natural landscapes are transformed to the maximum degree and a person is mostly isolated from nature. However, in post-industrial times the natural landscapes that have been 'denied' in every big city due to its growth and development are increasingly valued. The need to preserve natural landscapes within the city limits is complementary to the creation of a semi-artificial urban environment which not only tires the townspeople, but also negatively influences their physical and mental health.

St. Petersburg, with a population of c. five million people, is the largest city in the world in a natural zone of taiga (boreal forest), one of the least populated in the Globe. The city was created 'deliberately', as a new European capital of Russia, in a landscape that was little suited to sustain millions of people. For more than 300 years the territory of the city has increased a hundred times and at present the administrative borders lie at close to 1,400 km². Within the city, the specific landscapes of the lowlands of a taiga zone are present: woods, bogs and the coastal sites. In spite of the fact that practically all these landscapes at different periods

---

1   Translated by Yevg. Bonver.

were exposed to various human impacts (wood cuttings, drainage, agricultural use, etc.), the character of contemporary processes here is close to their natural form. The specificity of the formation of the Petersburg territory has been a constant increase in dry land from the partial filling of a shallow Neva bay in the Gulf of Finland.

This chapter represents the analysis of modern conditions of St. Petersburg's natural landscapes, which the authors understand as territories basically covered with natural vegetation (including the landscape created at the expense of water areas). Conclusions are based upon our own field research data, the analysis of maps and remote images from different times as well as State statistics. This chapter considers the authors' experience of applying the landscape-dynamic approach to the planning and management of protected natural territories in St. Petersburg.

Natural landscape pattern of Saint Petersburg

The territory of St. Petersburg, despite a seeming monotony, is rather diverse regarding landscapes. The central or historical part of the city occupies an extensive enlargement of the so-called Littorina terrace formed by loamy and sandy deposits of the Littorina sea, which was here 8,000-3,000 years ago. Altitudes of the terrace do not exceed 10 metres, the lowest levels have been periodically subjected to flooding; therefore construction on these sites has been made on piled ground with a thickness reaching five meters. Outside the central part of the city, the Littorina terrace basically presents a narrow strip along the coast of the Gulf of Finland. Non-developed sites of the Littorina terrace are covered with bogs and marshes, woods with a dominance of black alder, reed and willow thickets as well.

The next altitudinal level of Petersburg's landscape is created by lake and lacustrine-glacial terraces with heights up to 30 metres above sea level. The terraces are formed mainly by sandy and clay sediments. In the past these terraces were to a large extent boggy, then their greater part was drained and used for agriculture and building. On lake and lacustrine-glacial terraces are situated most of the new residential areas and industrial zones of Petersburg. The prestigious and picturesque Kurortnyi (Resort) district occupies mainly the sandy lacustrine-glacial terrace that is basically used for recreation and cottage building. Non-used areas of lake and lacustrine-glacial terraces are mainly covered with woods, with a prevalence of pine and birch.

Above a monotonous surface of lacustrine-glacial terraces, the low hills of various genesis (morainic, kame) tower; their altitude quite often exceeds 100 metres. They are less boggy and consequently have been populated and used for agriculture for a long time. However, the kame hills in the Kurortnyi district, formed by coarse-grained sand, have not been developed. Now these kames, covered mainly with pine woods, are actively used for recreation. The south-western part of St. Petersburg is located on Izhorskaya upland with altitudes of

c. 100 meters, formed by Ordovician limestone and covered with carbonate till. Owing to the high natural drainage ability and soil fertility, this territory has always been densely populated and deforestated. Here lies St. Petersburg's highest point (Duderhof heights), with an altitude of 176 meters.

As already noted, the natural landscapes of the taiga zone are woods, bogs and the coasts with marshes and grassy vegetation. We will consider next how these types of landscape are present in the territory of St. Petersburg.

Woods

Before the foundation of St. Petersburg in 1703 most of its present territory was covered with woods. Despite the presence of numerous settlements here, by the beginning of the eighteenth century forest occupied not less than 70 per cent of the considered territory. In the process of city building and agricultural development of the adjoining territories the area of woods, as a whole, decreased, though this process has not been linear. The analysis of different-time maps shows that, while woods on some sites were cut for building, other sites were overgrown by forest, especially former agricultural lands. This latter process has been especially intensified from the beginning of the twentieth century. As a result, St. Petersburg's tree area of at the beginning of the twenty-first century corresponds approximately to that of mid-nineteenth century, whereas at the beginning of the twentieth century the forest area was almost 20 per cent less.

At present the total area covered with wood and brushwood (without taking into account boulevards and small squares) makes up approximately 327 km², or about 24 per cent of the city's territory. However, 44 km² from this area are occupied by gardens and parks, which we do not consider here. Another 3 km² are covered with willow thickets with a prevalence of shrub species. Thus, approximately 280 km² (20 per cent of the city area) remain wooded, with a dominance of taiga tree species which we consider to be city woods.

It should be noted that many territories regarded as city parks are really parts of natural, sometimes even quite old, woods. Some woods are located on the lands which have no special status and, thus, do not enter into any accounts of wooded plantations. The total area of such woods can be estimated at around 20 km². As a rule, these are small-leaved woods, not very valuable, which have grown on former agricultural lands, wastelands and peat fields, and also on the territories prepared for building and later abandoned. However, the same category also includes valuable forest communities regarding their natural features, for example black alder woods.

The uncertainty of a legal status of the city woods was raised by the implementation of the new Forest Code and the General plan for the development of St. Petersburg. First, the status of woods within the city limits so far has not been brought into agreement with the legislation: they have not been allocated either to the federal forest fund, or to the city woods in the legal sense. Secondly, by comparing the General plan with the management plan for the city forest parks, it is found that only 140 km² of forest parks are included in the planning zone of 'woods and forest parks' – from a total of 240 km² (under 60 per cent). The remainder of the city forests falls on other planning zones where partial or full cutting is permitted. Such situation can lead to a sharp decrease in the city woods area.

In the process of the city's development both pine (*Pinus sylvestris L.*) and spruce (*Picea abies (L.) Karst.*) forests were replaced by small-leaved stands with a dominance of birch (*Betula pendula Roth, B. pubescens Ehrh.*). Now the greatest area of city woods has a dominance of pine (44 per cent) and birch (38 per cent), and the share of spruce stands is only 13 per cent of the forest area (Figure 17.1). In the past, apparently, the significant area, especially on coastal sites, was covered with black alder woods (*Alnus glutinosa (L.) Gaertn*); however these territories were built up first, therefore now black alder groves occupy just less than 1 per cent of the city woods.

The composition of woods differs essentially in various types of landscapes (Figure 17.2). So, the greatest share of the most valuable coniferous woods is present on the Kurornyi district's kame hills. On sandy terraces, mainly in the city's northern part, coniferous woods occupy two thirds of the forest area; on the clay lake terraces and Littorina terrace they take about half the area; and on morainic plains and flat hills coniferous forests cover somewhat more than one third of the forested area of the corresponding types of landscapes. In all types of landscapes, except the Littorina terrace, pine prevails among coniferous trees.

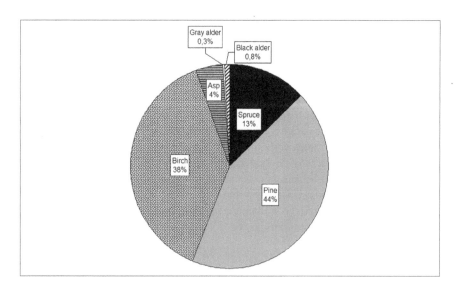

Figure 17.1 Distribution of the wooded area in Saint Petersburg by dominant tree species

In all types of landscapes the considerable part of woods has grown on abandoned agricultural lands. In such woods, small-leaved trees, basically birches, with participation of grey alders (*Alnus incana (L.) Moench*) usually dominate.

In recent decades the tree composition of the city woods has been subjected to fast and contradictory changes. So, in the isolated forest massifs surrounded by building areas pine have gradually been replaced by birches and other small-leaved tree species. This is the result both of pressure from recreational activities, and air and soil pollution. On the other hand, in areas where commercial timber cuttings have stopped after they were included in the city limits, there is a tendency to increased spruce share in small-leaved and pine woods which is general for the European taiga (Isachenko and Reznikov 1996).

Approximately 20 km² (6 per cent) of the city woods grow on peat-bogs. Oligotrophic and mesotrophic bogs in their natural conditions are covered with dwarfshrub-shagnous, cottongrass-sphagnous and sedge-sphagnous woods with a dominance of oppressed pine; the woods with a prevalence of birch and black alder are less common. On the drained peat-bogs and peat fields (former peat excavations) birch and pine-birch stands dominate, sometimes of quite a high growth class. In the cases where the drainage network continues to function, tree stands grow very fast. However, these sites are subjected to frequent peat fires, accompanied by full destruction of forest stands. In the event of damage to the drainage network secondary bogging progresses rapidly and this causes the waterlogging and subsequent death of trees.

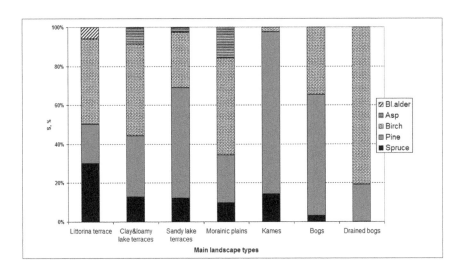

**Figure 17.2 Distribution of wood area with dominance of main tree species
by main natural landscape types in Saint Petersburg**

Source: Authors' estimates using *Ecological Atlas of St. Petersburg*, 1992, edited by D. Gorelik and V. Kuznetsov. St. Petersburg: Ecological union 'Monitoring' (Russian).

Bogs and marshes

At the beginning of the eighteenth century, not less than 40 per cent of the area of modern St. Petersburg was boggy. The area of peat-bogs (peat thickness over 0.5 meters) made about 15-20 per cent, the maximal peat layer in places exceeded 6 meters. Some bogs had been located in the city centre. Therefore the development and building of the new Russian capital was impossible without drainage, and quite often filling up, of bogs as well. The long-term struggle of government authorities and city dwellers against bogs resulted in a sharp reduction of their number and area.

From mid-nineteenth century to the end of the twentieth century, peat excavation has also been conducted in the present territory of St. Petersburg. In this period a total of not less than 45 km² of peat-bogs was excavated. About half the area of peateries has been built up, and filled up with mineral ground and prepared for building as well. The remaining peateries now present specific landscapes of partially watered peat quarries or levelled peat fields. The latter are partially used as agricultural lands.

Now the territory of St. Petersburg has more than 90 km² of boggy lands, including about three tenths peat-bogs whose total area exceeds 76 km². Most of these peat-bogs have been essentially changed by humans using drainage, full or partial excavation (Table 17.1).

Table 17.1 Change in the character of maintained peat-bogs in St. Petersburg

Conditions	Area, km²	Share of the sum, %	Share of the city area, %
Full or partial peat excavation	22	29	1.6
Drainage	37	49	2.6
Natural conditions (non-disturbed)	17	22	1.2
TOTAL	76	100	5.4

Source: Authors' estimates using *Peat deposits of the Leningrad Region* (1980), Moscow (in Russian).

The areas of peat-bogs and marshes on the St. Petersburg territory have no special status. The majority of the natural and drained bogs are situated on the lands of parks and forest parks. The drained bogs and peateries ever used in agriculture are considered lands for agricultural purpose, even if they are already overgrown by wood. Most peateries are seen as unused lands and are allocated by city authorities for industrial and residential building. As the new Water Code of Russia was brought into force, a legal conflict emerged: bogs are now considered not as lands, but as water objects, and, as is the case of all such objects, should be federal property. However, the city authorities are not interested in letting an important part of the land slip from their domain.

The attitude of the authorities and of the population of St. Petersburg towards bog protection is extremely ambiguous. While there is a certain consensus concerning the preservation of city woods, bogs are perceived by many people as 'bad lands', unsuitable either for building or for recreation. Nevertheless, under the insistence of scientists the main massifs of natural bogs and some sites of the drained peat-bogs are included in protected natural territories (PNT).

Among St. Petersburg's bogs and marshes the oligotrophic (raised) bogs prevail. In non-disturbed conditions the central part of such bog is occupied by open pine-dwarfshrub-sphagnous or pine-cottongrass-sphagnous communities, and the periphery is covered with oppressed pine. Most of raised peat-bogs are to some extent drained and completely covered with wood.

On the Littorina terrace, except for raised bogs, there are a lot of transitional (meso-oligotrophic and mesotrophic) bogs and eutrophic fens. The birch and pine-birch sphagnous woods are specific of transitional bogs, open sedge-sphagnous communities are less often. The main area of fens and marshes is located on the southern coast of the Neva bay and on Kotlin Island. Fens and marshes here are covered with black alder groves, willow and reed thickets. Fens and marshes are the most unattractive for people, and for this reason many species of waterfowl and shore birds nest and feed here, and some rare plant species occur here as well.

Coast of the Gulf of Finland

Within the limits of St. Petersburg there are about 200 km of a coastal line of the Gulf of Finland (without considering coastal line of internal water areas of ports and shipyards, dams, piers, etc.). The character of coasts is presented in Table 17.2.

The most picturesque are the coasts formed by sands and sandy-boulder deposits. The greatest extent of sandy beaches is in the Kurortnyi district. For each inhabitant of St. Petersburg there is approximately 1cm of beach line, and on hot summer days the density of users is at its highest.[2] As a result, the birds which nest on a mineral ground near the coast line are completely deprived of such opportunity and are close to disappearance. More favourable conditions for coastal animals and plants are created on the coasts densely covered with boulders and shingle that make them inconvenient for bathing.

---

2   The beaches of the Neva bay have largely lost their recreational value in recent decades because of the increase in water pollution due to the construction of the dike that should protect the city from floods.

Table 17.2 Types of coastal line in the territory of St. Petersburg

Index	Conditions	Extent, km	Share of the sum, %
	Natural coast, including:	*114*	*52*
N1	Sandy and sandy-boulder beaches	54	27
N2	Coast of low recreational value (willow, reed, bulrush thickets, fens, marshes)	50	25
	Artificial coast, including:	*95*	*48*
A1	Port and defensive constructions, quays, shipyards, warehouses, factories and other territories of limited access	44	22
A2	Unused (available to the public) former fortifications	12	6
A3	Embankments available to the public	8	4
A4	Artificial coastal line in vacant territories available to the public (mainly in-washed coast)	18	9
A5	Artificial or strengthened coast in parks of common and limited use	13	7
	TOTAL	199	100

Much less attractive are low coasts with fragments of fens and marshes, but it is here that the biotopes of specific coastal plants and animals are present. Some parts of artificial coast also are of value for their natural peculiarities. So, vacant in-washed territories are used by stints for nesting.

Now St. Petersburg's coastal landscapes are changing very rapidly. Owing to wash-out of coasts, the area of beaches has decreased and the area of sandy-boulder abraded coasts has increased. Plant overgrowth and the greater shallowness of the water area have led to an increase in the proportion of overgrown and bogged coasts at an expense of sandy beaches. Vacant sites on in-washed coasts have gradually built up. Besides, on some coastal sites port constructions have completely destroyed natural coastal landscapes.

The legislation of the Russian Federation stipulates a special legal status for coasts. In particular, on seashores the water-protecting zone is 500 meters wide, and also there is a coastal protective strip 30-50 meters in width, depending on the local relief. However, restrictions in these zones are aimed basically at protecting reservoirs, but not coastal landscapes. The only restriction in water-protecting zones, really applicable in city conditions, is a prohibition on the movement and parking of motor transport outside roads and sites without a firm covering that is essential in areas of intensive recreation. However this legal requirement is broken everywhere as well.

Protected natural territories of St. Petersburg

Basically, all natural landscapes on the St. Petersburg territory are now protected by law. Woods, parks, forest parks are protected by the city laws 'About the protection of green plantations' (2004) and 'About green plantations in common use' (2008). Moreover, city woods are under the purview of the Forest Code of the Russian Federation. Historical parks are included in lists of objects of cultural heritage and protected by the corresponding federal and regional legislation. Water bodies are protected by the Water Code of the Russian Federation.

The most valuable natural objects are regarded as protected natural territories (PNT). Their creation and functioning is defined by the federal law 'About especially protected natural territories' and by the city law 'About state natural sanctuaries and natural monuments of regional value in St. Petersburg' (2005).

To maintain and run the city's PNTs, a specialized institution, 'The Directorate of protected natural territories of St. Petersburg', has been created, which is subordinated to the Committee on Nature Use, Environmental Protection and Ecological Safety of the St. Petersburg Administration. The network of St. Petersburg's protected natural territories now includes 6 PNTs with a total area of 21.5 km^2 or about 1.5 per cent of the metropolitan area. Some coastal PNTs include parts of water area of the Gulf of Finland.

According to the new General Plan for the Development of St. Petersburg, by 2025, 21 new PNTs are meant to be created, with a total area of 236.5 km^2. Together with existing protected natural territories, by 2025 their total share should constitute about 20 per cent of the city area. However, more detailed analysis of the situation brings some amendments to this prospect. First, as specified above, it is necessary to exclude water areas from the mentioned figure. Secondly, the largest PNT (162 km^2) planned in the Kurortnyi district can be related to a PNT only formally, and will actually be developed as a resort zone. Having recounted the areas of the projected PNTs with regard to the stated reasons, we will re-estimate the dry-land area of the latter as 54.5 km^2, which together with existing PNTs will make up 74.5 km^2 (5.3 per cent of the city's territory).

Should the program to create new PNTs be implemented, the sites of all basic types of natural landscapes within the city limits will be taken under protection. Furthermore, a rather high degree of natural landscape protection will be provided (Table 17.3).

Table 17.3 The landscape structure of existing and planned St. Petersburg's protected natural territories (PNTs)*

Type of landscape	The area, km²	
	2009 year	**Planned for 2025 year**
Lake and lacustrine-glacial plains	6.0	25.2
Morainic plains and flat hills	1.0	6.8
Littorina terrace	2.9	19.3
Kame and fluvio-glacial hills and ridges	0.0	7.3
Morainic hills on carbonate till	0.7	0.7
Peat-bogs, fens and marshes (except for excavated)	7.5	29.8
Former peateries	0.3	0.6
Dunes	0.1	0.8
Landscapes of coasts of the Gulf of Finland (present terrace)	0.6	2.6
Valleys and floodplains of the rivers and streams	0.8	2.2
Technogenic surfaces, territories with building	0.4	1.2
Internal reservoirs, rives, streams	1.1	5.9
TOTAL	21.4	95.2

Note: * Without considering 'The Medical and health area of the Kurortnyi district'.

Landscape-dynamic approach in planning and management of the protected natural territories in St. Petersburg

A traditional conservational approach applied to natural reserves is obviously unsuitable for the protected natural territories in a metropolitan area. First, the landscapes of most protected territories are essentially transformed by human impacts (bog drainage, forest cuttings, agricultural use, etc.) and continue to change rapidly – both in the course of regenerative succession and as a result of continuing human influences. Secondly, any attempts at mutual isolation of city-dwellers and the remaining plots of wild or semi-wild nature have no prospects: natural landscapes must also carry out recreational functions. Mass visits of the townspeople 'to nature' with all possible negative consequences such as forest fires, heaps of garbage and damaged trees are specific manifestation of a love for nature – but such love can be destructive …

The model of landscape management of protected natural territories is the dynamic landscape concept, according to which characteristics of landscape are divided into attributes of *landscape site* (relatively stable characteristics of relief, soil structure and bedrock) and attributes of *landscape states* (more dynamical parameters related to vegetation and soil). Thus landscape dynamics is considered to be a change in landscape states within a constant 'framework' of landscape sites (Isachenko and Reznikov 1996; Isachenko 2007).

As regularity of change in landscape states due to various influences is most likely defined by characteristics of a landscape site and corresponding influence, the simulation of landscape dynamics and the design of management actions concerning PNT landscapes can be expediently carried out on the basis on landscape site pattern.

To realize what to do with protected natural territories in a big city, it is necessary to answer two questions:

- What are our targets in the estimated territory? It is necessary to define them, whether we want to make a given PNT more like a natural reserve, a refuge for protected plant and animal species, or a recreational area. A landscape state (or set of states), to the greatest degree corresponding to its functions in the structure of a PNT and consequently being the purpose of its management – we will name this a *target landscape state*.
- What processes are now being realized on the given PNT and how will they result in the preservation of present conditions? Processes of landscape change, caused by the present combination of natural factors and human impacts, we examine as *spontaneous processes (changes)*. Study of these changes allows us to predict the future state(s) of each local landscape.

The general decision-making scheme in the design of landscape management in the city's PNTs is presented in Figure 17.3. We will illustrate the contents of the blocks specified in this figure with the example of the natural monument 'Komarovskiy coast', located in St. Petersburg's Kurortnyi district (Komarovskiy coast 2002).

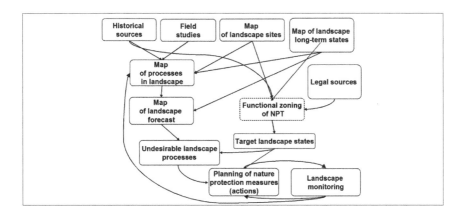

Figure 17.3 The algorithm of protected natural territories management based on the landscape-dynamic approach

(1) Using the results of field researches and high-resolution remote sensing imagery treatment, a detailed landscape map of the PNT, including layers of landscape sites and long-term landscape states (plant communities and soils), is made.

(2) The previous states of landscapes and their past changes are studied, in particular the results of human influences during the different historic periods. This research is based on analyses of historic sources, comparison of different-time maps and aerial photographs, and dendrochronology methods as well.

(3) The present processes going on in various landscape sites (e.g. replacement of birch by spruce, reforestation of abandoned meadows, bogging of woods, etc.) are revealed. The processes are defined from comparison of present landscape states with their previous states, and by materials of forest management and geo-botanical surveys. In taiga wood landscapes the directions of processes are revealed based on changes in stand composition or forming/destruction of a tree layer. Recreational digression is regarded as a separate direction of processes (Table 17.4).

To move from directions of processes to proper processes, it is necessary to correlate these directions with present landscape states. Such correlation is reported in Table 17.5, where the rows contain types of plant communities reflecting present landscape states, and the columns include directions of processes. The filled cells of Table 17.5 correspond to observable processes. Spatial distribution of the processes is reflected on the map (Figure 17.4).

(4) The forecast of spontaneous landscape dynamics for the period usually defined by the time of growth of one generation of forest stands (40-50 years) is made. So, if on a certain wood site spruce now started to replace birch, in 50 years the wood, most likely, will have a spruce dominance; if the meadows are now overgrowing with birch, a birch wood will be formed there, and so on. Based on the map of present processes in landscapes (Figure 17.4), we will obtain a map scenario of the spontaneous landscape change of the investigated area in 40-50 years (Figure 17.5).

Table 17.4 Directions of processes in landscapes of the protected natural territory 'Komarovskiy coast'

Index	The description of a direction of process
=	Maintaining the present state, composition and the phytomass of the vegetation
~	Dynamic balance (mainly in spruce forests)
C	Increase in phytocenotic role of pine
СБ	Joint increase in pine and birch tree stock
Е	Increase in phytocenotic role of spruce
Е–	Dying off spruce stands as a result of bogging
Б	Increase in phytocenotic role of birch
Оч	Increase in phytocenotic role of black alder
Ч	Increase in bird cherry and shrub phytomass
Д	Recreational digression in combination with local cuttings and forest fires

Table 17.5 Processes in landscapes of the protected natural territory 'Komarovskiy coast' (explanations for Figure 17.4)

Landscape state (types of plant communities)	Index	Directions of processes (indexes see Table 17.4)								
		=	~	Е	Е-	С	Б	Оч	Ч	Д
Woods with spruce dominance	Е	Е=	Е~	Е+	Е-			Е/Оч		
Woods with pine dominance	С	С=		С/Е		С+				С/Д
Woods with birch dominance	Б			Б/Е			Б+	Б/Оч	Б/Ч	
Woods with black alder dominance	Оч	Оч=		Оч/Е	Оч/Е-			Оч+	Оч/Ч	
Bird cherry thickets and shrub	Ч								Ч+	
Open peat-bogs and mires	Бол	Бол=								
Meadows	Л						Л/Б		Л/Ч	
Open dunes vegetation	Д									Д/Д
Sandy beach herbs	П									П/Д

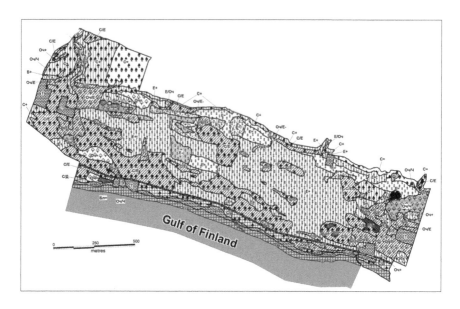

**Figure 17.4 Map of present processes in landscapes of the protected
natural territory 'Komarovskiy coast'**

Note: Explanations in tables 17.4 and 17.5.

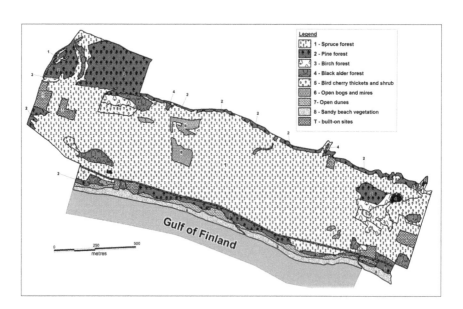

**Figure 17.5 Map scenario of the spontaneous landscape change of the
protected natural territory 'Komarovskiy coast' in 40-50 years**

Comparing the received map (Figure 17.5) with the map of processes in landscapes (Figure 17.4), it is easy to see that the landscape pattern of the territory will essentially change.

(5) On the grounds of inventory researches and a landscape map, the scheme of functional zoning of PNT is made (Figure 17.6).

(6) The approach to designing the target landscape states depends on a landscape site pattern and the present state of the landscape. So, for zones of strict protection the target state corresponds to the so-called climax condition of vegetation; for historical parks – to their primary design. For the recreational zones the optimum combination of availability and attractiveness with keeping of sanitary-hygienic criteria is chosen as the target state (Table 17.6).

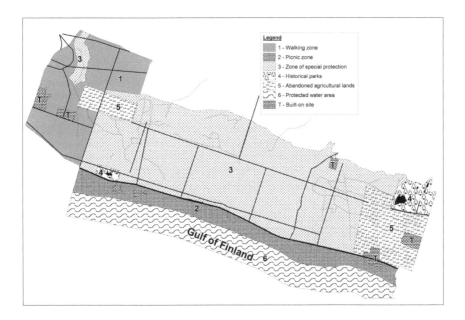

Legend

1 - Walking zone
2 - Picnic zone
3 - Zone of special protection
4 - Historical parks
5 - Abandoned agricultural lands
6 - Protected water area
T - Built-on site

Figure 17.6 Scheme of functional zoning of the protected natural territory 'Komarovskiy coast'

Note: For a detailed description of zones see Table 17.6.

**Table 17.6 Characteristics of functional zones and the target landscape
states for the protected natural territory 'Komarovskiy coast'**

Number on the map*	The name of a zone	Area, ha.	Basic use	Maximal recreation load, person/ha. per day	Target landscape state
1	Zone of moderate recreation (walking zone)	24.5	Walks, picking mushrooms and berries, sports and children's games	7-8	Dry pine and spruce forests
2	Zone of intensive recreation (picnic zone)	18.1	Picnics, bathing, walks on foot, roller skates and bicycles	50-60	Light pine forests, dunes with fixed ground, beaches with disperse vegetation
3	Zone of special protection	86.3	Lonely walks, picking mushrooms and berries, pass to the sea coast	2-3	Old-age spruce woods and black alder groves
4	Historical parks of the beginning of the twentieth century	5.0	Individual visits	2-3	Pine-spruce-broad-leaved park woods
5	Abandoned agricultural lands	16.1	Rare visits	<1	Spruce and black groves; objects of infrastructure

Note: * See Figure 17.6.

(7) As a result of comparing the map of target landscape states and the forecast of spontaneous landscape change, undesirable processes in landscape are defined, which prevent landscapes from reaching their target states. For the territory of interest there is, for example, the destruction of spruce due to damaged drainage, the growth of bird cherry thickets on former meadows and the erosion of a scarp of the Littorina terrace. On the basis of these data, the map of planned actions needed to prevent undesirable processes is made (Figure 17.7).

(8) To control changes in landscape states it is necessary to set up long-term landscape monitoring. This monitoring is carried out in two directions: 1) check of conclusions concerning trends of spontaneous processes in landscapes; 2) estimation of the results of nature protection actions. Joint processing of these two sets of observations will allow future correction of nature protection planning.

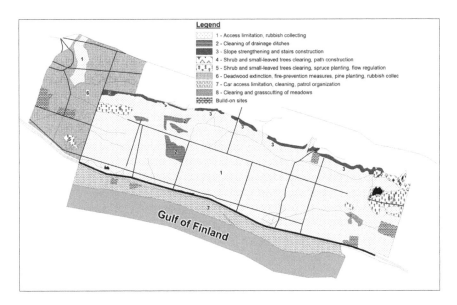

**Figure 17.7 Planning of nature protection actions in the protected
natural territory 'Komarovskiy coast'**

Conclusion

Saint Petersburg being the largest city in a zone of boreal forest has a considerable
variety of natural landscapes (woods, bogs and marshes, sea coasts). Until the
first half of the twentieth century the utilitarian approach to natural landscape as
an obstacle to city growth and development prevailed. At the end of the twentieth
century the need to preserve maintained natural landscapes was realized by
the common public and city authorities, and the creation of a protected natural
territories (PNT) network was started. According to the new General Plan for
the Development of St. Petersburg, by 2025 the share of PNTs should increase
substantially, and the sites of all basic types of natural landscapes within the city
limits will be taken under protection.

In 2007 the Directorate of Protected Natural Territories of St. Petersburg
started to implement the program of PNT landscape management developed by
the authors on the basis of the landscape-dynamic approach. At present, the first
results of this work have already been reached: clearing the territory of historical
park, reconstructing the pond system ('Komarovskiy coast'), building stairs on
the eroded slopes, planting trees and bushes on open places ('Duderhof heights'),
preventing the entry of motor transport on the PNTs, etc. Now plans for preserving
and restoring landscapes are being developed for all existing St. Petersburg's
PNTs.

Acknowledgements

Authors are grateful to the Directorate of Protected Natural Territories of St. Petersburg and personally to the Head of the Directorate, T.V. Kovaleva, for their comprehensive help with this work.

References

Isachenko, G.A. 2007. Long-term conditions of taiga landscapes of European Russia, in *Landscape Analysis for Sustainable Development. Theory and Applications of Landscape Science in Russia*, edited by N. Kasimov and A. Kushlin. Moscow: Alex Publisher, 144-155.
Isachenko, G.A. and Reznikov, A.Y. 1996. *Taiga of the European North-West Russia: Landscape dynamics*. St. Petersburg: Russian Geographic Society (in Russian).
Volkova, E.A., Isachenko, G.A. and Khramtsov, V.N. (eds) 2002. *Komarovskiy coast – complex natural reserve*, Saint Petersburg (in Russian).

Chapter 18

The Evolving Landscape of the Austin-San Antonio Corridor

Frederick A. Day and James W. Vaughan

Introduction

Today we are on the cusp of the regional coalescence of one of America's premier urban growth areas. Both Austin and San Antonio, Texas, have grown from relatively small cities in the last half-century to bustling million-plus population metropolitan areas in the present decade. Their rapid growth appears unabated as the region fills in the 90-mile corridor between these two nearby, yet quite different cities in the heart of Texas. Continuous inter-decadal population growth of 20 per cent to 40 per cent has been fuelled by high-tech, administrative and construction job growth (Handbook of Texas Online 2005). Both the physical and human landscapes have played curious, and critical, roles in the development of this burgeoning region. The corridor is literally bifurcated by the Balcones escarpment, a geologic divide along I-35, a major interstate highway, and has evolved into two distinctly different human landscapes (Figure 18.1).

When we came to the region over 20 years ago to be at Texas State University, we were immediately intrigued as to how the hilly west side of the I-35 divide was so favoured by the locals over the relatively flat, somewhat undulating eastern plains. Both more expensive as well as middle-class housing dominated the western portion of the urbanized area. This distinct division of the urban region into spatially affluent and poor is as pronounced as in any urban area of the country, and it has seemingly remained so up to the present. Real estate prices and agents have perpetuated the distinct differences in the cost of land and housing, and the average person, although perhaps not explicitly cognizant of the 90-mile length of this urban division, knows: the west side of the interstate is where developers build the more affluent housing sub-divisions, and the east side of I-35 ... well, that's where 'Bubba' (lower income whites) lives, or young families find 'starter homes', or the predominately poorer Hispanic communities are.

Over the years we have continued to follow these regional growth patterns in depth, both observing landscape change and pursuing local growth figures at quite detailed scales of analysis. In 1992 the senior author took over the introductory course to the Texas State Geography Masters program, one entitled *Geographic Analysis*. To uncover the differences in the two sides of the region, the new grad students were assigned a paper to research the geographic evolution of two

Figure 18.1 Hill Country and Blackland Prairie in the Corridor Region

small towns, one in the eastern portion of the area and one in the west. The small towns, some indeed hamlets or villages in size, were pre-selected to allow the students to use any analytical means to uncover the historical, socio-economic and physiographical differences affecting the development of the two places. The students have taken this somewhat unusual paper assignment quite seriously, doing extensive in-depth interviews of older residents, detailed archival work and intensive statistical analyses to uncover the reasons underlying differential growth of the two sections of our urban region.

On the last day of class we would discuss what we collectively learned. In over 15 years of papers, the insights have been intriguing and consistent: a historical transition from prosperity to stagnation in the eastern part of the Austin-San Antonio urban region, and a change from hardscrabble farms to middle and upper income exurban residences in the western Hill Country part of the region. In the Blackland Prairies there were stagnant, tiny towns full of atrophying decades-old historic houses nestled among sometimes modest, sometimes dilapidated homes of Hispanic Americans, in some ways more reminiscent of a small village in Mexico. In the Hill Country, resort, retirement and 'bedroom community' ranch-style homes (many being exceedingly large 'McMansions') threatened to submerge the historic small-town buildings, preserved mainly for retail and/or local image.

The central Texas corridor: Population growth

The central Texas urban corridor could be considered one of our most profoundly rapidly growing major metropolitan regions in the US. Both Austin and San Antonio have had inter-decadal population growth rates since the 1960s that have placed them in the top 20 urban growth areas of the US (US Bureau of the Census, 2000). The projections for the counties in the region are truly awesome. Between 2000 and 2008 two of the fastest growing counties in Texas were Hays (45.8 percent) and Comal (38.6 percent) (Texas Data Center 2008), both counties with much of their land area lying west of the Balcones Escarpment and I-35. The four largest metros in Texas are all in the nation's top ten in terms of adding on numbers to their populations, and Austin has been the fastest growing large metro in the country. Historical numbers bear out this point: the post-war growth of the region leave these urban areas perhaps only second in rapid growth to the Las Vegas urban area, a city of a seemingly more tenuous economic base (Table 18.1).

Table 18.1 Population growth rates for the corridor,
Texas and the USA, 1950-2000

	1950	2000	% Change
USA	151,325,798	281,421,906	86.0
Texas	7,711,194	20,851,820	170.4
Austin/San Antonio Corridor	813,126	2,842,146	249.5

Source: US Bureau of the Census, 2000.

The Austin – San Antonio corridor: One region, two landscapes

Differences in the physical environment are stark: the hardscrabble soils and rocky terrain of the Texas Hill Country in the west contrast vividly to the fertile soils of the flat Blackland Prairie to the east of Interstate 35 (Figure 18.2). From the early settlement in the 1840s to the present, these two distinct parts of the region have alternatively been the preferred places of human settlement, the fertile east being favoured early for its bountiful cotton crops, the west being favoured after World War II for its scenic, tree-covered, and hilly sites for homes.

The Texas Hill country, populated by immigrant Germans and western-pushing Americans from the Upland South (Tennessee and other hilly states), met with only marginal success in the beginning, with cattle and goat ranching ultimately proving best in the thin calciferous soils. To the east of the Balcones escarpment, farmers historically met with more success with crops, most notably cotton, responding well in the rich soils and generally flat lands of the Blackland Prairie. Indeed, migrants from the Old Deep South, notably Mississippi and Alabama, as

**Figure 18.2 Left = Typical Texas Hill country soils;
Right = Agricultural field in the Blackland Prairie**

well as German immigrants, transformed this area stretching to the eastern Texas borders into a white sea of cotton fields, by 1900 the largest cotton-growing region in the world (Handbook of Texas Online 2009).

The population growth in the Austin-San Antonio region picked up in the post-World War II era, most notably in the 1960s: San Antonio's growth partly related to a large presence of military bases and white-collar services; Austin, traditionally home to the state government and one of Texas' large flagship universities branched out into the high tech sector in the 1970s and 1980s. In both cities new growth most frequently spread to the northwest, into the environmentally sensitive Hill Country and the Edwards Aquifer Recharge Zone (Vaughan 2006). As suburban sprawl filled in the outlying areas, developers opted for Hill Country landscapes for new higher-end residential development. They almost completely overlooked the Blackland Prairie, which was generally cleared crop or ranch land, yet indeed cheaper for large-scale residential development. In the last five years the belated development of the rural landscape of the Blackland Prairie has picked up momentum with tract housing developments being built at a pace virtually unequaled elsewhere in the US. The homes being built, most by big name, nation-wide developers (Pulte, DR Horton, KB Homes, etc.) are targeted for the lower and middle middle-class, in a price range of about $110,000 to $160,000.

In this context this chapter seeks to understand the critical factors underlying the physical and human impressions on the region's growth. We suggest that geographic differences, including the region's importance as a transitional zone proximate to a diversity of resources, have played a surprisingly dominant role in the modification of the region's urban and rural landscapes during the evolution of one of America's most favoured locations for settlement in the early twenty-first century.

Spatial patterns of socio-economic change

Viewing the spatial patterns of socio-economic change in the region from 1980 to 2000, we can see the distinct inequality in development. It is almost uncanny to see the close identification of social and cultural characteristics of the two sides of the region with their physical landscape. For example, the region's census tracts with higher mean household incomes, or incomes more than 0.5 standard deviation above the region's median income, are almost all west of Interstate 35 in the Texas Hill Country (Figure 18.3), and this differentiation has increased over the years. This same pattern holds for house value (Figure 18.4). And many of these high-dollar homes in the Hill Country are very large as measured by housing units with four or more bedrooms (Figure 18.5).

Finally, the Hill Country tracts appear to be the last bastions for non-Hispanic whites. The isolation index is an 'exposure' index that measures the degree of potential contact between minority and majority group members. Most of the non-Hispanic white isolated census tracts are found in the Hill Country portion of the region (Figure 18.6). Isolated tracts to the east of I-35 are found in formerly small but desirable towns that are within commuting distance to Austin or San Antonio, and are evolving into exurban bedroom communities. Overall, the sharp divide in socio-economic characteristics is clearly marked by the interstate highway and the Balcones Escarpment which closely parallels the highway.

What has underlain this East-West, distinctly differentiated settlement pattern over the last five decades? It is not agricultural worth, as the Blackland Prairie land mollisols and clays are among the most fertile in Texas. Likewise, the flat topography has allowed large-scale crop farming and the possibility of much cheaper large-scale tract development of homes or businesses. In addition, its looser soils make excavation for septic systems, swimming pools, and foundations relatively easy. This is in stark contrast to the sometimes thick, hard limestone undersurface of most of the Texas Hill Country. Regardless, today the cost of raw rural land varies markedly: Blackland Prairie land may sell for one or two thousand dollars per acre, while somewhat comparable Hill Country land goes for $10,000 to $15,000 – approximately a ten-fold difference!

Discussion

Accessibility is similar for the two sides of the region: in fact it is perhaps somewhat easier and faster to get around in the eastern portion of the region, and easier to get to I-35, the very congested and dominant traffic artery that bifurcates the region and

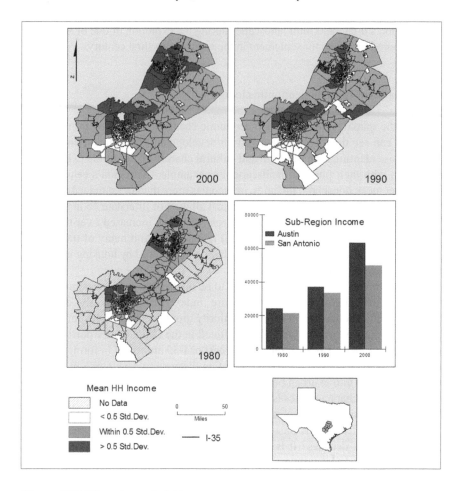

Figure 18.3 Mean household income
Note: Figure includes bar chart of income change by extended MSA. 'Extended MSA' includes census tracts located outside the Corridor's MSA designated counties but with strong economic ties to the primary city based on commuting patterns.

has an overly proportionate load of both local and inter-regional vehicular traffic. Indeed, this region has no other interstates, other than I-10 in the far south.

Truly, what has transformed the different rural landscapes of this region are local perceptions of the landscape. The Hill Country is seen as one of more beautiful views and secluded homesites nestled in evergreens and oaks (Figure 18.7). The Blackland Prairie lands are viewed as flat and unattractive. This view has been perpetuated without question by developers and the real estate industry. It is indeed a subjective opinion, and one that could be questioned. There are indeed gorgeous locations for housing in both sides of the region, and bleak locations

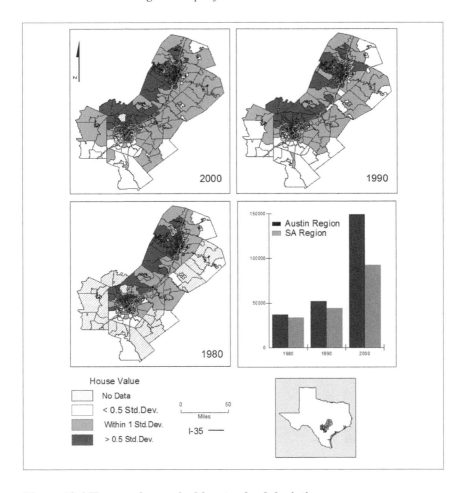

Figure 18.4 House value ranked by standard deviation

Note: Bar chart compares 1980, 1990, and 2000 house values for the Austin and San Antonio sub-regions.

as well. Indeed, most of the homes in the Hill Country have no view to speak of. However, this landscape perception has pushed up the price of land in one area, while depressing it in the other. In addition, of course, as more expensive housing and developments are confined to one region, land and housing values increase as well in a positive feedback loop. The images of luxury and grandeur are perpetuated as well by the real estate industry in advertising, dedicated magazines, and sales pitches such as describing the Hill Country as being '[…] celebrated for its undulating, cinematic beauty. Imagine topography not unlike a Scottish moor, vegetated with southwestern wild flowers, live oaks and lush grasses tall enough to tickle the underbelly of a mustang pony' (Figure 18.7).

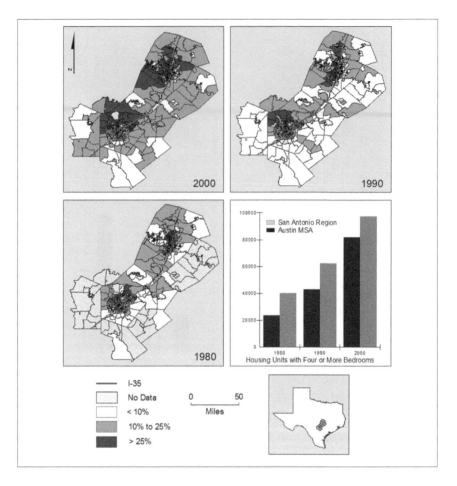

Figure 18.5 Percentage of housing units with four or more bedrooms by census tract

Note: Figure includes chart of total housing units with four or more bedrooms for each extended MSA.

Today it is hard to say how much of this trend will be reversed. The pure pressure of population growth and lack of suitable or affordable land has driven developers to the eastern portion of the region. Also, moving the new regional airport east of Austin in 1999 and the opening of Texas Highway 130, a new four-lane tolled bypass to the east of I-35, have set in motion rapid development of the Blackland prairie portion of the region. Several small towns, notably Lockhart, Hutto, and Pflugerville, have aggressively accommodated the new toll road. Indeed, the small town of Lockhart allowed the new toll road to pass close by with concomitant growth of new housing subdivisions. We are yet to know how this whole rural

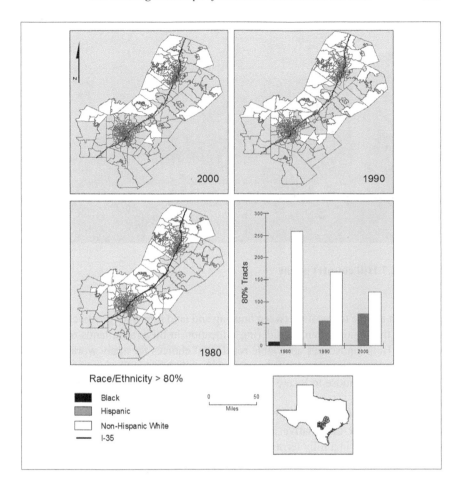

Figure 18.6 Isolated tracts in the corridor region
Note: Highlighted tracts are 80 per cent or more of one ethnicity or race. The bar chart illustrates the number of isolated tracts for each ethnicity or race.

landscape, set to undergo intense and rapid development, will be perceived, and indeed developed!

Future research directions

This study has suggested that the Austin-San Antonio Corridor is perhaps America's most distinctly bifurcated urban landscape. Intriguing questions still linger when landscape geographers ponder this post-World War II evolution, and several lines of inquiry might be pursued.

Figure 18.7 Hill country scene

First, what are the tradeoffs between housing and land prices for prospective home buyers in the region when considering a location in the prairie lands or the Hill Country? Both landscapes are viable residential choices for many workers in the region. For example, would it take a $40,000 or $60,000 difference in lot prices for someone to choose to locate in the less well-thought-of eastern portion of the region?

Second, it would be revealing to probe the attitudes of new residents in the region to uncover the relative importance of being nestled in the seclusion of junipers and oaks in a hilly terrain, relative to the importance of a 'grand' view of a mountain, lake or river. Likewise, it would be interesting to know how important it is that neighbors not be visibly close. Other similar questions can show how critical residential landscapes are in people's choices of where to live.

Third, with next year's 2010 Census, it could be revealing to work with census tract level data and see if the results of this study showing the notable socio-economic differences of residents of the two parts of the Austin–San Antonio urban region are persisting or changing. The decennial census data, along with the ongoing annual American Community Survey from the US Census Bureau, may allow us to suggest that the two halves of the landscape region are converging or indeed continuing to maintain distinct identities.

References

Handbook of Texas Online 2005. Urbanization. [Online] Available at http://www. tsha.utexas.edu/handbook/online/articles/AA/had3.html [accessed: 6 July 2005].

Handbook of Texas Online 2009. Cotton Culture. Available at http://www. tshaonline.org/handbook/online/articles/CC/afc3.html [accessed: 14 July 2009].

Texas State Data Center 2008. *Estimates of the Total Populations of Counties and Places in Texas for July 1, 2007 and January 1, 2008.* Institute for Demographic and Socioeconomic Research, College of Public Policy, University of Texas at San Antonio.

US Bureau of the Census 2000. Census Summary File 3. Washington, DC: US Department of Commerce, Data User Services.

Vaughan, J. 2006. *Growth and Change in a Paradigmatic Region: Is it Sustainable? Does Planning make a Difference?* PhD Dissertation: Texas State University-San Marcos.

Chapter 19

Land Use Changes as Basis for Environmental Protection: The Example of the South Moravian Region, Czech Republic

Hana Skokanová and Tereza Stránská

Introduction

Land use and land cover change can be a major threat to biodiversity as result of the destruction of the natural vegetation and the fragmentation or isolation of nature areas (Verburg et al. 2006). Therefore, a good understanding of land use development dynamics and processes that cause land use changes is essential.

Land use development dynamics can be studied by identification and analysis of trajectories of land use changes (e.g. Swetnam 2007; Käyhkö and Skånes 2006; Crews-Meyer 2004). These trajectories often describe the overall spatial-temporal transformations in land use/biotopes and can identify the core character of landscapes and thus can contribute to the management and conservation of valuable landscapes (see e.g. Käyhkö and Skånes 2006).

The focus of land use change trajectory analysis on change as a dynamic process is different from an approach where landscape patterns, i.e. states of a landscape, are observed and measured at any one moment in time. That is why the change trajectories cannot be determined without spatial intersection of data and at least three time steps are needed in order to build a change trajectory (Käyhkö and Skånes 2006).

Key biotopes are generally considered to support greater biodiversity than other biotopes. That is why these biotopes are often protected either as specially protected areas or are included in NATURA 2000 sites.

NATURA 2000 is a network of protected areas created by the EU countries on the basis of two key legal instruments – Council directive 79/409/EHS on the conservation of wild birds (the so-called Birds Directive) and Council directive 92/43/EHS on the conservation of natural habitats and of wild fauna and flora (the so-called Habitats Directive) (Chytrý et al. 2001). These two directives have been implemented in the Czech Republic within the 114/92 Act about landscape protection and nature conservation. The aim of the NATURA 2000 network is to ensure protection of the most significant sites of European nature. So-called bird areas and sites of community importance (SCI) are established on the basis of the above mentioned directives. This chapter deals with SCIs only.

There are two biogeographical regions, continental and Pannonian, in the Czech Republic. The continental region covers almost all the area of the Czech Republic; the Pannonian region covers South Moravia. Nine hundred and five SCIs, i.e. 9.6 per cent of the total area, were proposed so far and approximately 67 per cent of these sites coincide with already established special protected areas.

The aim of this chapter is to verify if NATURA 2000 sites have been managed in the same way for many centuries and thus yielded high diversity. The verification will be based on the analysis of land use maps from the last 180 years, with an emphasis on land use development and its dynamics.

Study site

South Moravian region covers 7192.7 km² and is situated at the borders with Slovakia in the east and Austria in the south (Figure 19.1). Both biogeographical regions – continental and Pannonian – can be found in South Moravia. There are 100 SCIs in the Pannonian region and 76 SCIs in the continental region (Havlíček 2004). Figure 19.2 shows the spatial distribution of these SCIs. Localities, which represent habitats of specific species, e.g. bats, fruit bats etc., where not included in the analyses because they are created mainly by the attics of chateaus, churches etc., and have point character.

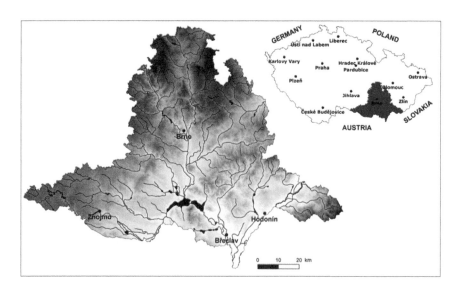

Figure 19.1 Delimitation of the study area

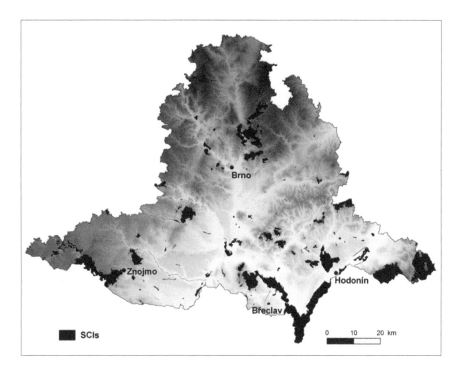

Figure 19.2 Spatial distribution of sites of community importance (SCIs)

Natural conditions

South Moravian landscape represents a typical agricultural region with very favourable natural conditions for this sector: the major relief forms are wide floodplains of the lower parts of the Dyje, Svratka, Svitava and Morava rivers, surrounded by hilly areas of geomorphological regions: Křižanovská vrchovina, Bobravská vrchovina and Drahanská vrchovina highlands in the north, Jevišovická pahorkatina hilly land in the west, Ždánický les forest, Litenčická pahorkatina and Kyjovská pahorkatina hilly lands and Bílé Karpaty mountains in the east and Mikulovská vrchovina highland in the south (Demek and Mackovčin 2006). The average elevation of these hilly areas is 380-500 meters a.s.l. Despite the relatively flat to hilly relief and thick geological substrate, the soils are much differentiated (Mackovčin et al. 2007).

The soils in the floodplains are very fertile with predominant chernozems and fluvisols. Luvisols and cambisols are typical for hilly areas. From the climatic point of view, the region belongs to the warmest areas of the Czech Republic with relatively high temperatures in the summer (average 18-20°C in July), mild winters and low precipitation ranging from 200 mm to 350 mm (Quitt 1970). The northern parts are characterized by lower temperatures (17-18°C in July) and more precipitation (350-450 mm). The relief together with land cover influence climate significantly on a meso-scale.

The main rivers that irrigate the study area are the Jihlava, Svratka, Svitava, Dyje and Morava, with their tributaries the Bobrava and Bítýška (both left tributaries of the Svratka), Oslava (left tributary of the Jihlava), Rokytná (right tributary of the Jihlava), Jevišovka (right tributary of the Dyje), Litava (left tributary of the Svratka), Trkmanka and Kyjovka (both right tributaries of the Dyje). Large diversity of flora and fauna is typical for the South Moravian region. The main factor is the intersection of three fytogeographic unions – Pannonian termofyticum, Carpathian mesofyticum and Czech-Moravian mesofyticum.

From the perspective of potential vegetation, termophilous oak woodlands, oak-hornbeam and lime-oak woodlands, alluvial woodlands, scree and ravine woodlands, herb-rich beech woodlands, acidophilous beech and silver fir woodlands, acidophilous woodrush-, silver fir-, birch-, and pine-oak woodlands and basiphilous perialpine pine woodlands occur in this region (Neuhäuslová et al. 2001).

Despite intensive agriculture, there are many unique biotopes with many protected species. Many of these biotopes are protected by Czech legislation – there are 215 small specially protected areas with the mean size of 0.35 km^2 that cover 75.1 km^2 and parts of four large specially protected areas (protected landscape areas Bílé Karpaty, Pálava and Moravský kras and the national park Podyjí) that cover 151.1 km^2, which is about 4.7 per cent of the study area.

Economic conditions

Thanks to these natural conditions, South Moravia has been cultivated since the Neolithic agricultural revolution. The main agricultural focus is on growing crops, vegetable, fruits and wine. The main political and economic centre is Brno in the north, with a population of 370,592 (in 2009). Other big cities are Břeclav and Znojmo, which are situated in the south. From the economic point of view, the region can be divided into two parts – the north with a concentration of industry, especially in Brno and its surrounding area, and the south with a concentration of agriculture. The main industrial sectors used to be machinery, textile, rubber and food industry. Commerce was also an important part of the economy. Nowadays, the significant industries are machinery and the electronic industry. Other main economic sectors are commerce and tourism.

Materials and methods

Assessment of the land use development

The analysis of land use changes, or generally of cultural landscape changes, is based on a variety of sources, including topographic and old (historical) maps, aerial and satellite photographs, land registers with geodetic survey maps and land plot records as well as various statistical and archival data (Bender et al. 2005).

Five map datasets in medium to large scale, mainly from military surveys, were used for the derivation of land use data. They represent periods of 1830s, 1880s, 1950s, 1990s and 2000s. Their overview together with their positional error is stated in Table 19.1.

Table 19.1 Map sources and their characteristics

Period	Name	Date of creation	Scale	Positional error
1830s	2nd Austrian military survey	1836-1852	1:28 800	11-30 m
1880s	3rd Austrian military survey	1876-1880	1:25 000	13-30 m
1950s	Czechoslovak military topographic maps	1952-1955	1:25 000	10-15 m
1990s	Czechoslovak military topographic maps	1988-1995	1:25 000	10-15 m
	Czechoslovak topographic base maps (ZABAGED 2)	1982-1996	1:10 000	5-10 m
2000s	Czech topographic base maps (ZABAGED)	2002-2006	1:10 000	5-10 m

All topographic military maps were scanned. Maps from the first two periods were rectified in a MATCART program (Brůna et al. 2002; Čada 2006); maps from 1950s and 1990s were rectified according to control points represented by corners of map sheets created for the military topographic survey. Topographic base maps were obtained in their digital form. Czechoslovak topographic base maps from the 1990s were used as a complementary source to the military topographic maps as these significantly underestimate area of permanent grassland.

On the basis of the captions to these datasets, nine land use categories were identified: 1 = arable land; 2 = permanent grassland; 3 = orchard; 4 = vineyard and hop-field; 5 = forest; 6 = water area; 7 = built-up area; 8 = recreational area; and 0 = other area.

The maps were manually vectorized in the ArcGIS software using onscreen method. Vector data, which were derived from the maps, were overlaid. As a result, a GIS database for further analyses was created.

To analyse land use changes dynamics a method called stability mapping was used. This process systematically identifies those areas which have been most prone to land use change, and is based on calculation of three indices that altogether distinguish six classes of land use change trajectories (Swetnam 2007).

The three indices identifying class of land use change trajectory are similarity, turnover and diversity. Similarity captures information about the dominance of any one category at a particular location throughout the period. Turnover records how many changes occurred between adjacent pairs of years. And diversity is simply the number of different categories recorded for the five time steps (Swetnam 2007). The six land use change trajectory classes derived from the combination of these indices are: Stable, quasi-stable, stepped, cyclical, dynamic and no constant trend (NCT). Stable class records the same land use category in each of time steps. Quasi-stable class reflects the dominant trend with only one change of category. Stepped class indicates those locations where there had been one point of change between two dominant categories. Frequent change between just two categories indicates cyclical class, while frequent change among more than three categories is classified as dynamic class. The last class with no constant trend (NCT) means that the land use categories changed several times during the researched period but this change has been variable.

*Classification of SCIs for the purpose of historical
land use development analyses*

For the purpose of historical land use development analyses, the SCIs were divided according to two criteria. The first criterion was proportion of forest. Three types were established (Table 19.2):

- forested sites (more than 50 per cent of the area is covered by forest, as the example of Figure 19.3);

Figure 19.3 Example of a forested site: National Park Podyjí
Source: Stránská 2005.

- non-forested sites (forest covers less than 30 per cent of the area – Figure 19.4);

**Figure 19.4 Example of a non-forested site: Rock habitats in
          the Rokytná river valley**
Source: Stránská 2006.

- transitional sites (forest covers 30-50 per cent of the area – Figure 19.5).

**Figure 19.5 Example of a transitional site:
          Protected Landscape Area Bílé Karpaty**
Source: Havlíček 2009.

Table 19.2 Number and proportion of SCIs according to the proportion of forest

Type	Number	Proportion of the SCI area (%)
Forested	57	73.3
Non-forested	111	3.3
Transitional	8	23.5

The second criterion was size of the site. According to this criterion three types were established as well (Table 19.3):

- small sites (with an area under 10 ha);
- medium sites (with an area between 10 and 100 ha);
- large sites (area over 100 ha).

Table 19.3 Number and proportion of SCIs according to the size

Type	Number	Proportion on the SCI area (%)
Small	74	0.3
Medium	60	3.5
Large	42	96.2

If we consider the relationship between these two criteria, we can say that forested SCIs are represented mainly by large sites, and non-forested SCIs by small sites: From 57 forested SCIs, only 5 sites are small (with the total area c. 19 ha), 19 sites are medium (with the total area c. 840 ha) and 33 sites are large – their total area is c. 44,000 ha; in the case of non-forested SCIs, small sites prevail (68 sites with the total area of 213 ha), followed by medium sites (38 with the total area of 1,271 ha) and only four large sites with the total area of 548 ha. As for transitional SCIs, they are the least represented type with prevailing large size sites: only one site is small (8 ha), two sites are medium (total area being 28 ha) and five sites are large (total area being 14,345 ha).

Concerning biotopes (according to the NATURA 2000 classification – see Chytrý et al. 2001), we can find valuable biotopes represented by termophilous oak woods, Pannonian termophilous oak woods, oak woods on sand, herb-rich beech forests, ravine forests or ash-alder alluvial forests in forested SCIs. Biotopes found in non-forested SCIs are valuable vegetation formations with high biodiversity and many protected and endangered species. The most common biotope is represented by steppe narrow-leaved dry grasslands, which were used as pastures and nowadays have distinctive high biodiversity. Beside steppe biotopes, other common biotopes are extensively used herb-rich meadows, a typical element of the Bílé Karpaty mountains. Other valuable biotopes situated in South Moravia are connected with wetlands, water areas or cliffs but their occurrence is rather patchy.

Results

Land use development of the whole region

If we consider the proportion and development of individual land use categories in the whole South Moravian region, we can say that the prevailing category is arable land, which exceeds 55 per cent of the total area throughout the researched period. It peaked in the 1950s and had the lowest area at the beginning of the twenty-first century. The second most widespread category is forest with the average proportion of 22 per cent. The proportion of this category was smaller in the first two periods and then it increased and remained more or less stable.

The biggest changes occurred in the category of permanent grassland. A significant drop in the area was noted in the first half of the research period, i.e. 1840-1950, with the decrease of more than 10 per cent. From the 1950s to the 1990s the drop continued but was not so remarkable because almost all permanent grassland was already destroyed. A slight increase in 2002-2006 was a result of land abandonment.

The proportion of vineyards decreased in the 1950s as a consequence of wine diseases and economic situation, and increased again in the 1990s, as a result of the systematic management of both cooperative farms and few private farmers. There has been a systematic increase in the proportion of orchards since the 1950s which can be attributed to both favourable conditions for growing fruit, especially warm-requiring sorts like peaches or apricots, and the development of fruit industry in the region.

Quite a steep increase in the proportion of built-up area in the second half of the twentieth century reflected mainly economic development, i.e. construction of industrial plants, residential housing etc. While in the 1950s the increase in this category was at the expense of permanent grassland, in the 1990s it was at the expense of arable land. With the economic development of the region, an increase in the other area was a clear result as lot of construction material was needed for factories, houses etc. The recreational area occurred to a larger extent in the 1950s and its proportion has increased since then.

A decrease in the proportion of water area at the end of the nineteenth century was caused by draining several lakes and ponds. Increase in the second half of the twentieth century was a result of the re-establishment of several ponds, namely Pohořelické rybníky and Jaroslavický ponds, and building new water bodies of Brněnská přehrada lake and Nové Mlýny water body (Skokanová et al. 2009).

Land use development of the NATURA 2000 sites

The character of the NATURA 2000 sites determines to some extent land use which is found in the sites. Logically forested sites will then be covered mainly by forests. Non-forested sites are covered by other categories, mainly by permanent grassland and/or arable land. The proportion of forest in transitional sites will be less pronounced than in the forested sites but more than in the non-forested sites. The development of forest shows a general trend of increasing proportion in all types of sites (Figure 19.6). Similarly a general trend, only in the opposite direction, can be identified in the development of permanent grassland in the forested and transitional sites. A decline in the proportion of permanent grassland in the non-forested sites occurred only between 1830s and 1950s; it rose in the 1990s as a consequence of the establishment of special protected areas with special management and then decreased again at the detriment of water area and forests.

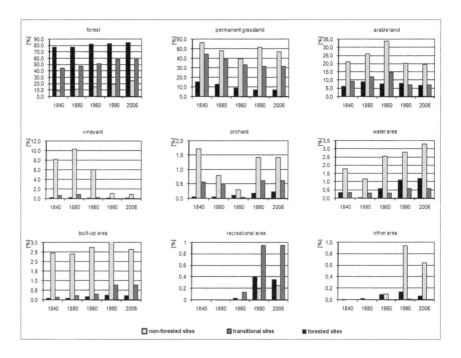

Figure 19.6 Development of land use categories in the forested,
non-forested and transitional sites in the period 1836-2006

Unlike the whole South Moravian region, arable land has never been a prevailing land use category, even though it covered a significant part of non-forested sites where it reached peak in the 1950s, which coincides with the whole region and also with transitional sites. The peak of arable land in forested sites was noted earlier – in the 1880s, when the peak of agricultural revolution was typical for the whole country (Bičík et al. 2001).

Orchards and vineyards are typical for non-forested sites, with the proportion of orchards being the highest in the 1990s and 2000s; in the case of vineyards the peak was noted in the 1880s.

Concerning the water area, again the largest proportion was noted in non-forested sites. All sites experienced a decrease in this category in the 1880s and an increase since then.

The proportion of built-up area was very low in both forested and transitional site (maximum 0.8 per cent). While the built-up area remained more or less the same in forested sites, there was a very slight increase documented in transitional sites. Non-forested sites display rather larger proportion of this category in comparison with the other two types and there was a slight increase in the proportion from the 1880s to the 1990s. 'Other area' and 'recreational area' are not very significant because both categories cover only up to 1 per cent of the total area of the sites, at the most. Recreational area occurred since the 1950s and can be found mainly in transitional and forested sites. Other area is situated especially in non-forested sites and has been present to a larger extent in the 1990s and 2000s.

From the perspective of size, large sites are covered mostly by forest, while small sites are covered mainly by permanent grassland and arable land which dominated only in the 1950s. The prevailing category in medium sites was represented by permanent grassland during the nineteenth century and then forest took over.

Forests gradually increased their area from the 1840s to the present in both medium and large sites which was also seen in the case of forested and transitional sites. In small sites there was a slight decrease in the forest area in the 1880s.

While the area of arable land in small and medium sites reached a peak in the 1950s and then declined (as was seen also in non-forested and transitional sites), in large sites the peak of this category was in the 1990s.

A very different trend in the development of the area of permanent grassland can be seen in all three types of sites (Figure 19.7). In the case of small sites, there was a decrease in the area until the 1950s, which was followed by an increase lasting until the present. Medium sites also showed a decrease until the 1950s and then an increase in the 1990s but this was again followed by a decrease in the 2000s. This corresponds to the development of permanent grassland in non-forested sites. Large sites experienced a decline in this category until the 1990s and then a slight increase.

Orchards and vineyards occurred in a larger extent only in small and medium sites but their higher proportions were identified in different time steps. Vineyards were typical for the first half of the studied period, i.e. from the 1830s till the 1950s, and orchards for the second half of the period, i.e. from the 1950s till the 2000s.

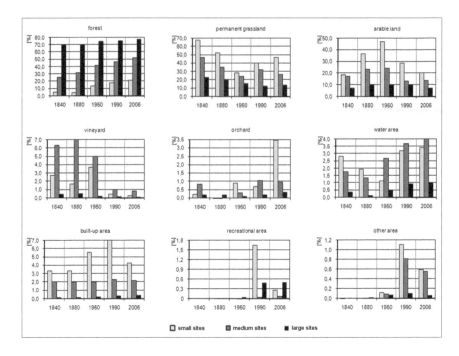

Figure 19.7 Development of land use categories in small, medium and large sites in the period 1836-2006

Small and medium sites also show bigger proportion of water area than large sites. Proportion of this category declined in the 1880s in medium and large sites and then it rose to its maximum in the 2000s. In the case of small sites the decline lasted until the 1950s. Since then it experienced the same development as the other two types.

Built-up area covers mostly small sites and its proportion peaked in the 1990s, which is similar to the non-forested sites. This category in the medium sites showed a more or less steady trend during the researched period and the same trend was also characteristic for large sites. Recreational area was identified in small and large sites; in small sites a relatively big proportion was noted for the 1990s, in large sites the proportion remained the same during the 1990s and 2000s. Other area was also most pronounced in the 1990s and 2000s but it was detected mainly in small and medium sites.

Land use change trajectories

The South Moravian region can be considered stable, with the stable area covering over 74 per cent of it. Nearly 61 per cent of these stable plots were used as arable

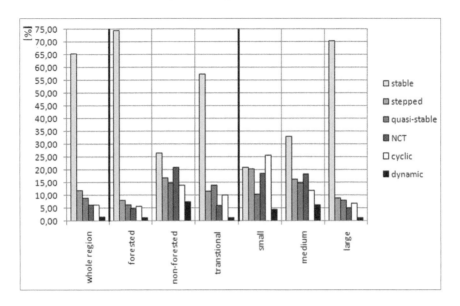

Figure 19.8 Land use change trajectories and their proportion (%) of the total area in the whole region, forested, non-forested, transitional, small, medium and large sites

land, 34 per cent were used as forests and only 2.8 per cent were used as built-up area and 1.4 per cent as permanent grassland respectively.

A stepped class, second most common change trajectory, was identified for 11.7 per cent of the whole region (Figure 19.8). The prevailing changes were from permanent grassland to arable land and from arable land to built-up area. Also changes between permanent grassland and forest were of some significance. The changes were detected especially between the 1880s and 1950s or 1950s and 1990s.

The proportion of quasi-stable class was 6 per cent and included changes mainly from permanent grassland either to arable land or forest at the end of the nineteenth century, from arable land to permanent grassland or forest (at the end of the nineteenth century and beginning of the twenty-first century respectively) and from forest or vineyard to arable land again at the end of the nineteenth century.

Cyclic class included above all changes between forest and permanent grassland or arable land and concerned 5.5 per cent of the whole region. The remaining plots were classified either as dynamic or with no constant trend (NCT). Changes in these plots did not show any clear pattern or occurred very frequently. They applied above all to land use categories of permanent grassland, arable land and forest.

Forested sites show the highest proportion of stable class (74.5 per cent), followed by transitional sites (57.3 per cent). Both types have dominant land use category of forest (94.4 per cent for forested sites and 69.6 per cent for transitional sites respectively) but in the transitional sites a proportion of permanent grassland of the total area of the class is also high (28 per cent). Non-forested sites have only 26.4 per cent of stable areas with the stable categories being permanent grassland, arable land, forest, built-up area, water area and vineyard.

While in the forested site the second most common class is stepped change (as is in the whole South Moravia), which is represented by transition mainly between permanent grassland and either forest or arable land, in the non-forested sites the second most common class is NCT (mostly changes between permanent grassland, arable land, forests and vineyards) and in the transitional sites it is quasi-stable class (changes from arable land to permanent grassland at the end of twentieth century or changes between permanent grassland and forest in both the 1880s and 2000s).

Stepped change in non-forested and transitional sites was recorded at nearly 17 per cent and 12 per cent of the total area respectively. In transitional sites, changes occurred between permanent grassland and/or arable land. This finding is similar to changes detected in forested sites (see above). In non-forested sites changes were also found between vineyards and permanent grassland or arable land.

Unlike the stepped class where change can occur any time during the researched period, in the quasi-stable class change occurs only at the beginning or at the end of the period. This situation means that if the start or end of the researched period is shifted, the class can change into a stepped one. The quasi-stable class in non-forested sites covers 14.7 per cent of the total area which was the highest proportion of all three types. Changes were mainly between permanent grassland and forest or vice-versa and between arable land and permanent grassland. The same changes were also recorded for forested sites where the quasi-stable class was found at 6.2 per cent of the total area (the third most common type of change).

Changes between forest, permanent grassland and arable land within cyclical class are typical for both forested and transitional sites. However while in forested sites forests dominate, in transitional sites the dominant categories are permanent grassland and arable land. This is in agreement with land use development (see above). Non-forested sites display typical rotational agricultural management with changes between arable land, permanent grassland and vineyards.

As was mentioned before, plots where no constant trend occurred were typical especially for non-forested sites. These sites also had a large proportion of dynamically changing land uses. Dynamic changes in forested and transitional sites covered only up to 1 per cent of their area. Whereas in forested sites these changes were recorded among permanent grassland, arable land and forest, in transitional sites also vineyards, orchards and recreational areas participated in the change.

From the viewpoint of size, we can say that large sites show similar proportions and pattern as forested sites, i.e. large proportion of stable areas (70.4 per cent) followed by stepped and quasi-stable changes (8.8 and 8 per cent, respectively), with the dynamic class and class with no constant trend being the smallest (1.1 and 5.1 per cent, respectively). This similarity reflects the fact that most forested sites belong to large areas (7.3 per cent). Yet there are some differences between these two types and these are clearly reflected in the proportion of stable categories within the stable class: in both types forests cover the majority of the type (94.4 per cent for forested sites and 69.6 per cent for large sites) but the second most common categories are different (arable land in forested sites vs. permanent grassland in large areas) and their proportion is also different (2.9 per cent vs. 28 per cent).

Another similar pattern in the proportion of land use change trajectories can be identified between non-forested and medium sites; non-forested sites are the most common and also cover more than 50 per cent of the total area of medium sites. Again, there are some differences in the character of classes. Significant class from the perspective of area in the case of medium sites is forest even though there are only 18 forested sites out of 60. Still, they represent continual forest complexes and thus their influence on land use change trajectories is significant. Also permanent grassland and arable land are very common, especially in non-forested and transitional sites where they create a distinctive landscape mosaic, sometimes also with patches of vineyards and orchards. Concerning cyclical change, in medium sites there is also a quite significant proportion in rotation between forests and arable land, unlike the case of non-forest sites. Also in stepped changes forests play an important part. On the other hand, the role of vineyards is not as significant as in the case of non-forested sites. This refers especially to the dynamic and NCT changes.

Small sites show very different patterns from all other sites. This is due to their size – they have area up to 10 ha and that is why they are more susceptible to changes than larger sites. The prevailing class is cyclic change (25.6 per cent of the total area) with the rotation between permanent grassland and arable land, followed by stable class (20.8 per cent) and stepped change (20.3 per cent). Permanent grassland dominates the stably used plots. Other stably used plots are those of arable land, built-up area and forest. Transitions from permanent grassland to forest or arable land are the main changes within the stepped class. Changes among these three land use categories but also among orchards and water areas are typical for NCT class. Quasi-stable class shows quite big proportions in changes from permanent grassland to arable land and vice versa (24.9 and 19.2 per cent of the class, respectively) but also from water area to permanent grassland. Finally, the dynamic class records the biggest variety in the changes among land use categories from all types.

Conclusions

European sites of community importance within the NATURA 2000 network represent biodiversity islands in the cultural landscape of South Moravia. To find out if these islands were present in the past or were created recently, which can reflect in their biodiversity, it is useful to analyse land use/land cover on the basis of old maps and other sources.

In total, 177 sites of community importance were researched in South Moravia with the aim to determine relationships between stability of the sites, presumably connected with higher biodiversity, and land use changes.

The researched sites were divided according to two basic criteria – site size and proportion of forest. Analyses of land use development and subsequent land use change trajectories helped verify a hypothesis of dependence of intensity of changes on these criteria.

The results show that while small sites are generally less stable and more susceptible to land use changes, sites larger than 100 ha constitute stable biotopes in the form of large forest complexes are not easily changed. This is also a result of the fact that small-scaled sites are composed of mosaic of permanent grassland, vineyards, arable land and orchards where the change in land use is rather easy. Moreover, a large part of these sites has been abandoned and consequently turned into extensive biotopes of narrow-leaved dry grasslands or herb-rich meadows.

Analyses only of land use development and land use change trajectories cannot determine whether a site has high biodiversity but can indicate parts where this biodiversity is likely to be found due to both land use and management practises.

Acknowledgements

This study is a part of a research project on sources and indicators of biodiversity in cultural landscape in the context of its fragmentation dynamics which has been carried out at the Department of Landscape Ecology, Silva Tarouca Research Institute for Landscape and Ornamental Gardening, Czech Republic.

References

Bender, O., Boehmer, H.J., Jens, D. and Schumacher K. 2005. Using GIS to analyse long-term cultural landscape change in Southern Germany. *Landscape and Urban Planning*, 70(1-2), 111-125.

Bičík, I., Jeleček, L., and Štěpánek, V. 2001. Land-use changes and their social driving forces in Czechia in the 19th and 20th centuries. *Land Use Policy*, 18, 65-73.

Brůna, V., Buchta, I. and Uhlířová, L. 2002. Identifikace historické sítě prvků ekologické stability krajiny na mapách vojenských mapování. *Acta Universitatis Purkynianae – Studia Geoinformatica II.*, 81. Ústí nad Labem.

Čada, V. 2006. Hodnocení polohové a geometrické přesnosti prvků II. vojenského mapování lokalizovaných v S-JTSK, in *Historická geografie, Supplementum I.*, edited by Šimůnek, R. Praha: Historický ústav, 82-105.

Chytrý, M., Kučera, T. and Kočí, M. (eds) 2001. *Katalog biotopů České republiky.* Praha: AOPK ČR.

Crews-Meyer, K.A. 2004. Agricultural landscape change and stability in northeast Thailand: historical patch-level analysis. *Agriculture, Ecosystems & Environment,* 101(2-3), 155-169.

Demek, J. and Mackovčin, P. 2006. *Zeměpisný lexikon. Hory a nížiny.* Praha: AOPK ČR.

Käyhkö, N. and Skånes, H. 2006. Change trajectories and key biotopes – Assessing landscape dynamics and sustainability. *Landscape and Urban Planning,* 75(3-4), 300-321.

Mackovčin, P., Jatiová, M., Demek, J., and Slavík, P. 2007. *Brněnsko* (Chráněná území ČR Edition). Praha: AOPK ČR a EkoCentrum Brno.

Neuhäuslová, Z., Blažková, D., Grulich, V., Husová, M., Chytrý, M., Jeník, J., Jirásek, J., Kolbek, J., Kropáč, Z., Ložek, V., Moravec, J., Prach, K., Rybníček, K., Rybníčková, E., Sádlo, J. 2001. *Mapa potenciální přirozené vegetace ČR.* Praha: Academia. 341.

Quitt, E. 1970. *Mapa klimatických oblastí ČSSR.* Brno: Geografický ústav ČSAV.

Skokanová, H., Stránská, T., Havlíček, M., Borovec, R., Eremiášová, R., Rysková, R. and Svoboda, J. 2009. Land use dynamics of the South Moravian region during last 170 years. *Geoscape,* 4(1), 58-65.

Swetnam, R.D. 2007. Rural land use in England and Wales between 1930 and 1998: Mapping trajectories of change with a high resolution spatio-temporal dataset. *Landscape and Urban Planning,* 81(1-2), 91-103.

Verburg, P.H., Overmars, K.P., Huigen, M.G.A., de Groot, W.T. and Veldkamp, A. 2006. Analysis of the effect of land use change on protected areas in the Philippines, *Applied Geography,* 26(2), 153-173.

Chapter 20

Assessing the Sustainability Impact of Land Management with the Ecosystem Services Concept: Towards a Framework for Mediterranean Agroforestry Landscapes

Tobias Plieninger and Thanasis Kizos

Introduction

The concept of sustainability is used to estimate the state and potential of an area or an activity to achieve 'sustainable development'. The Millennium Ecosystem Assessment (MA) operationalizes sustainability from a human-centred ecosystems perspective, with indirect driving forces (e.g. population, technologies, and lifestyles) translating into direct drivers (e.g. land use changes) that put pressure on ecosystem services, which directly affect human well-being. All interactions act at different scales, from the local to the global. Ecosystem services, 'the benefits people obtain from ecosystems' (Millennium Ecosystem Assessment 2003: 53), are the core of the framework, grouped into 21 single services within four broad categories of provisioning, regulating, cultural, and supporting services. The MA puts a special focus on the identification of trade-offs and synergies between different ecosystem services that depend on specific management practices in an ecosystem.

For land use practices, two spacial scales that can be considered are the small (management) scale and the large (landscape) scale. At management level, practices express the complex sets of attitudes, implicit rules, knowledge, experiences, etc. of a group of land users in a specific area. They shape the appearance of the land, the architecture of farming constructions and the social distribution of labour in and among farms. They imprint linkages between economic, social, technological, ecological and political dimensions upon the landscape. As such, management practices determine the services that farm ecosystems provide in the MA framework. At landscape level, management practices are 'summed up'; they translate into quantifiable area-related parameters and determine the overall sustainability state of the area. For such scales the impacts of the management practices can be assessed with the use of area level indicators (e.g. land cover change, soil and water quality, etc.). These two levels should not be viewed as separate and distinct but as a continuous spectrum band from 'action' to 'result' across which the impacts of management practices can be assessed.

In the Mediterranean, land management practices have been shaped throughout millennia of farm development and ecosystem responses (Grove and Rackham 2001), which have spawned a diversity of land uses and wildlife. Landscapes that have slowly evolved via the interaction of management practices and the natural environment over long periods and that 'form a characteristic structure reflecting a harmonious integration of abiotic, biotic and cultural elements' (Antrop 1997: 109) have been termed 'traditional cultural landscapes'. 'High Nature Value (HNV) Farming Systems' (Cooper et al. 2007) are common in these landscapes.

The aim of this chapter is to explore the conceptual framework and propose some indicators for the sustainability assessment of two Mediterranean land use systems over different spatial scales with the use of the ecosystem services concept. We specifically seek to investigate the differences that arise from the use of different spatial scales (the small scale, i.e. the operating range of the farmer or the landholder, and the large scale, i.e. that of the landscape) in the assessment of sustainability for the same type of ecosystems, land uses and their changes. These different scales correspond roughly with the local and regional approaches of sustainability impact assessment discussed above.

The case of Mediterranean agroforestry systems

Agroforestry systems stand for, according to the World Agroforestry Center (ICRAF), a 'collective name for land-use systems and technologies where woody perennials (trees, shrubs, palms, bamboos, etc.) are deliberately used on the same land management unit as agricultural crops and/or animals, either in some form of spatial arrangement or temporal sequence. In agroforestry systems, there are both ecological and economic interactions between the different components' (Gordon et al. 1997: 1). Agroforestry systems not only provide agricultural and forestry commodities, but also deliver services such as soil conservation, reduction of surface runoff (Ceballos et al. 2002), carbon sequestration (Pandey 2002), biodiversity conservation (McNeely and Schroth 2006), and landscape beauty.

In the Mediterranean, agroforestry systems date back to the Neolithic with a variety of trees and management systems. Some of the typical tree species used in agroforestry systems are olive (*Olea europaea*), oak (*Quercus sp.*), chestnut (*Castanea sativa*), carob (*Ceratonia siliqua*), apple (*Malus sp.*), almond (*Prunus dulcis*) and date (*Phoenix dactylifera*). Geographically, some species and systems spread throughout the region and others are limited to particular locations.

A large-scale agroforestry system of the Mediterranean is the *dehesa* on the Iberian Peninsula – the cultivation of oaks combined with grazing of sheep, goats or pigs and cultivation of arable crops (Grove and Rackham 2001). *Dehesas* are characterized by the complementary relationship between pastoral, agricultural and forestry components, resulting in a high resource use efficiency and important interactions with different natural ecosystems. Management activities result in a large assortment of commodities (amongst others meat, wool, cork, charcoal, milk, ham, honey, and acorns) and non-commodities (e.g. soil conservation or regulation of the hydrological cycle). The tree species range from holm oak (*Quercus ilex*) to cork oak (*Quercus suber*). Pig husbandry is tightly related to *dehesa* management as hogs provide the best use of acorn mast by producing highly priced acorn ham. Forest management is another integral component of the *dehesas* with regular clearing and pruning. Rotational ploughing was a common management strategy for crop cultivation and for the control of shrub encroachment. Cultivation in *dehesas* historically had long rotation cycles of four to seven years, but since the 1980s the extent of arable farming has strongly decreased. *Dehesas* have become renowned Europe-wide for supporting outstanding levels of species and habitat diversity and a rich diversity of plants and animals.

Another widespread system comprises the cultivation of olive trees and grazing of the understorey by goats and sheep. Low-input traditional plantations (against intensified 'modern' or abandoned plantations) are the ones that can be considered an agroforestry system. In Beaufoy's frame (2000) these have potentially the highest natural value of all types of plantations, due to features such as terraces, old trees, stone walls and high floral diversity; and the most positive effects (e.g. in regard to water management in upland areas) as well as the least negative effects on the environment. At the same time they are the least viable in economic terms and hence most vulnerable to abandonment (Kizos and Koulouri 2010).

Methods

The approach followed here attempts to integrate different scales and concepts in assessing the sustainability impact of land use changes. At the small scale, the analysis of intangible management practices can provide the type of services and the type of impact, while at the regional scale information on the visible landscape effects of these individual practices determines the overall impacts. Since at the small scale farms with diverging practices and therefore different impacts exist side by side, the trade-offs and synergies of these impacts have to be assessed at the large scale.

Indicator selection was performed on the basis of Cooper et al.'s (2007) core criteria for high nature value farming: intensity of land use, presence of semi-natural features, and presence of a land use mosaic, adapted for the specifics of agroforestry systems. An initial exhaustive list of indicators was reduced to those features that proved most relevant for both systems and important for ecosystem services provision.

The indicators used for the small scale are concerned with specific management practices and are adapted to the particular systems. Therefore, small scale assessment is performed according to specific research on the particular systems and our personal experience when the evidence is non-conclusive or absent. On the contrary, the landscape scale indicators used are more 'conventional' in the sense that they are derived from existing European approaches (e.g. Wascher 2000; EEA 2005).

The type of relationships between management practice and ecosystem service follow Peco et al. (1999): (i) Positive relationship, where a change in the value of the management practice reflects a change in the same direction of an ecosystem service; (ii) Negative, where a change in the value of the management practice reflects a change in the opposite direction to the ecosystem service; and (iii) Bell-shaped, where intermediate values of the management practice show an optimum ecosystem service while the two extremes show minimum ecosystem services.

For the management scale, we propose a list of 19 indicators grouped into four categories: (i) changes in overall farm management with nine indicators (Table 20.1); (ii) livestock grazing with seven indicators (Table 20.2); (iii) crop cultivation with three indicators (Table 20.3), and (iv) forest management with four indicators (Table 20.4) to grasp management changes and their ecosystem impact. We tried to select non-redundant indicators, but, as agroforestry systems are characterized by a multitude of interactions between their components, there are various interdependencies (e.g. the abandonment of transhumance would not have been possible without external feed inputs). For each of these broader categories the practices and the indicators are classified into two distinct classes: extensification and intensification indicators. Extensification is understood as a reduction of input levels of capital, labour, energy, and/or agrochemicals, while intensification refers to a rise of the level of inputs per area of land. For each indicator, six different types of services are analysed in four categories: provisioning (with two subcategories: livestock and crop production); regulating (with two subcategories: forest regeneration and erosion control); cultural that here is used as potential for heritage/eco-tourism services; and supporting that include biodiversity. Overall, a matrix is filled for all the indicators and all the services for the two systems. These indicators are not exhaustive and they are adapted to the particular systems, as we have chosen to focus on change and the distinction between older and novel practices in order to monitor continuity or rupture in the type of ecosystem services that result.

The indicators for the landscape scale are divided into four broad themes: land cover – land use; intensity of management; landscape elements; and incomes – local products. For each indicator there is an overall assessment of the impact of the increase of its values on the overall environmental services in the area (Table 20.5). Moreover, the indicator is characterized with regard to the type of process it entails, with three types: intensification, extensification and abandonment. Again, these indicators are not exhaustive and their purpose is to capture as best as possible the complexity of the situation at the small scale and translate at the large scale the major impacts and trends in the area.

The material that is used comes from published research (see next section for details) and from our own personal experience, including interviews with farmers and managers, data from aerial photographs or satellite imaging and observations in the field. The overall approach is exploratory and aims more at providing a conceptual framework than arriving at final assessments.

Changes at the management scale and ecosystem services

At farm level, eight indicators are selected for the overall landscape and farm structural features, four refer to extensification and four to intensification: the extensification indicators include indicators for the abandonment of elements such as stonewalls, stone farm structures, footpaths and terraces and an indicator for land abandonment. For land abandonment, the services are mixed with some positive, many negative and some bell-shaped effects (Table 20.1). The intensification indicators include elements such as watering ponds (positive for livestock and biodiversity services), roads (positive for provisional services, negative for the rest), wire fences (negative) and irrigation facilities (again positive for provisional services but negative for the rest). Most of the abandonment processes do not affect provisioning or regulating services to a large extent, but they definitely impact cultural services and in some respect biodiversity in a negative way (Plieninger and Schaich 2006). On the other hand, modernized structures facilitate livestock and crop production and can modify regulating services, but most seem incompatible with cultural services and biodiversity.

Table 20.1 Impact of overall farm management on ecosystem services

	Provisioning services				Regulating services				Cultural services		Supporting services	
	Livestock		Crops		Forest regeneration		Erosion control		Heritage/ Eco-tourism		Biodiversity	
	1	2	1	2	1	2	1	2	1	2	1	2
Extensification indicators												
Abandonment of stonewalls and other stone farm structures	0	0	0	0	0	+	0	0	–	–	–	+
Abandonment of traditional footpaths	0	0	0	0	0	0	0	0	–	–	0	0
Abandonment of terraces	a)	0		0		0		–		–		–
Outright abandonment/ transition to 'natural' woodland/shrub	–	–	–	–	+	+	?	?	–	–	∩	∩

Table 20.1 continued

	Provisioning services				Regulating services				Cultural services		Supporting services	
	Livestock		Crops		Forest regeneration		Erosion control		Heritage/Eco-tourism		Biodiversity	
	1	2	1	2	1	2	1	2	1	2	1	2
Intensification indicators												
Construction of watering ponds	+	+	0	0	0	0	0	0	0	0	+	+
Construction of roads	+	+	+	+	0	0	–	–	–	–	–	–
Introduction of wire fences	0	0	0	0	0	0	0	0	–	0	?	0
Construction of irrigation facilities	0	0	+	+	–	–	?	?	0	0	–	–

Note: 1 = dehesas; 2 = olive groves; a) terraces are rarely found in dehesas.

For the livestock grazing theme, six indicators are selected, three for extensified and three for intensified systems: for the former, the abandonment of animal husbandry (negative for provisional and some cultural services, positive for regulating and supporting ones), the abandonment of seasonal livestock movements and other forms of traditional grazing management (negative for almost all services) and the reintroduction/maintenance of indigenous breeds (mostly positive as these breeds are better adjusted to local conditions) are included. For intensified systems, the increase of livestock density and the introduction of external inputs are considered negative for all services except some provisional ones, while the replacement of sheep from cattle that is encountered in dehesas is considered as a practice with mixed and some unknown yet results (Table 20.2). Next to the increase in numbers, the livestock sector experienced the loss of genetic diversity, i.e. the decline of many indigenous livestock breeds such as the Spanish Merino negro sheep, and also a loss of complex rangeland and grazing management techniques.

For the crop cultivation theme, three indicators are selected, one for extensification and two for intensification. The abandonment of crop cultivation is considered negative for all services except forest regeneration. The introduction of external input (negative for all services except some provisional ones) and the shortening of fallow cycles (again mostly negative) are considered intensification indicators (Table 20.3), as the former land use mosaic of cultivated, grazed and shrubby parts has been simplified. Questions remain in regard to impact of this on the ecosystem services. On the land where crop cultivation has remained, the fallow cycles have been shortened to two years or less, and external inputs (herbicides, mineral fertilizer) have been introduced.

Table 20.2 Impact of livestock husbandry practices on ecosystem services

	Provisioning services				Regulating services				Cultural services		Supporting services	
	Livestock		Crops		Forest regeneration		Erosion control		Heritage/ Eco-tourism		Biodiversity	
	1	2	1	2	1	2	1	2	1	2	1	2
Extensification indicators												
Complete abandonment of animal husbandry	–	–	–	–	+	+	+	0	–	0	∩	?
Abandonment of seasonal livestock movements and other forms of traditional grazing management	+	+	0	0	–	–	–	–	–	0	–	–
Reintroduction/ maintenance of indigenous breeds	+	+	0	?	?	0	?	0	+	+	+	+
Intensification indicators												
Increase of livestock density	+	+	0	0	–	–	–	–	0	0	–	–
Introduction of external inputs (feed, high performance breeds)	+	+	0	0	?	–	–	–	–	–	–	–
Replacement of sheep by cattle	+	0	0	0	?	0	–	0	0	0	?	0

Note: 1 = dehesas; 2 = olive groves.

Finally, for the forest management theme, four indicators are selected, three for extensification and one for intensification: the abandonment of practices such as pruning and charcoal production (neutral for most services, positive for forest regeneration); the neglect of natural regeneration of trees (negative) and the brushing with heavy machinery (mostly negative except for provisional services). Artificial regeneration is considered a positive practice in general (Table 20.4). The forest management sector has been largely dominated by neglect (Díaz et al. 1997). Oak stands are often overaged and lack regeneration, which is a feature of both traditional and modernized systems (Pulido and Díaz 2005). This may have consequences on regulating services (Moreno-Marcos et al. 2007), biodiversity (Díaz et al. 2001), cultural services (Plieninger 2007), and even on crop productivity (Moreno et al. 2007). A modern practice to confront this issue is afforestation, introduced by the EU schemes. Olives are less affected from neglect, as they may be cultivated with limited management practices, while harvesting can be negotiated 'at leisure' (weekends and afternoons). At the same time, olive oil always kept a considerable market value that can cover household needs and finally, plantations represent important investments.

Table 20.3 Impact of crop cultivation practices on ecosystem services

	Provisioning services				Regulating services				Cultural services		Supporting services	
	Livestock		Crops		Forest regeneration		Erosion control		Heritage/ Eco-tourism		Biodiversity	
	1	2	1	2	1	2	1	2	1	2	1	2
Extensification indicators												
Abandonment of crop cultivation	–	–	–	–	+	–	+	–	–	–	–	–
Intensification indicators												
Introduction of external input (pesticides, mineral fertilizer, farm machinery)	0	–	+	–	–	–	–	–	–	–	–	–
Shortening of fallow cycles	0	0	+	+	–	–	–	–	0	0	–	–

Note: 1 = dehesas; 2 = olive groves.

The combination of abandoned traditional practices and intensified modernized practices leads to a simplification of the land use systems discussed here. As many historical features persist over long periods, this simplification may not be apparent in the landscape at first glance. However, they have become relics with a lost functionality. Today they fulfil other ecosystem services (mainly cultural services) than in the times when they were created. Only agroenvironmental schemes provide some new incentives for the supply of these new services. However, most *dehesas* and olive groves, even if they have probably lost some of their nature value, should continue to be considered high nature value farmland.

Landscape scale changes and ecosystem services

Overall, 16 indicators are selected, in four broad themes. The first theme is the land cover – land use indicators that can also cover issues of land use change when monitored over time. Here four indicators are included, the 'area of native tree plantations' (positive for the overall environmental value of the area); the 'area of dry pasture or cropland' for the rest of the grazing areas and former agroforestry

Table 20.4 Impact of forest management practices on ecosystem services

	Provisioning services				Regulating services				Cultural services		Supporting services	
	Livestock		Crops		Forest regeneration		Erosion control		Heritage/ Eco-tourism		Biodiversity	
	1	2	1	2	1	2	1	2	1	2	1	2
Extensification indicators												
Abandonment of pruning and charcoal production	0	0	0	0	–	0	0	0	–	0	?	?
Heavy machinery brushing	+	+	+	+	–	–	–	–	0	0	–	–
Neglect of natural regeneration	–	–	–	–	–	–	?	–	–	–	–	–
Intensification indicators												
Artificial regeneration	–	–	+	0	+	+	+	+	+	+	?	?

Note: 1 = dehesas; 2 = olive groves.

systems (negative for the overall environmental value); the 'area of abandoned systems' (negative for environmental value) (Pinto-Correia and Mascarenhas 1999); and the 'area of natural (unmanaged) forest and shrubland' (of bell-shape concerning the environmental value).

The second theme concerns the intensity of management at the farm level, as an average or aggregated total, with five indicators: the 'area of irrigated land' (negative in environmental value terms); the 'grazing density' (negative, Shakesby et al. 2002); the 'tree density' (bell-shaped in environmental value terms); the external inputs for fertilisers and animal feeding stuff (both negative, indicate intensification).

The third theme is the landscape elements indicators that cover the change of some specific elements, with four indicators: the 'share of stone structural elements in the landscape' (positive for the environmental value, Plieninger and Schaich 2006); the 'average quality of stone structural elements' that reflects the condition of these elements; the 'maintenance of terraces' (positive); and the 'watering ponds density' (positive).

The fourth theme is for indicators of incomes and local products, which cover issues of value and incomes distribution in the area that can help extensification of management practices and provide additional incomes for farmers and land managers. Three indicators are included: the 'number of traditional or specific character products in the area' (positive, signifies the extent in producing a commodity that can produce added value); the 'number of agri-tourists in the area' (positive for the same reasons); and the 'number of agri-tourism holdings' (also positive, if practised at a relatively small scale).

For each indicator there is an overall assessment of the impact of the increase of its values to the overall environmental services in the area (Table 20.5) and a characterization with regard to intensification, extensification and abandonment. The availability of data is also discussed, with three classes: available data from official sources published and readily used; merely available data, have to be calculated from aerial photography – satellite imagery or collected from local sources; and unavailable data, have to be collected from primary research.

Considerations for future research

The development of the particular framework for assessing sustainability at two different scales reveals that the relationships between land use and land cover changes at different scales remain poorly understood. There are well-established frameworks, indicator sets, and reliable data for regional scale assessments. However, such higher-scale assessments detect only coarse land management changes, e.g. the transition from forest to grassland. They are often generalized and thus unable to assess qualitative, gradual, and more subtle changes in land management. But small-scale structural modifications within a land cover category may be as significant for the provision of ecosystem services as land cover change itself. Moreover, changes in management practices (e.g. the uprooting of trees or the abandonment of crop cultivation) may trigger regional landscape changes (e.g. reduction of tree cover, increase of natural shrubland) that become visible only at a later date. This study demonstrates that both the *dehesa* and the olive agroforestry systems of the Mediterranean have experienced significant land management changes that tend to be ignored in large-scale assessments.

An issue that emerges from this approach is the integration of small scales in the regional scale. At the small scale, practices can and indeed do vary significantly from farms lying side by side. This creates a complex mosaic of different practices and diverging services and landscapes (e.g. abandoned farms next to intensified or extensified ones in a random way). Regional scale indicators can only make average assessments and not really integrate all these different patches of the mosaic into a coherent system. Despite some existing approaches (e.g. landscape cohesion from Opdam et al. 2003 and landscape metrics Botequilha Leitão and Ahern 2002, among others), to our knowledge this integration is still an open issue.

Another important issue that emerges concerns the 'summing' of the impacts of the individual indicators vertically and horizontally along the tables. In most cases this is not possible without a deeper understanding of the interactions between the different processes and changes. Thus an overall estimation is very difficult at the small or the regional scale for more than one service for the same indicator or even

Table 20.5 List of landscape scale indicators and their relationship to overall environmental value

Theme	Indicator	Overall environmental value	Process	Data availability
Land cover – land use indicators	Area of native tree plantations	Positive	–	A
	Area of dry pasture or cropland	Negative	In	A
	Area of abandoned systems	Negative	Ab	B
	Area of natural (unmanaged) forest and shrubland	Bell-shaped	Ex - Ab	B
Intensity of management indicators	Area of irrigated land	Negative	In	A
	Grazing density	Negative	In	A
	Oak/olive density	Bell-shaped	–	B
	External inputs (fertilisers)	Negative	In	B/C
	External inputs (animal feeding stuff)	Negative	In	B
Landscape elements indicators	Share of stone structural elements in the landscape	Positive	Ex	B
	Average quality of stone structural elements	Positive	Ex	C
	Maintenance of Terraces	Positive	Ex	C
	Watering ponds density	Positive	In	B
Incomes – products indicators	Number of traditional or specific character products in the area	Positive	Ex – In	A/B
	Number of tourists for agri-tourism	Positive	Ex	B/C
	Number of agri-tourism holdings	Positive	Ex	B

Note: Process: In – Intensification, Ab – Abandonment, Ex – Extensification; Data availability: A – available (published from official sources), B – merely available, may have to be calculated from aerial photography - satellite imagery or by collecting data from local sources, C – unavailable, have to be collected from primary research).

Source: Adapted from Peco et al. 1999.

for a single service over different indicators. The construction of a single index therefore is not possible without further research on the small scale interactions.

In conclusion, with this approach, we propose that small scale should be taken into account in sustainability assessment of land use change, as these changes and the ecosystem services relationships take place exactly at this level. The regional scale indicators used here cannot integrate small scale diversity and therefore this approach is exploratory. It is also tailored to the specific land uses and ecosystems, a fact that is inherent to all such approaches that deal with specific localities and land uses.

References

Antrop, M. 1997. The concept of traditional landscapes as a base for landscape evaluation and planning. The example of Flanders Region, *Landscape and Urban Planning*, 38, 105-117.

Beaufoy, G. 2000. *The Environmental Impact of Olive Oil in the European Union: Practical Options for Reducing the Environmental Impact.* [Online] Available at http://ec.europa.eu/environment/agriculture/pdf/oliveoil.pdf [accessed 08 November 2010]. Peterborough, Madrid: European Forum on Nature Conservation, Pastoralism and the Asociación para el Análisis y Reforma de la Política Agro-rural.

Botequilha Leitão, A. and Ahern, J. 2002. Applying landscape ecological concepts and metrics in sustainable landscape planning, *Landscape and Urban Planning*, 59, 65-93.

Ceballos, A., Cerdà, A. and Schnabel, S. 2002. Runoff production and erosion processes on a dehesa in western Spain, *Geographical Review*, 92, 333-353.

Cooper, T., Arblaster, K., Baldock, D., Farmer, M., Beaufoy, G., Jones, G., Poux, X., McCracken, D., Bignal, E., Elbersen, B., Wascher, D., Angelstam, P., Roberge, J.-M., Pointereau, P., Seffer, J. and Galvanek, D. 2007. *Final Report for the Study on HNV Indicators for Evaluation. Contract Notice 2006-G4-04.* London: Institute for European Environmental Policy.

Díaz, M., Campos, P. and Pulido, F.J. 1997. The Spanish *dehesas*: A diversity in land-use and wildlife, in *Farming and Birds in Europe. The Common Agricultural Policy and its Implications for Bird Conservation*, edited by D.J. Pain and M.W. Pienkowski. London: Academic Press, 178-209.

Díaz, M., Pulido, F.J. and Marañón, T. 2001. Diversidad biológica en los bosques mediterráneos ibéricos: Relaciones con el uso humano e importancia para la sustentabilidad de los sistemas adehesados, in *Beneficios comerciales y ambientales de la repoblación y la regeneración del arbolado del monte mediterráneo*, edited by P. Campos and G. Montero. Madrid: CIFOR-INIA, 269-296.

EEA 2005. *Agriculture and Environment in EU-15: The IRENA Indicator Report.* [Online] Available at http://reports.eea.europa.eu/eea_report_2005_6/en/

EEA_report_6_2005.pdf [accessed 25 June 2008]. Copenhagen: European Environment Agency.

Gordon, A.M., Newman, S.M. and Williams, P.A. 1997. Temperate agroforestry: An overview, in *Temperate Agroforestry Systems*, edited by A.M. Gordon and S.M. Newman. Wallingford: CAB International, 1-8.

Grove, A.T. and Rackham, O. 2001. *The Nature of Mediterranean Europe: An Ecological History*. New Haven, CT. London: Yale University Press.

Kizos, T. and Koulouri, M. 2010. Same land cover, same land use at large scale, different landscapes at the small scale: Landscape change in olive plantations on Lesvos Island, Greece *Landscape Research*, 35, 449-467.

McNeely, J.A. and Schroth, G. 2006. Agroforestry and biodiversity conservation – traditional practices, present dynamics, and lessons for the future, *Biodiversity and Conservation*, 15, 549-554.

Millennium Ecosystem Assessment 2003. *Ecosystems and Human Well-Being: A Framework for Assessment*. Washington, DC: Island Press.

Moreno, G., Obrador, J.J. and Garcia, A. 2007. Impact of evergreen oaks on soil fertility and crop production in intercropped dehesas, *Agriculture Ecosystems & Environment*, 119, 270-280.

Moreno-Marcos, G., Obrador, J.J., García, E., Cubera, E., Montero, M.J., Pulido, F.J. and Dupraz, C. 2007. Driving competitive and facilitative interactions in oak dehesas through management practices, *Agroforestry Systems*, 70, 25-40.

Opdam, P., Verboom, J. and Pouwels, R. 2003. Landscape cohesion: An index for the conservation potential of landscapes for biodiversity, *Landscape Ecology*, 18, 113-126.

Pandey, D.N. 2002. Carbon sequestration in agroforestry systems, *Climate Policy*, 2, 367-377.

Peco, B., Malo, J.E., Oñate, J.J., Suárez, F. and Sumpsi, J. 1999. Agri-environmental indicators for extensive land-use systems in the Iberian Peninsula, in *Environmental Indicators and Agricultural Policy*, edited by F.M. Brouwer and J.R. Crabtree. Wallingford: CAB International, 137-156.

Pinto-Correia, T. and Mascarenhas, J. 1999. Contribution to the extensification/ intensification debate: New trends in the Portuguese montado, *Landscape and Urban Planning*, 46, 125-131.

Plieninger, T. 2007. Compatibility of livestock grazing with stand regeneration in Mediterranean holm oak parklands, *Journal for Nature Conservation*, 15, 1-9.

Plieninger, T. and Schaich, H. 2006. Elementos estructurales del paisaje adehesado tradicional en Monroy y Torrejón el Rubio (Cáceres) y su importancia para la conservación de la naturaleza y el desarrollo rural, *Revista de Estudios Extremeños*, 62, 414-484.

Pulido, F.J. and Díaz, M. 2005. Regeneration of a Mediterranean oak: A whole cycle approach, *Ecoscience*, 12, 92-102.

Shakesby, R.A., Coelho, C.O.A., Schnabel, S., Keizer, J.J., Clarke, M.A., Contador, J.F.L., Walsh, R.P.D., Ferreira, A.J.D. and Doerr, S.H. 2002. A ranking methodology for assessing relative erosion risk and its application to dehesas and montados in Spain and Portugal, *Land Degradation & Development*, 13, 129-140.

Wascher, D.M. 2000. *Agri-Environmental Indicators for Sustainable Agriculture in Europe*. Tilburg: European Centre for Nature Conservation.

Chapter 21

Claiming Territorial Identity and Local Development: From Wishes to Deeds

Zoran Roca, José António Oliveira and Maria de Nazaré Roca

Introduction

Claims for the (re)affirmation of local identities as a vital condition for promoting the competitiveness of socio-economically lagging places and regions on the global markets of goods, services and ideas has been an integral part of the development policy discourse in the European Union since the nineties. Likewise, in Portugal, a highly integrated country in the global economy and culture, but with still sizable lagging areas, a commonplace in academic, political and media discourse has been the claim to combat *descaracterização*, that is, in Portuguese, the loss of authenticity,[1] at local and regional levels. *Descaracterização* has indeed become a pressing problem in places and regions, mostly rural, which thrive on the promotion of local identity as a development resource. It has been argued that people, especially the young, should not only know more about, but also take greater care of, the natural environment, cultural heritage and other authentic specificities of the geographic space of their residence, work, or leisure.

The pro-identity claims assume the likelihood of the resilience of globally forged local identities (Amin and Thrift 1994), the importance of the sense of place (Massey 1991; Rose 1995) and, especially, of topophilia, or 'the affective bond between people and place, or setting' (Tuan 1990: 4). Actually, it is assumed that *descaracterização* and topophilia are the inversely correlated antonyms: the stronger *descaracterização*, the weaker topophilia, or, in other words, territories where topophilia is weak, vanishing, or completely gone, are more susceptible to *descaracterização* than those where high levels of topophilia are shared among locals.

In spite of the quests to strengthen topophilia, *descaracterização* has affected virtually all spheres of the economy and culture in Portugal. Notorious instances at local and regional levels are, for example: the adulterations of traditional rural landscapes in the North, where the long-established terraced vineyards and mixed

---

1   The Portuguese word *descaracterização* is kept here for it effectively encapsulates the effect of abandonment, profanation, mistreatment, degradation, waste and other processes that harm and eliminate, often irreparably, the originality, authenticity, uniqueness, typicality, etc., of valuable natural, social, cultural, economic and other identity features in a given territory.

cropping are increasingly being substituted by technologically more advanced, generally subsidized and, thus, more competitive monoculture production; the rising acceptance of the consumer models that favour concentration of commercial activity over traditional retailing; the replacement of traditional social values, such as those related to community safety-nets and inter-generational solidarity, with a growing evidence of solitude and indeed exclusion as commonly established (post)modern social patterns; the loss of demographic vitality through ageing and emigration, and, consequently, the reconfiguration of original spatial distribution of the population and settlements; the visual pollution caused by the abandonment of what is traditional and the indiscriminate adoption of 'modern' construction materials, shapes and colours (Roca 2004).

Why such a gap between the pro-identity/development claims and the anti-identity reality, as evidenced in the fading topophilia and the spreading *descaracterização*? How to determine which territorial identity features are relevant for sustainable development? Is topophilia sufficient as a barrier to *descaracterização* and as a lever to local development?

The objective of this chapter is to offer answers to the above questions. To this end, the results of an exploratory research using a newly introduced concept, named terraphilia, are presented. Defined as the 'affective bond between people and territory that encourage local development intervention' (Oliveira et al. 2010), the concept of terraphilia[2] essentially complements the concept of topophilia as its 'pro-developmental extension'.[3] The exploration of the topophilia-terraphilia interface was carried out by applying the Identerra Model, a conceptual-methodological framework for the study of territorial identity as a development resource (Roca and Roca 2007).

The Identerra Model

According to the Identerra Model, territorial identities reveal the uniqueness of a geographic space in terms of landscape and lifestyle-related qualities. Landscapes are constituted by spatial fixes, defined in the Model as the sum of permanently and temporarily rooted and anchored elements of the natural heritage, population and

---

2   The Greek word *topos* (place, location) was substituted by the Latin word *terra* (land, earth) in order to emphasize more explicitly the focus on the development of concrete territorial entities, such as landscapes, places and regions. In Portuguese and other languages, the notion of *terra* is often synonymous to the birthplace, homeland, social roots, or ancestral livelihood setting. In fact, *terra* has been an intrinsic identity feature of rurality and, increasingly, a nostalgia motive among urbanites.

3   The notion of development assumes an *a priori* positive relation between economic and cultural change, on one side, and, on the other, populations' well-being. However, every 'development intervention' driven by terraphilia can bring about positive or negative effects for different stakeholders and/or environmental, economic, social or cultural sphere(s). What actually matters is that positive effects prevail over negative ones and that power-relations among the development stakeholders are as symmetrical and consensual as possible.

human-made economic and cultural heritage in a geographical area. Lifestyles, understood as the patterns of use and management of spatial fixes, are constituted by the activities, relations and meanings within horizontal and vertical networks and systems that determine the functioning of Nature, Society, Economy and Culture.

Two basic aspects of territorial identity – the objective and subjective – are distinguished in the Identerra Model. The objective territorial identity combines spatial fixes and flows that can be identified and assessed on the basis of macroscopic analyses of secondary and remote data sources and images of landscape- and lifestyle-related facts. The subjective territorial identity can be studied from the point of view of two basic sets of spatial fixes and flows: those experienced in real life and that can reflect topophilia, and those that are only sought and can reflect terraphilia. Both can be assessed by means of participatory research aimed at recording primary data and images.

Macroscopic analyses and participatory research are complementary: the former enables an overall contextualization of landscape and lifestyle-related identity features and the selection of locations and methods for the latter. The findings of the participatory studies should point to the discrepancies and/or synergies between the subjective (experienced and sought) and objective identities that might be relevant for local development policies striving to strengthen the topophilia-terraphilia interface.

Macroscopic analysis

The research was carried out in the Oeste region (a NUTS III, northwest of the Lisbon Metropolitan Area), where the official pro-identity rhetoric encouraging competitiveness has been placed high on the development agenda. The macroscopic analysis, performed in order to test the hypothesis of the Identerra Model, i.e., to detect landscape- and lifestyle-related components (spatial fixes and flows) of the objective territorial identity of the Oeste region, was undertaken in the following phases:

i.   exploration of available statistics from the latest inter-Census period 1991-2001 at the level of all 121 parishes (smallest statistical units) of the 12 counties (basic local administration unit) of the Oeste region;[4]
ii. selection of indicators and justification of their capacity to define and measure different landscape- and lifestyle-related components (i.e., spatial fixes and flows) hypothesized in the Identerra Model;

---

4   Statistics at parish level are relatively scarce, so a full coverage of spatial fixes and flows defined by the Identerra Model was not possible, especially regarding the natural environment.

iii. the application of the 'principal components' multivariate analysis to the data matrix constituted by the selected indicators for 121 cases (parishes);
iv. the application of a numeric taxonomy to the matrix of scores obtained in the previous phase.

The selected quantitative indicators and justification of their relevance for macroscopic analyses of landscape and lifestyle-related territorial identity features are presented below.[5]

Natural heritage

- Percentage of forest and bush areas with no crops, over total farm area, 1999 (As this indicator includes spontaneous and sub-spontaneous forests and bushes, i.e., not only industrial forests, it is the closest possible indirect approximation to the illustration of the state of 'natural' landscapes.).

Economic heritage

- Percentage of area with permanent crops over used agricultural land of the farm units, 1999 (This indicator is a possible approximation to the extent of the Oeste region's traditional rural landscape where fruit production is very important.);
- Percentage of individual farmers with completed upper secondary education, 1999 (This indicator illustrates the potential for modernization and integration of the agricultural activity into the market economy and more competitive farming systems.);
- Percentage of individual farmers who dedicate up to 25 per cent of their time to agriculture, 1999 (This indicator of the farmers' pluriactivity reflects non-agricultural income generation that can result from the introduction of new activities.);
- Percentage of seasonal or second home dwellings over total number of dwellings, 2001 (The second home phenomenon is a consequence of territorial attractiveness – environmental, cultural and other amenities – that can have important impacts on local economic and cultural change. It is a powerful driving force of landscape and lifestyle change especially when occurring inside very traditional settings.);

---

5   Ideally, the macroscopic study of the topophilia/terraphilia interface could be carried out using other types of indicators besides the ones presented here, reflecting the types and levels of *descaracterização* (e.g. land use anarchy, break up of community solidarity bonds, preference for fast food, etc.), or of pro-development initiatives (e.g. strengthening of local economic structure and job creation, valorization of local know-how, land use planning efficiency, etc.). Such data, however, are not readily available at county or parish levels.

- Percentage of population that works or studies outside the county of residence, 2001 (Indicator of territorial interactivity, showing functional dependencies between home, education and working places.);
- Rate of change of population who works and studies outside the county of residence, 1991-2001 (Increased social and economic integration and economic dependence. An indicator of the improvements in infrastructure and transportation services.);
- Percentage of commuting workers and students using private car, 2001 (Indicator of specific lifestyles conditioned by the nature of road infrastructure and public transportation policies and the level of functional dependence between the main urban centres.);
- Rate of change of commuting workers and students using private car, 1991-2001 (Increased motorization rate induced by higher living standards and by the improvement of road infrastructure and public transportation policies and the level of functional dependence between the main urban centres.).

Cultural heritage

- Percentage of foreigners over total population, 2001 (Indicator of territorial attractiveness related to local and regional economy, such as job opportunities, or to environmental and cultural amenities, such as fixation of retired foreigners.);
- Rate of change of single-parent families, 1991-2001 (Indicator of change from traditional to modern lifestyles related to urbanization and post-industrial social values.);
- Rate of change of seasonal or second home dwellings, 1991-2001 (Indicator of landscape change in terms of built environment, as well as, indirectly, of probable cultural 'adaptation' of residents through increased interaction with newcomers.);
- Percentage of families with more than five members over total number of families, 2001 (Persistence of traditional socio-economic, cultural and other determinants of fertility.);
- Percentage of individuals aged 65+ in single person families, 2001 (Extent of new family relations (e.g., break-up of family ties) associated with rural to urban migration and increased life expectancy that can result in social exclusion, as well as in new demand for health and social care services.).

Population/society

- Population density (km^2), 2001 (Illustrates the level of the population pressure (demand) on territorial development resources and of the urbanization levels.);

- Rate of change of the total population, 1991-2001 (Indicator of change in the availability of human capital (production and consumption) and changing pressure on territorial development resources.);
- Percentage of population with completed upper secondary education over total population, 2001 (Availability of qualified human capital potential. NB: National average is very low, i.e., in 2001 less than 7 per cent of the population completed upper secondary education.);
- Rate of change of population with completed upper secondary education, 1991-2001 (Changing availability of qualified human capital potential.);
- Aged/child ratio, 2001 (Indicator of population ageing.);
- 15-39/40-64 years old population ratio, 2001 (Indicator of the vitality level of the population in the active age.).

The application of the 'principal components' multivariate analysis to the final matrix of 20 indicators and 121 cases revealed that four components could explain 51.69 per cent of the common variance. Consequently, a numeric taxonomy was applied by using (i) the scores of the first factor as a measure of distance among cases and (ii) the 'nearest neighbour' as the clustering strategy. This analysis resulted in three major regional clusters, labelled as 'Urban Fixes with Agitated Flows'; 'Rurban Fixes with Intensified Flows', and 'Rural Fixes with Crystallized Flows' (Figure 21.1). Their description, based on the relative contribution of each indicator to the clusters' homogeneity, is summarized as follows:

- Cluster 1, 'Urban Fixes with Agitated Flows', consists of 37 parishes where the county seats are located, or are in their immediate fringe. Compared to the entire Oeste region, these parishes are characterized by higher levels of urbanization. The ways in which the main statistical indicators contribute to define this cluster are: (i) the highest rate of change of inter-county commuters and the highest population densities; (ii) an increased presence of second homes, mostly in parishes where their presence was below average for the Oeste region in 2001; and (iii) a weak presence of part-time farmers, which indicates low importance of agriculture in such dynamic urbanized parishes. This situation of 'urban agitation' is corroborated by a relatively higher youthfulness of the population and a higher weight of single-parent families (an indicator of change from traditional to modern lifestyles related to urbanization and post-industrial social values).
- Cluster 2, 'Rurban fixes with intensified flows' is composed of 27 parishes, most of which are in the coastal areas of the Oeste region, or in the neighbourhood of the county seats, and its main aspects are: (i) a lower proportion of bush and forest areas (without any crops); (ii) a higher 15-39/40-64 year-old population ratio; (iii) a higher percentage of part-time farmers; (iv) a weaker presence of farmers with completed upper secondary education; and (v) a lower proportion of families constituted by only one elderly person. The variation pattern of these indicators highlights

the strengthening of the urban sprawl and its effects on traditional rural landscapes and lifestyles. Furthermore, in relation to the Oeste region as a whole, this cluster has a higher increase in the motorization rate, certainly due to increasing inter-county commuting flows.

- Cluster 3, 'Rural fixes with crystallized flows' contains 57 parishes, characterized by: (i) a higher proportion of bush and forest land (without any crops); (ii) a lower rate of change of second homes; (iii) lower levels of education in the population; (iv) a lower rate of change of inter-county commuting; (v) a lower presence of foreigners in the total population. These parishes correspond to the economically and socially less dynamic, even depressed, rural areas. While permanent crops such as vineyards and orchards, and also forests, occupy an important proportion of agricultural land, this part of the Oeste region is also marked by the permanence of original natural landscapes (e.g. the Environmentally Protected Area of the Montejunto Mountain) and the resurgence of 'natural' landscapes on the abandoned lands, covered by spontaneous and sub-spontaneous forests and bushes.

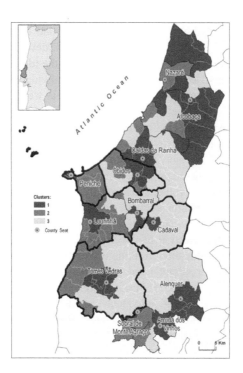

Figure 21.1 Landscape and lifestyle-related typology of territorial identities of the Oeste region

Participatory analysis

On the basis of the obtained intra-regional taxonomy of spatial fixes (landscapes) and flows (lifestyles) of the Identerra Model, illustrating the objective dimension of territorial identity, the counties of Óbidos, Peniche, Torres Vedras and Cadaval of the Oeste region were selected for the study of types and levels of topophilia and terraphilia (i.e., the subjective dimension of territorial identity). Every county incorporates a balanced proportion of parishes that belong to each of the three clusters, so that all four counties together are a representative sample of the Oeste region in terms of diversity of landscape- and lifestyle-related territorial identity features, i.e., from the economically and culturally vibrant, increasingly diversified, highly urbanized Atlantic coast to the still traditionally rural interior, including the Environmentally Protected Area of the Montejunto Mountain.

KAP workshops

The detection and assessment of the subjective territorial identity dimension in terms of topophilia (experienced territorial identity) and terraphilia (sought territorial identity) was based on the recordings of local knowledge, attitudes and practice (KAP) regarding qualities of landscape and lifestyles. To this end, a participatory method, labelled KAP Workshop, was developed and applied in each of the four counties. Following the principles of the Identerra Model, a KAP Workshop is structured so as to enable recordings of individual and group opinions based on the sense of belonging to a specific territory, as well as of diagnostic and prospective appraisals of actual (experienced) and claimed (sought) qualities of the constituents of the natural environment, society, economy and culture.

A KAP Workshop is based on phased collection and processing of primary information obtained from individual and group statements, closed and open discussions, and of group-based prioritizations and proposals for change. The initial recording of the participants' profiles and their definition of the concrete local identity features, such as those of natural and built environment, social customs and habits, arts and crafts, is followed by retrospective and prospective diagnostics consisting of individual and group selection of the 'most important' positive and negative features and of their classification in terms of duration ('traditional' vs. 'recent'), stability ('vanishing' vs. 'resistant') and the participants' feelings ('optimism' vs. 'pessimism') about their evolution. Furthermore, local, regional, national and/or supranational institutional and/or individual responsibilities are attributed to the changing qualities of all territorial identity features. After these diagnostics have been shared among the groups of participants, the KAP Workshop comes to its final stage – the formulation of proposals based on the previous retrospective and prospective diagnostics – in which the participants (i) define priority issues of territorial identity as a local development resource; (ii) propose concrete policy solutions and actions in the field, aimed at the (re)affirmation of the desired territorial identity features; and (iii) suggest key responsible agents of change.

Participants

The socio-demographic characteristics and work experience of the participants of the KAP workshops held in the four selected counties of the Oeste region guaranteed a collection of data based on a high level of familiarity with landscape- and lifestyle-related territorial identity and local/regional development issues. A total of 47 participants evenly representing the counties of Peniche (13), Torres Vedras (12), Cadaval (12) and Óbidos (10) included 23 men and 24 women with ages between 21 and 74 (median age: 40 years). All of them both lived and worked in the Oeste (40 per cent for at least 30 years and 80 per cent for at least 10 years), mostly (70 per cent) within the same county; 50 per cent were born in the same county and 15 per cent in another county of the Oeste region. A majority in Peniche, Torres Vedras and Óbidos had tertiary education (69 per cent in Social Sciences and Humanities) and in Cadaval, upper secondary education. Their professional experience ranged from administrative and technical functions (including decision-making and coordination), through commercial entrepreneurship and professional training, to the management of the natural and cultural heritage. The so-called 'third sector', such as local/regional development associations (26 per cent) and private charities (17 per cent), as well as local government entities (21 per cent), predominated in terms of the participants' institutional affiliation.

Individual responses

In responding to the question 'Which are the most important elements that characterize your county?', the participants specified a wide range of landscape- and lifestyle-related elements, i.e., from those that are part of the objective traditional (e.g. windmills, gastronomy and built heritage) and emerging identity (e.g. new orchards, urban settlements and tourism industry), to the subjective experienced/sought identities (e.g. distressed/improved urban environment, traditional/modern social relations, and fragile/competitive economy).

The participants were requested to identify two aspects that, in their opinion, most positively and most negatively affect their county of residence, and to classify them as 'traditional' or 'recent' and as 'resistant' or 'vanishing'. They also had to declare their pessimism or optimism regarding the future evolution of these aspects. The open-ended responses were classified and their frequencies recorded in accordance with the landscape- and lifestyle-related territorial identity features (Nature, Society, Economy and Culture) of the Identerra Model, as shown in Figures 21.2 and 21.3.

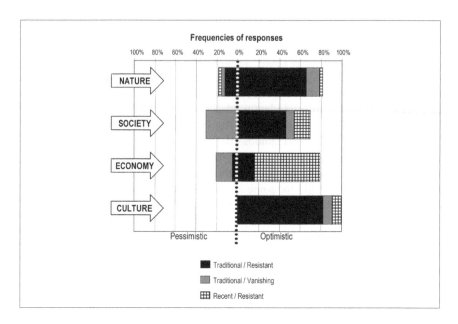

Figure 21.2 Positive identity features: Retrospective and prospective diagnosis

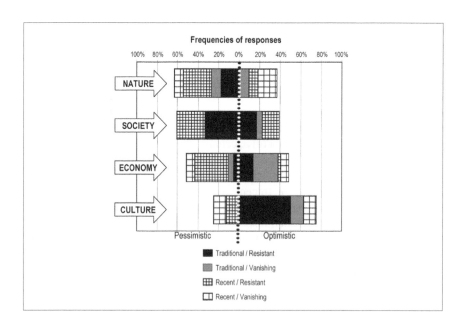

Figure 21.3 Negative identity features: Retrospective and prospective diagnosis

The participants' definitions and assessments of priority positive and negative identity features of their counties are synthesized thus:

- the responses are more unanimous about positive identity features, while the negative ones are very diverse and hard to define in concrete terms, which is a general indication of a high degree of topophilia shared by the participants;
- the assessment of the natural environment is predominantly positive, although some threats are differentiated between the more urbanized counties (e.g. pollution, car traffic) and the more rural ones (e.g. waste depository in the Cadaval county); this should be considered by future local development policies;
- the social issues, frequently referred to as stemming directly from the activity of local agents, clearly emerge on the positive side when related to community cooperation and assistance networks, but also on the negative side when related to the rural settings marked by strong social control (e.g. small communities' resistance to some aspects of social modernization);
- the economy is, no doubt, the identity element subject to strong individual and collective disagreement; on the one hand, the weak bases of local economies emerged as negatively assessed features from the point of view of both unemployment and low quality of the entrepreneurial structures; however, on the other hand, the components of economic infrastructure, such as those that improved accessibilities and potentials for the development of tourism industry, are positively assessed;
- regarding cultural features, such as, first, the attachment to the legacy of the rural *milieu*, intrinsic qualities of local people and gastronomic tradition, and, second, the elements of built heritage which in every county constitute important spatial fixes, the former are assessed rather negatively though with some hesitance, while the latter are eulogized in such a manner that the high level of topophilia mentioned above is actually confirmed.

Group responses

Experienced territorial identity and development features:
Priorities and responsibilities

The consensual group responses revealed that the common denominators of the positive priority territorial identity and development features of the participants' counties are (i) landscapes and historic heritage; (ii) natural landscapes; (iii) quality of life; and (v) rurality. As negative features, the groups consensually prioritized (i) economic development; (ii) social development; (iii) infrastructure and public services; (iv) human capital development and traditional productive activity; (v) social and economic development; and (vi) associativism.

The content analyses have shown that there is a high level of correspondence between the experimented subjective identity (topophilia) and the objective identity established on the basis of the indicators used in the macroscopic analysis. Qualities of the landscapes and of the cultural-historical heritage and the preserved rurality that amalgamates specificities of natural and cultural landscapes and lifestyle patterns are the most prominent among the positive territorial identity aspects. Among the negative territorial identity aspects, weaknesses of the development process, low levels of human capital development, and inadequate accessibility to public services are highlighted.

Sought territorial identity and development features:
Priorities and responsibilities

The components of the experienced territorial identity features, reported as consensual priorities by the participants (tables 21.1 and 21.2) were object of further group discussions aimed at the formulation of concrete action proposals for the maximization of positive and minimization of negative territorial identity aspects. Every proposal for action was accompanied by the groups' suggestions as to which development agent should be involved. The responses referring to action proposals and development agents were classified according to the character of the envisaged action and the institutional framework, respectively. The results of this classification are presented synthetically hereunder, as follows: in Table 21.1 all consensually defined positive and negative territorial identity and development priority issues are intersected with the groups' proposals for actions; in Table 21.2, the proposed actions are intersected with the groups' suggestions regarding the agents to be engaged in the concretization of such actions. This entire exercise (i.e., the KAP Workshop and data processing) enabled detection of the incidence and nature of terraphilia amongst the participants.

Considering that the sense of terraphilia increases with the capacity to formulate proposals to solve the weaknesses and to maximize the defined strengths, Tables 21.3 and 21.4 can have two-fold readings and interpretations. First, reading by the rows enables assessment of the intensity of terraphilia based on the numbers and kinds of priority issues and on the typology of suggested actions for the solution of these issues. Second, reading by the columns enables delineation of actions considered most pertinent/relevant for resolving the weaknesses and maximizing strengths, as well as the identification of agents that should implement these actions and could, at the same time, become targets of specific actions aimed at raising their levels of terraphilia (e.g. activities of territorial marketing, or at least the sensitization for their involvement in some actions through pointing to their specific problem solving capacities).

Table 21.1 Proposed actions in favour of territorial identity and development priority issues (group responses)

Positive and negative territorial identity and development priorities issues	Human capital development	Employment creation	Infrastructure and public facilities	Valorization of natural environment	Promotion of collective and public services	Dissemination and awareness creation	Promotion of tourism industry	Restoration and valorization of built heritage	Associations (management and organization)	Spatial planning and management	Total n° of references
Economic development (-)	19	8			3					4	34
Landscapes and historic heritage (+)				10		12		4			26
Social development (-)	8	7			8						23
Natural landscapes (+)	5			11		4				2	22
Quality of life (+)		7			6		7				20
Rurality (+)			4				7	5			16
Infrastructure and public services (-)			15								15
Human capital development and traditional productive activity (-)	12		3								15
Social and economic development (-)	9	5									14
Associativism (-)									7		7
Total n° of references	**53**	**27**	**22**	**21**	**17**	**16**	**14**	**9**	**7**	**6**	**192**

	Main priorities		Main actions		Mainstream terraphilia

Also in tables 21.1 and 21.2, the darkest shaded area on the intersection between rows and columns can be interpreted as the clearest and strongest (mainstream) course of terraphilia, which could be directly used as a framework for designing an Action Plan for the Affirmation of Territorial Identity as a Development Resource.

Table 21.2 Proposed agents to carry out actions in favour of territorial identity and development priorities (group responses)

Proposed actions	Local administration	Local associations and charities	Central administration	Schools and training centres	Firms and entrepreneurs	Others	Individuals	Economic associations	Mass media	Total n° of references
Human resources development	8	5	10	10	5	4	6	1	2	51
Employment creation	3	2	7	7	2	1	1	3		26
Infrastructure and public services	9	2	3	4	2		1	1		22
Valorization of natural environment	6	6	3		2	2	1		1	21
Dissemination and awareness creation	6	1	1	2	1	1	1	1	2	16
Promotion of collective and social services	3	9				3		1		16
Promotion of tourism industry	3	6	1	1	2			1		14
Restoration and valorization of built heritage	3	1	2			2	1			9
Associations (management and organization)	2	2	1		1	1				7
Spatial planning and management	2	2			1			1		6
Total n° of references	**45**	**36**	**28**	**24**	**16**	**14**	**11**	**9**	**5**	**188**

	Main priorities		Main actions		Mainstream terraphilia

Conclusions

The concept of terraphilia brought forward in this chapter is an attempt to enrich Tuan's concept of topophilia with a pro-developmental approach to the subjective (experienced) dimension of landscape- and lifestyle-related features of territorial identity. It was shown that the conversion of terraphilia from a normative concept into an analytical (operational) category could be achieved in the framework of the Identerra Model, which enables empirical delimitation and systematization of the state (spatial fixes) and change (spatial flows) of natural, economic, societal and cultural features of territorial identity.

Given that it embodies the sought (claimed) element of the subjective dimension of territorial identity, the operationalized concept of terraphilia is the lever between the subjectively experienced (topophilia) and objective (factual) territorial identity of the Identerra Model. Specifically, the materialization of terraphilia necessarily results in enhancements (development) of landscape- and lifestyle-related features of the objective territorial identity. Furthermore, it was shown that the operationalization of the concept of terraphilia facilitates the recording of development agents' (i) knowledge and assessment of manifestations of territorial identity; (ii) claims toward change; and (iii) capacity to envisage viable policies and actions that promote affirmation of territorial identity as a local development resource.

The promotion and activation of terraphilia is important in the process of identifying and assessing strategically relevant elements of local and regional development, such as (i) the sense of belonging to a territory, which reflects the levels of satisfaction with the environmental, social, economic, cultural and other conditions provided by that territory; and (ii) territorial attractiveness, which can be decisive for the fixation of new economic activities and for the introduction of social innovation in that territory.

It is worth stressing that to explore development agents' knowledge, attitudes and practice in a given territory in the framework of the topophilia and terraphilia concepts may be fundamental for the orientation of every planning process. This is especially important in Portugal, where public participation has been explicitly reinforced in the planning legislation only recently, since 2007. In this context, the Identerra Model can facilitate the systematization of diverse components that constitute one's affection to specific territorial settings (topophilia), or one's keenness to promote territorial development (terraphilia). This, in turn, can yield tangible results that can help more efficient focusing and implementation of development policies, particularly those related to landscape change.

References

Amin, A. and Thrift, N. 1994. Living in the Global, in *Globalisation, Institutions and Regional Development in Europe*, edited by A. Amin and N. Thrift, N. Oxford: Oxford University Press, 1-22.

Massey, D. 1991. A Global Sense of Place, in *The Making of the Regions, Open University, D103 Block 6*. Milton Keynes: Open University Press, 12-51.

Oliveira, J.A., Roca Z., Leitão, N. 2010. Territorial Identity and Development: From Topophilia to Terraphilia. *Land Use Policy*, 27(3), 801-814.

Roca, Z. 2004. Affirmation of Regional Identity between Rhetoric and Reality: Evidence from Portugal, in *Cultural Uniqueness and Regional Economy*, edited by E. Boneschansker et al. Leeuwarden, Ljouwert: Fryske Akademy, 29-52.

Roca, Z. and Roca, M.N.O. 2007. Affirmation of Territorial Identity: A Development Policy Issue, *Land Use Policy*, 24(2), 434-442.

Rose, G. 1995. Place and Identity: A Sense of Place, in *A Place in the World? Places, Cultures and Globalisation*, edited by D. Massey and P. Jess. Oxford, Oxford: University Press/The Open University, 87-132.

Tuan, Y.F. 1990. *Topophilia – A Study of Environmental Perception, Attitudes and Values*. New York: Columbia University Press/Morningside Edition.

Chapter 22

How Can the Different Functions of Gardens in Suburban Areas be Evaluated? Towards a Framework for an Integrated Method of Evaluating Aesthetic, Ecological, Environmental and Economic Functions of Gardens in Suburban Areas

Athanasia Mavridou and Thanasis Kizos

Introduction

Urbanization is a worldwide trend today (UNFPA 2006; UNDP 2000), not always connected with population growth (Robinson et al. 2005). This development can be quick and results in the fragmentation of former rural and/or agricultural landscapes (Armstrong 2004), causing land cover and land use changes (Marzluff and Ewing 2001; McKinney 2006), and creating 'urbanized villages', the so-called suburbs (Alston and Richardson 2006).

The continuation of this process requires more and more land – including areas which are environmentally valuable – to be converted into housing sites (Kaplan and Austin 2004). People move out to the suburbs, as they seek low-density residential areas, freedom of choice as well as expression and access to nearby natural areas (Frumkin 2001; Kaplan and Austin 2004; McKinney 2006). Rising incomes, falling commuting time and cost and dependence on extensive automobile use assist this development (Robinson et al. 2005; Wasilewski and Krukowski 2004). The lack of green areas in inner cities creates a need for attractive, green settings and favourable location (Luttik 2000), as idealized with a house with a garden. These settings are important for the quality of life, offer recreation, mental and physical health and enjoyment of the natural environment (Kaplan 1984; Armstrong 2000; Beer et al. 2003; Syme et al. 2004).

The garden scale may be small in size compared to other semi-natural land uses or un-built environments, but the extensive spread of gardens in suburban areas has important overall impacts as it modifies the functions of space. This chapter aims to review the literature on the evaluation of the aesthetic, ecological, environmental and economic functions of gardens in suburban areas, present the findings of an empirical study in the town of Mytilini, on Lesvos Island, in Greece, and sketch an integrated method of evaluating these functions. Since gardens in suburban areas usually replace agricultural landscapes, these functions will be compared against the functions of the agricultural landscape. In the next section, the functions of the agricultural landscape are presented, followed by those of gardens.

Multifunctionality of agricultural landscapes

The agricultural landscape can be understood as a complex of ecological, economic, and cultural qualities on which human and other life forms depend (Wasilewski and Krukowski 2004) and therefore is by definition multifunctional, as multifunctionality consists of the integration of different functions in a given spatial and/or temporal unit at a given scale (Guiomar and Fermandes 2007). There is a sizable literature that documents this multifunctionality of agriculture and its landscapes and the OECD (1998) recognizes that agriculture's primary function is to supply food and fibre, but it can also shape the landscape, provide environmental benefits such as land conservation, the sustainable management of renewable natural resources and the preservation of biodiversity, and contribute to the socio-economic viability of many rural areas. Several authors have explored the production relationships between commodity and green outputs (Gatto and Merlo 1999) and have discussed and evaluated agriculture's productive and non-productive functions (OECD 2001). The key elements of multifunctionality are: i) the existence of multiple commodity and non-commodity outputs that are jointly produced by agriculture; and ii) the fact that some of the non-commodity outputs exhibit the characteristics of externalities or public goods, with the result that markets for these goods do not exist or function poorly (OECD 2001). This plurality of outputs is an admitted fact in European agriculture (Brunstad et al. 2001) and the notion of multifunctionality is used in close relation with sustainable development or rural development (Kroger and Knickel 2005). The European Union (EU) has embraced the multifunctionality of agriculture, incorporating it into a 'European farming model' through which quality products are produced in a living countryside, conserving biodiversity, the landscape and the 'rural heritage' of Europe (CEC 1999).

In this context, agriculture becomes less about producing food and more about non-commodity outputs and functions (e.g. resources and biodiversity protection, leisure and open space) which fulfil additional private or societal needs related to the use of land and landscapes (Wiggering et al. 2006; Slee 2007).

Multifunctionality of gardenscapes

Today research on topics that concern gardens are divided into two large and well discernible approaches: for the first, gardens are viewed from the perspective of ecology and for the second from the perspective of landscape architecture. A holistic and integrated view of gardens in suburban areas is absent from the literature, as most studies tend to ignore environmental functions of gardens and/or deal mostly with gardens in urban areas. Historically, most studies concerning gardens referred to their aesthetic functions, but in the last decades there is a growing concern for the ecological functions of gardens especially in urban spaces (Breuste 2004).

The functions of gardens in general can be aesthetic-symbolic, ecological, environmental and economic: (a) gardens are made to decorate and enhance the space around the house and they are important for a variety of quality of life variables such as avoidance of stress, recreation and personal and social identity (symbolic functions); (b) gardens consist a natural or semi-natural habitat for a variety of flora and fauna species (ecological functions); (c) owners can use the whole or a part of their garden, for producing food for sale or for home consumption (economic functions); and (d) gardens also conserve resources and create microclimates (environmental functions).

Landscape architects, who traditionally design outdoor spaces incorporating plant materials, have developed their own design theories regarding the aesthetic and psychological benefits provided by plants (Thayer and Atwood 1978; Tsalikidis 2008). Plants are used in open spaces either for their aesthetic- visual characteristics in a presentational level such as: form, flowering effect, colour, texture, complexity, and other qualities or/and in a referential or representational level on which plants are perceived in terms of recognized function, symbolic associations, or other 'attachable' meanings (direction, screening, shade, wall covering, barrier, fruit benefiting, wind or sound protection) (Acar 2007; Tsalikidis 2008; Zagorski et al. 2004). Plants can increase pleasure by adding, subtracting, or interacting with other landscape elements, and they may accomplish this by altering either the presentational or referential characteristics of the stimulus field, or more likely both, and they can also play a positive role on human psychology (Thayer and Atwood 1978) and benefit physical or mental health (Kaplan 2001). Landscape preferences and perceptions of urban natural areas are linked with the desire for and benefits of having access to nearby natural areas (Frumkin 2001; Ozguner and Kendle 2006).

There is a growing body of recent research on ecological functions of gardens and parks, especially in urban areas (e.g. Breuste 2004). The biodiversity value of gardens is an issue of debate for many ecologists (Gaston et al. 2005a; Thompson et al. 2003, 2004) and two main groups of researchers are encountered: one supports that gardens contribute to biodiversity conservation in urban and suburban areas (Adams and Dove 1989; Cannon et al. 2005; Gaston et al. 2005a; Rudd et al. 2002; Thompson et al. 2005; Whelan et al. 2006; Helfand et al. 2006; Acar et al. 2007; Mathieu et al. 2007; Ozguner et al. 2007), while others consider gardens as threats to native flora and fauna (Reichard and Hamilton 1997; Hodkinson and Thompson 1997; Czech et al. 2000; Reichard and White 2001; Baskin 2002; Raloff 2003; Moffatt et al. 2004; Arevalo et al. 2005; DeStefano et al. 2005; Smith et al. 2006; McKinney 2006; Alston and Richardson 2006; Duguay et al. 2007).

In more detail, the first approach to the link between gardens and biodiversity recognizes that private gardens represent the largest single proportion of greenspace in many urban areas (Gaston et al. 2005b). The yard scale, the property occupied by a single residential dwelling, is relatively small, but constitutes a substantial part of the vegetated space within a city, and a mosaic of environmentally beneficial gardens can contribute to ecological health (Mathieu et al. 2007) as it can be a valuable tool regarding detecting and monitoring urban landscape biodiversity and cultural change (Acar et al. 2007). Gardens contribute to the biological integrity of the city by enhancing the survival of wildlife (Gaston et al. 2005a), by increasing species richness, by providing sources of food and shelter for wildlife (habitats for insects, birds and small mammals) (Ozguner et al. 2007) and they are considered as important refuges and food sources for indigenous species (Mathieu et al. 2007). They can also act as corridors between habitats (Adams and Dove 1989) and thus they are important contributors to a wider biological network which can enhance connectivity between vegetation communities and support the dispersion or survival of meta-populations (Mathieu et al. 2007). Within this context, many research findings indicate that the above functions are best served when native plants are used in gardens (Whelan et al. 2006; Helfand et al. 2006), as native fauna is best adapted to utilize native plant and fauna communities (Batten 1972).

Within the other approach, researchers support that the floras of private gardens are among the most unusual forms of botanical assemblages (Arevalo et al. 2005; Smith et al. 2006; Duguay et al. 2007). Therefore, when compared to most naturally developing floristic communities, domestic garden floras can threaten local species (Hodkinson and Thompson 1997; Moffatt et al. 2004; Smith et al. 2006). Such harmful effects due to non-native species are now regarded as one of the greatest threats to biological diversity worldwide (IUCN 2000) and ornamental plants comprise more than 40 per cent of widespread invasive plant species, far exceeding the share of plants introduced for other purposes (Reichard and White 2001; Reichard and Hamilton 1997; Baskin 2002; Raloff 2003; Alston and Richardson 2006).

The conclusions of The Wildlife Society for Suburban, rather than Urban, Landscapes suggest that the type of development currently dominant in the western world (single-family homes with the support services such as roads, power, water, sewerage) qualifies these areas as landscapes that mix the built environment with remnant wildlife habitats and newly created habitats such as backyards. These habitats attract or retain many wildlife species that can lead to high rates of human-wildlife conflicts, which demand large amounts of attention, time, and resources from local natural resource management agencies (Destefano et al. 2005).

Concerning the environmental functions of gardens, there are two different dimensions: the first dimension refers to the use of resources for the maintenance of gardens (water, fertilisers and plant protection from diseases products). Obviously, the type of garden and the type of plants used in it affect the amount and the type of resources used at gardens (Templeton et al. 1999; Helfand et al. 2006). One of the most important of these resources used is irrigation water, with studies reporting that as much as 56 per cent of total domestic usage of water takes place outside the house (e.g. on lawns, gardens or swimming pools) in semi-arid climates such as the Mediterranean (Loh and Coghlan 2003). The second dimension refers to the microclimates (temperature, humidity, noise) that gardens create (Tsalikidis 2008; Morancho 2003).

Finally, concerning the economic functions of gardens, some owners use parts of their garden for producing food for sale or for home consumption (Daniels and Kirkpatrick 2006; Mavridou and Kizos 2007).

A conceptual framework that can cover land use and landscape changes between the agricultural landscape and gardenscapes can be based on the identification of a hierarchy of the different functions of each landscape. While in agricultural landscapes the most important function is the economic followed by ecological-environmental and aesthetic functions, in gardenscapes the aesthetic function is considered the most important, followed by ecological and economic functions (Figure 22.1).

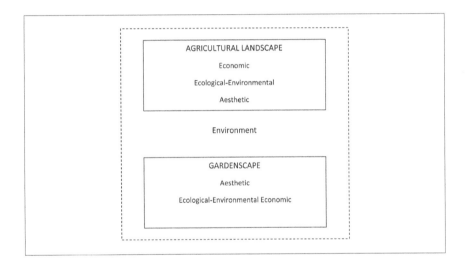

Figure 22.1 Hierarchical significance of functions between agricultural landscapes and gardenscapes

The use of this framework can assist the analysis of differences between the two types of landscapes, but at the same time, the literature review reveals that differences among gardens can also be very important. These differences are explored with more detail with an example from the suburban area of the Mytilene town on Lesvos Island in Greece.

An example: The gardens of the suburban area of Mytilene

The town of Mytilini is the capital of the Prefecture of Lesvos on Lesvos Island, located in the North Aegean Region, close to the Asia Minor coast (Figure 22.2). The climate of the town and the island is typically Mediterranean with mild winters and dry summers and an average of 725 mm of annual precipitation (typically from September to May). Over 40,000 people reside in the town and the nearby suburbs out of the 90,000 inhabitants of the island. The suburbs are of two kinds: either former small settlements close to town, which are today intensively developed and spread in the surrounding countryside (former olive plantations), or scattered buildings in olive plantations, some older country houses of old wealthy local families and some recent houses and villas. After the 1980s, the town spread very fast in the surrounding fields with either isolated houses or more often with blocks of two-storey houses (Figure 22.3). All types of houses that are built on former fields have smaller or bigger gardens.

For the example presented here, initially 56 gardens of suburban areas were surveyed and the topography (elevation, topography, geomorphology, size) along with the natural and artificial elements of the gardens were recorded. After the first phase was completed, some characteristic cases from the gardens of the first

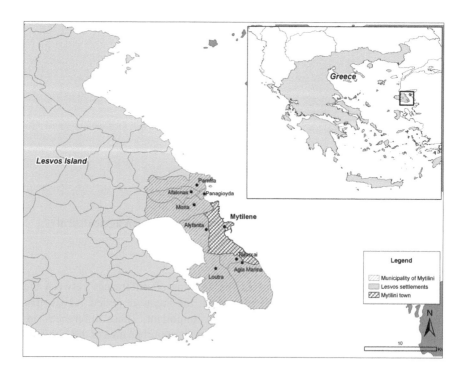

Figure 22.2 Map of Lesvos Island, with Mytilini

Figure 22.3 Fragmentation of the landscape

phase were chosen according to their size, and their owners were interviewed with the use of a semi-structured questionnaire in order to collect quantitative and qualitative data regarding gardening practices and personal views on gardens and their management. Simultaneously, professional gardeners and landscapers were interviewed to investigate their practices and the advice they offer (for more details see Mavridou and Kizos 2007).

There are four categories of gardens according to their size: small gardens ($<$50 m^2, 16 per cent of the total), medial (50-500 m^2, 44.6 per cent of the total), large (500-1000 m^2, 21.4 per cent) and very large gardens (1000-2000 m^2, 17.8 per cent). Garden size is very important for the characteristics found in the garden. First of all, the size of the garden and the type of building (villa, modern or traditional building) are related and the relationship is statistically important as villas are found only in large size gardens ($>$500 m^2). Moreover, the size of the garden and the ratio of natural and artificial elements in the garden are also related, with fewer natural elements in small gardens and more as the size increases. Finally, the size of gardens and their use for decoration and/or production correlate (Spearman's $r = 0.449$, $p = 0.01$), with all nine small gardens being exclusively decorative, while all the exclusively productive ones are large ones. This indicates that environmental pressures per m^2 are expected to be higher in smaller gardens, but could be important in medium and large gardens as well.

Concerning the flora encountered (Table 22.1), the most common plants recorded are bushes (in 84 per cent of gardens). Exotic species are also very common (in 73.2 per cent of gardens and in all size types) with species such as *Phoenix canariensis, Cycas revoluta, Araucaria heterophylla, Pittosporum tobira, Albizzia julibrissin, Thuja occidentalis, Euonymus fortunei, Hedera helix, Bougainvillea spectabilis, Ligustrum vulgare* etc. Ornamental evergreen trees such as *Pinus sp.*, *Cupressocyparis leylandii, Thuja occidentalis, Araucaria heterophylla*, follow with a presence in 42.9 per cent of the gardens and then natural fences with plants (in 41.1 per cent). Coverage with lawn (in 39.3 per cent of the gardens) appears mostly in medium-size gardens, while its management is costly due to dry summers, requiring frequent irrigation, fertilization and protection. Regarding the artificial environment of the gardens, the corridors (in 81 per cent of the total) and the parking area (in 73.2 per cent) are the commonest elements in all types of garden size. Swimming pools are mostly found in large gardens (Figure 22.4).

Table 22.1 Natural and artificial elements of the sample of gardens per size type

Land cover/elements	Size of gardens					% total (56 gardens)
	<50 m²	50-500 m²	500-1000 m²	1000-2000 m²	Total	
Natural elements						
Bushes	7	21	11	8	47	83,9
Exotic species	6	18	10	7	41	73,2
Ornamental evergreen trees	3	8	7	6	24	42,9
Fences of plants	3	12	5	3	23	41,1
Lawn	3	9	8	2	22	39,3
Fruit bearing trees	2	5	8	3	18	32,1
Ornamental deciduous trees		4	9	3	16	28,6
Olive groves		1		6	7	12,5
Vegetables			3	2	5	8,9
Artificial elements						
Corridors	6	20	11	9	46	82,1
Garage or parking area	6	17	10	8	41	73,2
Pergolas	2	2	4	2	10	17,9
Doghouse	1	4	1	2	8	14,3
Swimming pool		1	1	3	5	8,9
Barbeque		2		2	4	7,1
Playground		2	2		4	7,1
Other animal's houses			1	2	3	5,4
Courts		1		1	2	3,6
Storehouse			1		1	1,8

Figure 22.4 An aspect of a garden with view of the town of Mytilene

After the completion of the first phase, eight owners were interviewed on gardening practices and personal views on gardens and their management. Five from these eight owners had professional help in the design and construction of their garden and they continue to manage them with professionals as their free time is 'limited' and the garden is 'too big to manage'. All of them say that they can afford such professional assistance and are satisfied with the result. The other three have constructed and managed their gardens themselves as they enjoy gardening. These gardens are smaller and therefore easier to manage.

Gardening practices are similar whether by professionals or the owners: gardens are irrigated on a daily basis in summer, while in winter four owners do not irrigate. The quantity of irrigation water depends on the size of the garden and the type of plants, with lawns demanding more water (it is noteworthy that none regarded irrigation as an actual input or an environmental pressure, but something 'natural' and in-exhaustive). Regarding fertilization, practices vary: some declare their ignorance on the fertilization practices of professionals, while others may use manure annually, two fertilize twice a year with commercial fertilizers. For plant protection, one does not use chemical products at all, while the rest may use home materials (such as soap or alcohol solutions) for common plant diseases such as *Brachycaudus helichrysi*, *Aphis fabae*, and sulfur solutions for *Oidium sp.*, or commercial products from professionals.

Four owners declared that the usage of their gardens is exclusively for decoration and pleasure, while the others use their gardens not only for decoration but also for producing food in order 'to know what we and our children eat'.

To the question 'Do you think your garden is natural?' almost all of them answered 'yes', as they believe it has to do with the non-use of chemicals and fertilizers and the existence of 'green, flowers, and fruits'. Native plant species were not mentioned by anyone, although they seem to recognize native from exotic species and most of them prefer the exotic ones ('because I like them' is a typical answer). Likewise, they use aromatic native plants of Greek flora such as *Thymus vulgaris, Laurus nobilis, Origanum vulgare, Rosmarinus officinalis* for cooking and not for ornamental reasons.

Professionals seem to be well informed about the management practises and the pressures they inflict on the environment. In many cases, though, they fail to suggest friendlier environmental practises or better adapted species, but rather agree with the clients' desires for exotic species which are more expensive and demand more practises.

Discussion

The review presented here demonstrates that gardens remain one of the least studied and understood habitats in urban and suburban areas. In part this omission has been due to the difficulties inherent in obtaining ecological data on gardens and the lack of a methodology for classifying and analysing garden data (Mathieu et al. 2007; Smith et al. 2006). Moreover, unifying concepts and methodologies that integrate the different research approaches are also missing. Most of the approaches deal with either the aesthetic functions or the ecological functions of gardens. Also, most studies refer to urban green spaces such as parks and private gardens and not to suburban green spaces that are currently more in number and rapidly increasing (exceptions are Marzluff et al. 2001 and Destefano et al. 2005).

This chapter has investigated current approaches to the study of the different functions of gardens. The literature review reveals that an integrated method in which aesthetic, ecological, economic and environmental functions of gardens will be included is needed to evaluate the changes that take place in the urban fringes of modern cities in the western world. Especially in semi-arid climates such as the Mediterranean basin, such as approach will be very useful, as on the one hand suburban gardens increase in number and the space they cover, while, on the other, issues of the resources that the management of different land uses require and ecological functions of space are gaining ground.

The basic finding in our opinion is that the introduction of an integrative framework should take into account a double approach: (a) a comparison of function between gardens, as the example presented here reveals that the differences between gardens in terms of composition and practices can be significant; and (b) a comparison between gardens and former land uses.

For the former, the exploratory study presented here highlights that gardens can be categorized according to their ornamental and productive use and their size. This distinction is relevant for the presence of exotic species and in some degree for practices and environmental pressures. Additionally, the impacts from each garden may not be very important, but the sum of inputs from all gardens in an area can have very important impacts on resources and the environment. More research is nevertheless required in order to cover the theoretical and research gaps.

For the latter, the theoretical and research gaps are even wider. For example, in the case of Mytilini, gardens have replaced and continue to replace olive groves and some forests which are habitats for native flora and fauna. On the contrary, gardens are highly artificial biotopes of exotic species that may be unsuitable or even hostile for native species. But, a thorough analysis of the different functions and a method to compare these functions remains underdeveloped. Such an approach should take into account two different dimensions (i.e. the inter-garden dimension and the comparison with former land uses) and at least four different functions (i.e. economic, ecological, environmental and aesthetic).

References

Acar, C., Acar, H. and Eroglu, E. 2007. Evaluation of ornamental plant resources to urban biodiversity and cultural changing: A case study of residential landscapes in Trabzon city (Turkey). *Building and Environment*, 42, 218-229.

Adams, L.W. and Dove, L.E. 1989. *Wildlife Reserves and Corridors in the Urban Environment. National Institute for Urban Wildlife.* Columbia, MD.

Alston, K.P. and Richardson D.M. 2006. The roles of habitat features, disturbance, and distance from putative source populations in structuring alien plant invasions at the urban/wildland interface on the Cape Peninsula, South Africa. *Biological Conservation*, 132, 183-198.

Arevalo J.R., Delgado J.D., Otto, R., Naranjo, A., Salas, M. and Fernandez-Palacios, J.M. 2005. Distribution of alien vs native plant species in roadside communities along an altitudinal gradient in Tenerife and Gran Canaria (Canary Islands). *Perspect Plant Ecol Evol Syst*, 7, 185-202.

Armstrong, D. 2000. A survey of community gardens in upstate New York: Implications for health promotion and community development. *Health & Place*, 6, 319-327.

Armstrong, H. 2004. New Forms of Green for Mega-Cities: Peri-and inter-urban Agricultural Space. AILA, Australia. [Online] Available at http://www.aila.org.au/ONLINE/2004R/PAPERS/Armstrong.pdf [acessed].

Baskin, Y. 2002. The greening of horticulture: New codes of conduct aim to curb plant invasions. *Bioscience*, 52, 464-471.

Batten, A.L. 1972. Breeding bird species diversity in relation to increasing urbanization. *Bird Study*, 19, 157-166.

Beer, A.R., Delshammar, T. and Schildwacht, P. 2003. A changing understanding of the role of greenspace in high-density housing – a European perspective. *Built Environment*, 29, 132-143.

Breuste, J.H. 2004. Decision making, planning and design for the conservation of indigenous vegetation within urban development. *Landscape and Urban Planning*, 68(4), 439-52.

Brunstad, R.J., Gaasland, I. and Va°rdal, E., 2001. Multifunctionality of agriculture: An inquiry into the complementarity between landscape preservation and food security. *77th EAAE Seminar/NJF Seminar No. 325*, 17-18 August 2001, Helsinki, 12.

Cannon, A.R., Chamberlain, D.E., Toms, M.P., Hatchwell, B.J. and Gaston, K.J. 2005. Trends in the use of private gardens by wild birds in Great Britain 1995-2002. J. *Appl. Ecol*, 42(4), 659-671.

CEC 1999. *Agriculture, Environment, Rural Development: Facts and Figures.* Luxembourg: CEC.

Czech, B., Krausman, P.R. and Devers, P.K. 2000. Economic associations among causes of species endangerment in the United States. *BioScience*, 50, 593-601.

Daniels, G.D. and Kirkpatrick, J.B. 2006. Comparing the characteristics of front and back domestic gardens in Hobart, Tasmania, Australia. *Landscape and Urban Planning*, 78, 344-352.

Destefano, S., Deblinger, R.D. and Miller, C. 2005. Suburban wildlife: Lessons, challenges, and opportunities. *Urban Ecosystems*, 8, 131-137.

Duguay, S., Eigenbrod, Z.F., Fahrig, Z.L. 2007. Effects of surrounding urbanization on non-native flora in small forest patches. *Landscape Ecol*, 22, 589-599.

Frumkin, H. 2001. Beyond toxicity: Human health and the natural environment. *Am. J. Prev. Med*, 20, 234-242.

Gaston, K.J., Smith, R.M., Thompson, K. and Warren, P.H. 2005a. Urban domestic gardens (II): Experimental tests of methods for increasing biodiversity. *Biodiver. Conserv*, 14(2), 395-413.

Gaston, K.J., Warren, P.H., Thompson, K. and Smith, R.M. 2005b. Urban domestic gardens (IV): The extent of the resource and its associated features. *Biodiver. Conserv*, 14(14), 3327-3349.

Gatto, P., Merlo, M. 1999. The economic nature of stewardship: Complementarity and trade-offs with food and fibre production, in *Countryside Stewardship: Policies, farmers and markets*, edited by G. van Huylenbroeck and M. Whitby. Oxford: Pergamon Elsevier, 21-46.

Guiomar N. and Fermandes J.P. 2007, Multifunctionality of landscapes – rural development, landscape functions and their impact on biodiversity in 25 Years of Landscape Ecology: Scientific Principles in Practice. *Proceedings of the 7th IALE World Congress – Part 1*, edited by R.G.H. Bunce, R.H.G. Jongman, L. Hojas and S. Weel, Wageningen, The Netherlands, July 2007.

Helfand, G.L.E., Park, J.S., Nassauer, J. and Kosek, S. 2006. The economics of native plants in residential landscape designs. *Landscape and Urban Planning*, 78, 229-240.

Hodkinson, D.J., Thompson, K. 1997. Plant dispersal: The role of man. *Journal of Applied Ecology*, 34, 1484-1496.

IUCN (International Union for the Conservation of Nature and Natural Resources) 2000. IUCN guidelines for the prevention of biodiversity loss caused by alien invasive species. [Online] Available at http://intranet.iucn.org/webfiles/doc/ SSC/SSCwebsite/Policy_statements/IUCN_Guidelines_for_the_Prevention_ of_Biodiversity_Loss_caused_by_Alien_Invasive_Species.pdf [accessed: 12 February 2010].

Kaplan, R. 1984. Impact of urban nature: A theoretical analysis. *Urban Ecol.*, 8, 189-197.

Kaplan, R. 2001. The Nature of the View from Home: Psychological Benefits *Environment and Behavior*, 33, 507-542.

Kaplan, R. and Austin M.E. 2004. Out in the country: Sprawl and the quest for nature nearby, *Landscape and Urban Planning*, 69, 235-243.

Knickel, K.H. and M. Kröger, 2005. *Evaluation of policies with respect to multifunctionality of agriculture. Summary report of the workpackage 6.* Multagri project, 6th framework programme, global change and ecosystem.

Loh, M., Coghlan, P. 2003. *Domestic Water Use Study: Perth, Western Australia,* 1998-2001. Perth, WA: Water Corporation.

Luttik, J. 2000. The value of trees, water and open space as reflected by house prices in the Netherlands. *Landscape and Urban Planning*, 48, 161-167.

Marzluff, J.M. and Ewing, K. 2001. Restoration of fragmented landscapes for the conservation of birds: A general framework and specific recommendations for urbanizing landscapes. *Restoration Ecology*, 9, 280-292.

Mathieu, R., Freeman, C. and Aryal, J. 2007. Mapping private gardens in urban areas using object oriented techniques and very high-resolution satellite imagery, *Landscape and Urban Planning*, 81, 179-192.

Mavridou, A. and Kizos, T. 2007. Analysis, Typology and Evaluation of gardens in suburban areas; Evidence from Mytilene suburban area, Lesvos, Greece, *8th Pan-Hellenic Geographical Conference of Hellenic Geographical Association*, Athens.

McKinney, M.L. 2006. Urbanization as a major cause of biotic homogenization *Biological Conservation*, 127, 247 –260.

Moffatt, S.F., McLachlan, S.M. and Kenkel, N.C. 2004. Impacts of land use on riparian forest along an urban-rural gradient in southern Manitoba. *Plant Ecol.*, 174, 119-135.

Morancho, A.B. 2003. A hedonic valuation of urban green areas, *Landscape and Urban Planning*, 66, 35-41.

OECD 1998. *Agriculture in a Changing World: Which Policies for Tomorrow?* Meeting of the Committee for Agriculture at the Ministerial level, Press Communiqué, Paris, 5-6 March.

OECD 2001. Multifunctionality Towards an analytical framework. [Online] Available at www.oecd.org/dataoecd/43/31/1894469.pdf [accessed: 12 February 2010].

Ozguner, H., Kendle, A.D. and Bisgrove, R.J. 2007. Attitudes of landscape professionals towards naturalistic versus formal urban landscapes in the UK *Landscape and Urban Planning*, 81, 34-45.

Raloff, J. 2003. Cultivating weeds: Is your yard a menace to parks and wildlands? *Science News*, 163, 15-18.

Reichard, S., Hamilton, C.W. 1997. Predicting invasions of woody plants introduced into North America. *Conserv. Biol.*, 11, 193-203.

Reichard, S.H., White, P. 2001. Horticulture as a pathway of invasive plant introductions in the United States. *Bioscience*, 51, 103-113.

Robinson, L., Newell, J.P. and Marzluff, J.M. 2005. Twenty-five years of sprawl in the Seattle region: Growth management responses and implications for conservation. *Landscape and Urban planning*, 71, 51-72.

Rudd, H., Vala, J., Schaefer, V. 2002. Importance of backyard habitat in a comprehensive biodiversity conservation strategy: A connectivity analysis of urban greenspaces. *Restorat. Ecol.*, 10, 368-375.

Slee, B. 2007. Social indicators of multifunctional rural land use: The case of forestry in the UK. *Agriculture, Ecosystems and Environment*, 120, 31-40.

Smith, R.M., Thompson, K., Hodgson, J.G., Warren, P.H. and Gaston, K.J. 2006. Urban domestic gardens (IX): Composition and richness of the vascular plant flora, and implications for native biodiversity. *Biological Conservation*, 129, 312-322.

Syme, G.J., Shao, Q., PO M., Campbell, E. 2004. Predicting and understanding home garden water use. *Landscape and Urban Plannin*, 68, 121-128.

Templeton, S.R., Yoo, S.J. and Zilberman, D. 1999. An economic analysis of yard care and synthetic chemical use: The case of San Francisco. *Environ. Resource Econ.*, 14, 385-397.

Thayer R.L. and Atwood, B.G. 1978. Plants complexity and pleasure in urban and suburban environments. *Environmental Psychology and Nonverbal Behavior*, 3(2), 67-76.

Thompson, K., Austin, K.C., Smith, R.M., Warren, P.H., Angold, P. and Gaston, K.J. 2003. Urban domestic gardens (I): Putting small-scale plant diversity in context. *Journal of Vegetation Science*, 14, 71-78.

Thompson, K., Colsell, S., Carpenter, J., Smith, R.M., Warren, P.H., Gaston, K.J. 2005. Urban domestic gardens (VII): A preliminary survey of soil seed banks. *Seed Sci. Res.*, 15(2), 133-141.

Thompson, K., Hodgson, J.G., Smith, R.M., Warren, P.H. and Gaston, K.J. 2004. Urban domestic gardens (III) Composition and diversity of lawn floras. *J. Vegetat. Sci.*, 15, 373-378.

Tsalikidis, J. 2008. *Landscape Architecture: Introduction to Theory and Practice*, Thessaloniki: Epikentro.

UNDP (United Nations Development Programme, United Nations Environment Programme, World Bank & World Resources Institute) 2000. *World Resources 2000-2001: People and Ecosystems – the fraying web of life*. Amsterdam: Elsevier Science.

UNFPA United Nations Population Fund. *Population Issues: Meeting development goals* [Online] Available at http://www.unfpa.org/pds/urbanization.htm [accessed: 10 December 2006].

Wasilewski, A. and Krukowski, K. 2004. Study of Urbanization Around Warsaw and Olsztyn, Poland. *Environmental Management*, 34(2), 291-303.

Whelan, R.J., Roberts, D.G., England, P.R. and Ayre, D.J. 2006. The potential for genetic contamination vs. augmentation by native plants in urban gardens *Biological Conservation*, 128, 493-500.

Wiggering, H., Dalchowa, C., Glemnitz, M., Helming, K.,, Muller, K., Schultz A., Stachowa, U. and Zander, P. 2006. Indicators for multifunctional land use – Linking socio-economic requirements with landscape potentials, *Ecological Indicators*, 6, 238-249.

Zagorski, T., Kirkpatrick, J.B. and Stratford, E. 2004. Gardens and the bush: Gardeners' attitudes, garden types and invasives, in Daniels, G.D. and Kirkpatrick, J.B. 2006. Does variation in garden characteristics influence the conservation of birds in suburbia? *Biological Conservation*, 133, 326-335.

PART IV
Landscape Research and Development Planning

Chapter 23

Research Supporting Landscape Policy: A Challenge

Daniel Terrasson

Introduction

We have been educated with two simple ideas. On the one hand, research produces knowledge and it is assumed that scientific results are naturally disseminated into society following a top-down process that leads to the application of knowledge. On the other hand, public action takes place within a well-established process in which decision-makers arbitrate between public and private interests.

Moreover, research is interrogated by society to justify its very existence and its cost as well as to allay severe public anxieties: AIDS, global warming, mad cow disease … In this context, one can often read in the strategic plan of research institutes, in the justification of their research projects, that they are aiming to help public decision-making, producing tools for action, etc. From the research world as well as from the policy world, voices have claimed that the relations between science and society were probably not as satisfactory as this model suggests, and that it was necessary to avoid such simple ideas. In consequence a renewed interest in the science/society interaction arose during the last decade. In particular, we would mention the works of Scott, Gibbons, Nowotny and others in the sociology of science, *The new production of knowledge*, who have identified the emergence of 'mode 2 research', the works of Michel Calon and Bruno Latour, the working group 'Bridging the gap' launched by the European Commission, or the initiatives taken under the former German presidency of the European Union under the frame of 'Science meets society', etc.

If we want to investigate the possibilities for research to support landscape public action we have to answer three questions:

- What are the features of the decision-making system?
- What are the specificities of landscape as public action is concerned?
- What is lacking in the decision-making system and is research able to deliver appropriate inputs?

These three questions will provide the framework for this chapter. We will start by a short description of the evolution in public decision-making. Then we will discuss the specificity of landscape, and for that we will mainly rely on the outputs of the French research program 'Landscape and public policies'. Finally, we will see what has been done to help public decision.

Evolution in the decision-making system

We need to bear in mind two main points that have consequences on the advisory function (Duran 2008):

The extent of the decision making system and its polycentrism

Formerly, the role of the national government was preeminent in the definition of public interest. Nowadays, we are facing institutional plurality and interweaving. This is a consequence of the assertion of Europe and the growth of regionalism and decentralization. As a matter of fact, we have several levels of political responsibility and each level of government has its own political legitimacy and its own view of public interest according to its own agenda. This is a problem known as 'God's war': no one is right and everyone is right …

Furthermore, public action does not rely only on public institutions, but also on a number of private actors. There is no *one* decision-maker, there is a decision-making system with fuzzy limits.

The consequential revolution

Formerly, policies were mainly thought of in a logic of quantitative production. They were aimed at building new equipments (roads, schools, hospitals …) or improving the production of goods (tons of cereals, iron or coal). Nowadays a number of political issues are not quantitative, and are not sector-driven. They are qualitative and are crossing economic sectors: health, environment, security, quality of life … In this case, it is no longer public action itself that is important (the number of roads or hospitals …), but the consequences of the action (how the environment is improved by new instruments of the agricultural policy, for example).

We can first conclude that these two trends are leading to an increased complexity of public action. In this context, the question is not only to coordinate the activities of the various branches, but to organize cooperation between institutions that have all their own legitimacy. The problem is to manage collective action while overcoming four main difficulties:

- The growing complexity of public problems that cannot be solved by simple technical solutions;
- The differentiation of public arenas in which public problems are debated;

- The overloading of public services, exacerbated by the loss of their staff's technical competency, as they are in the same time required to reduce public deficit and fiscal pressure;
- The changing competency of staff in these institutions and the insufficiency in new required knowledge: from technical problems to the management of collective action.

The second conclusion is that these evolutions are increasing the need for the evaluation of public policies.

Specificity of landscape in public action

The trends reported above apply whatever the objective of the policy. Anyone would agree that what has been said on the public decision-making system is valid with respect to landscape: a qualitative issue crossing branch concerns, an object transformed in consequence of the action of numerous, uncoordinated private and public actors, a question concerning various institutional levels (Council of Europe, countries, regions, numerous localities, a large number of associations and citizens).

But landscape has several specificities that particularly justify a call for supporting public decision-making.

First of all, landscape is a relatively new concern for public policy. Even though regulations aiming at the protection of landscape in some specific cases already exist in a number of countries, they did not constitute a policy in the full meaning of the word (an identified problem, explicit objectives and appropriate means to attain them). In France, for example, the first laws are as old as 1906 but the landscape law (which is not itself a policy) was adopted only in 1993 (Cadieu et al. 1995). Moreover, the models on which landscape policies are based have changed over time (Rumpala 1998) and developed in three directions:

- Going from a 'don't touch it' conception that aims to protect landscape heritage in its current state, to a managerial conception that aims to adapt to change;
- At the same time, the landscape's status has changed from a public asset that must be safeguarded according to a heritage-preservation approach, to the status of a resource that must be valued by the market, and which as a result must be at least partially appropriated by stakeholders;
- Finally, policies no longer deal only with aspects that are out of the ordinary; they address the landscape on a day-to-day basis.

What may be mentioned here, however, is that, first, one model does not replace another; instead, these models are superimposed and generate a wide range of understandings by the players during implementation; second, this change has taken place from a regulatory viewpoint but has not yet been completely enacted on the ground, where the 'don't touch it' concept continues to dominate. The main consequence is that we are lacking long experience to conceive such public actions and to imagine their effects;

The second features are that the objectives of a landscape policy cannot be precisely defined at a global scale, and that the concept itself is difficult to translate into legal terms (Rousso 1996). Therefore, landscape is a good example of this new kind of public problem, for which regulation only defines legal participatory procedures, the precise objective of public action being decided locally. Under these conditions, local implementation is guided by how the concerned players interpret their purposes. Five concerns with five categories of corresponding actions can be identified:

- Aesthetics: making the surroundings more aesthetically pleasing;
- Heritage: preserving something that belongs to everyone, like putting it in a museum;
- Social: procuring well-being;
- Economic: highlighting the value of a resource, contributing to development and employment;
- Ecological: managing biodiversity (habitats, ecological structures, etc.).

Such a range of interpretations may cause conflicts among different groups of players, but this can also occur within most institutions: among individuals, within departments, between different levels of responsibility. As a result, there is a great deal of diversity when it comes to how public policies are applied at local level. This brings into sharp focus the question of the landscape culture of the main people in charge of implementing policy (elected officials, decentralized administrations, etc.).

Moreover, if the objective of landscape can vary, the status of the landscape may also take several distinct forms during public initiatives. These different sorts of status may be combined in some cases.

The landscape may be precisely the *focus* of the action undertaken. In this case, one observes what action is intended in formal and standardized landscaping terms: the initiative is part of a purely aesthetic or heritage-based approach. In this case it seems that applications are very limited.

The landscape may also be a *means* to reach a local or sector-wide development objective (especially agricultural). The landscape has social importance through the economic resource it is likely to become in order to bring value to a product (cheese, for example), or to a touristic activity. It is then an attribute of the territory and the landscape becomes the focus of negotiation because it represents an important issue in the development process.

The landscape may also be a *negotiation tool* when developing a territorial project. The landscape is then seen as intimately connected to a social situation and human activities. It is a way for the players to position themselves with regard to one another and to create their development trajectory. It refers to other goals that policies aim at: the organization of an area, the living environment, etc.

Finally, the landscape may become only an *alibi* used by certain players in order to reach other objectives, not always made explicit, in other fields. This is potentially the case, at times, in certain sector-wide actions (forestry development, agricultural diversification, etc.). It is even more so for projects that can be likened to unacknowledged social engineering.

In this context, where the status of the landscape is not always made explicit, and where there is a lack of genuine reflection about the substance and the objectives of landscaping action, applications tend to be focused on reproducing stereotypical models: rebuilding farmland enclosed with groves of trees, the struggle to keep the landscape from being closed off, planting flowers, etc. Such models are conveyed by certain groups of players and in particular the specialists (landscape architects, technicians from public services, scientists, etc.) who at this level play a significant role. Resorting to stereotypical models, however, tends to overshadow taking into account the particular features of the local context and the inhabitants' concerns.

This observation suggests that:

- It is urgent that the concerned players become aware of this phenomenon;
- There is a need for properly-adapted training and information.

And finally, the evaluation of landscape public policies is raising several difficulties. The first is that landscape policies often have a limited effect when compared to other policies. The policies with the greatest impact on landscape and people's mentalities were by far development policies with a significant influence on the surroundings and activities. There are many different direct and indirect effects, immediate and delayed effects, temporary and permanent effects. It is then useful to ask whether landscape policy has only a demonstrative character that drives social practices, or whether it reveals a regulatory framework that is poorly adapted to expectations.

Secondly, landscape policy does not provide for systems to evaluate their effects and effectiveness (lack of success indicators). The need to provide for the conditions of evaluation from the very creation of public policy is not specific to the landscape issue, but as objectives are not clearly stated, evaluation becomes problematical. Evaluating actions implies resituating their creation, implementation and the players involved in a specific context.

The sustainability of regulatory measures is poor when compared to the time steps of the social and natural dynamics of the landscape.

Evaluation is challenged by the very nature of the landscape, which includes material and immaterial dimensions. The immaterial dimension is poorly addressed by evaluation methods, but even the materiality is often not properly taken into account. Methods tend to split the landscape into spatial objects whose relevancy is justified both by the prospects they offer to facilitate dialogue and because they are the main target for policy action (low walls, hedges, paths, etc.). Restoring the uniqueness of the landscape is not settled by this twist, which moreover does not show what is meaningful in the material evolution of the landscape.

How can research support landscape public action?

Usually it is recognized that research can support public action in four different ways (Bureau 2008).

First of all, research is asked to organize access to existing knowledge and data (education of staff, data bases, knowledge transfer ...). This cannot be reduced to a simple publication of scientific results coming from research works conceived according to pure academic issues. The offer has to be adapted to the particular expectations and former knowledge of public actors. This is the second part of the so-called 'double translation process' (Duran 2008): translation of a problem into research questions, and interpretation of the results into principles for action. Moreover, an appropriation process of the final product is always necessary.

Secondly, research can produce expertise. Researchers can be mobilized as experts, either occasionally (for a crisis), or more regularly (as government advisors, member of an administrative committee ...). But research also has a role to play on its own initiative in the anticipation of problems and crisis. A difficult point concerns the competency of experts, as landscape is crossing academic fields and as landscape schools are more oriented towards gardening than landscape issues at a wider scale.

Third, research programmes or projects focused on public policies can be launched by public institutions, either prior to the conception of public action, or as support to its implementation. Several outcomes can be expected: generic knowledge on the process, dynamics and functioning of systems, diagnosis of system failures, tool-kits and methods for the implementation of public decision, evaluation of public policies ... Research supporting public action implies a complex interdisciplinary approach, but also an experience of dialogue between scientists and actors engaged in public action (engineers, policy-makers): specific competency and know-how are needed.

Finally, some long term cooperation between scientists and decision-makers can be organized, following specific agreements.

As regards landscape, we have seen that this is a relatively new concern for public policies, and that the objectives of the policies are poorly defined. The European Landscape Convention recognizes that landscapes policies are at an experimental stage (article 8) and that there is a need to share good experiences and practices. That means that research has a particular role to play. Several countries have launched national research programmes targeted at the implementation of landscape policies (Austria, France and Switzerland). In France, a first national research programme explicitly dedicated to the support of public action and called

'Landscape and public policies'[1] was launched by the Ministry of Ecology, Energy, Sustainable Development and Land Management in 1998. A new programme called 'Landscape and Sustainable Development'[2] followed in 2005 and is still under way.

The programme 'Landscape and Public Policy' has mainly helped to understand what is happening when landscaping action is undertaken. Who are the actors involved and who is not? What are their objectives and their respective roles? What is really at stake behind the discourse? How does the process evolve? Which tools and knowledge are used in the debate? The programme also shows the multiplicity and diversity of the initiatives that were undertaken at local level. And therefore, we have now a better understanding of the critical points in the implementation process of landscaping. But the programme has not really been able to evaluate the impact of public policies on landscape, as they were too recently implemented. Landscape is not an object that can automatically or easily be translated into legal terms, and it is not surprising that the implementation of policies that appeared even recently raises many questions. Unlike regulations anchored in the long term such as that on the sites, to date public landscape initiatives are still in the experimental phase. Nevertheless, observations about the impact of these policies on society, on land values and on the ability to take into account inhabitants' expectations draw particular attention to the need to be vigilant with regard to the ethics that must underpin landscape initiatives.

Finally, the programme has also created an interest in landscape policy into the research community, and one significant output is that a number of researchers have developed a competency on this theme, have been identified by the Ministry of Ecology and can now been mobilized as experts.

In terms of education and training, two points should be noticed. First, there is a very broad need to improve the training of private experts and elected officials. One of the most crucial questions for the future of policy in this field may be that

---

1 The national research programme 'Public policies and landscape: Analysis, assessment, comparisons' was funded by the Research Department of the French Ministry of Ecology, Sustainable Development and Sea (MEDDM). It started in 1998, with a five-year proposed duration, but was extended until 2005. Its aim was to provide the scientific foundations for the inclusion of landscape in public policy definition and implementation, as well as for the assessment of its effects. Twenty four projects were funded. The consolidated budget has been estimated at €4 million, to which the Ministry contributes €1.6 million. More information at http://www.ecologie.gouv.fr/Politiques-publiques-de-paysages.html.

2 The national research programme 'Landscape and sustainable Development' is funded by the Research Department of the French Ministry of Ecology, Sustainable Development and Sea (MEDDM). It started in 2005 and will end in 2010. Its aim is to give scientific evidence of the links between landscape and sustainability in the context of the implementation of the European Landscape Convention. Sixteen projects were funded and all of them are involving European collaborations. The consolidated budget is estimated at €5.5 million, to which the Ministry contributes €1.7 million. For more information see http://www.ecologie.gouv.fr/article.php3?id_article=5663.

of training competent consultants who are able to grasp landscape issues, both through the analysis of their evolving materiality and through the symbolic ties between populations and the area they live in. It is essential that expertise methods cease referring to old and outdated conceptions of landscape heritage, which they often agree with today, by limiting themselves to localizing the picturesque aspect of sites or breaking down the landscape into formal units without any ecological, social or historical depth. Finally, it is crucial for expertise to be understood as an essential contribution to opening up the debate involving the concerned players and leading to joint projects. The elected officials' cultural background and knowledge of associated policies is also limited when it comes to landscape issues. Therefore, as is the case for the experts, there is a problem with respect to training and making suitable information available. This is a matter of giving the key players the ability to grasp landscaping challenges but also the ability to develop a critical eye when it comes to schools of thought that are accepted thanks to their power of persuasion. But these targets are difficult to reach for several reasons: lack of convenient training structure, lack of time for complementary training, lack of motivation … On the contrary, training officers in public services has been strongly developed in recent years, but all of them are susceptible to change position quite quickly and the training needs are continuously renewed.

Moreover, the links between research and education in landscape schools is under progress but still relatively low (Luginbühl 1996). The graduated landscape architects are educated in schools whose origins are related to horticulture and gardening art or to architecture. In both cases the culture is based on artistic creation more than the analysis of natural and social interactions. The need for a closer link between education and research is still not seen as crucial by all school responsible.

Conclusion

Developing links between research and policies is always a challenge. It is difficult to harmonize the time-scales of scientists and policy makers and to overcome their mutual incomprehension. To get a real dialogue going is never easy. Some difficulties arise from the functioning of science (scientist's evaluation mainly based on peer-review, lack of criteria for the evaluation of applied research work). And several failure situations exist between scientists and decision-makers: incomprehension, research answers out of the scope of actors' questions, lack of translation of research products into understandable results, lack of clarity in their expectations …

But landscape is also a challenge for research as it involves complex interactions between social and natural processes. The development of interdisciplinarity has already produced useful results. Progresses are now needed in two directions:

- Methodologies for the evaluation of landscape public action taking in account the diversity of impacts (social external effect, fitting with social expectation in an evolving context ...) and guidelines for a set of evaluation criteria;
- Conditions of sustainability for landscaping actions.

References

Bridging the Gap. [Online] Available at http://ec.europa.eu/research/environment/newsanddoc/article_4055_en.htm [accessed: 12 February 2010].

Bureau, D. 2008. *La relation recherche-décision publique*, séminaire Appui à la décision publique, Antony, 10 January 2008.

Cadieu P., Corot D., Le Roy R. and Trapizine R. 1995. La loi 'paysages': Dossier expert – *La lettre du cadre territorial*.

Calon, M. and Latour, B. (eds) 1991. *La science telle qu'elle se fait*. Paris: La Découverte.

Duran, P. 2000. *Connaissance et action publique; l'évaluation comme savoir pratique*. Séminaire politiques publiques et paysages, Actes du séminaire d'Albi, 28-30 March 2000, 33-43.

Duran, P. 2008. *Evolution de la décision publique*, séminaire 'Appui à la décision publique', Antony, 10 January 2008.

Gibbons, M., Limoges, C., Nowotny, H., Schwartzman, S., Scott, P. and Trow, M. 1994. *The new production of knowledge: Dynamic of science in contemporary societies*. London: Sage.

Luginbühl, Y. 1996. Le paysage aujourd'hui et son enseignement. *L'information géographique*. 1, 20-29.

Rousso, A. 1996. Le droit du Paysage, un nouveau droit pour une nouvelle politique, *Cahiers de l'Environnement de l'INRA*, 26, 29-42.

Rumpala, Y. 1998. Les ambiguïtés d'une intervention publique dans la préservation des paysages: Retour sur les labels 'paysages de reconquête'. *Natures, Sciences, Sociétés*, 6(3), 39-44.

Chapter 24

'Landscape' as a Sign: Semiotics and Methodological Issues in Landscape Studies

Tor Arnesen

Introduction

In printed English the word landscape appeared in 1603, and had origins in Middle Dutch (landscap), Old Norse (landskap) and the German word *Landschaft*, meaning a restricted piece of land. A previous formation in English was 'landskip'. As is often discussed, e.g. by Olwig (1996), any confusions in research entailing the common word 'landscape' is not a novel affair. Hartshorne made a reflection on this in 1939 (apud Olwig 1996): Landscape is 'perhaps the most single important word in the geographic language', and he pointed to the confusion the word created. Hartshorne observes that 'the English word is [...] used in an aesthetic way to refer to 'appearance of land as we perceive it' and 'the section of the earth's surface and sky that lies in our field of vision as seen in perspective from a particular point' (Hartsthorne 1939; apud Olwig 1996: 152). This, Harsthorne argues, enable users to shift 'from the use of the same word to mean, on the one hand, a definitely restricted area and, on the other, a more or less definitely defined aspect of an unlimited extent of the earth's surface' (Hartsthorne 1939; apud Olwig 1996). These shifts between area and aspect are interesting – but I find it interesting in a slightly different manner: 'Landscape' is all about aspects of an area, and the relation between an area and an aspect is a vital core of what 'landscape' is. The relation between area and an aspect of the area is clearly something cognitive, thus it must irreducibly imply a logical third, someone for whom the relation between an area and an aspect exists. Semiotics is well equipped to analyse this triad of relations, between area, aspect and someone.

A semiotic approach to landscape studies

This approach relies on semiotics in a post-Peircean tradition (Charles Sanders Peirce 1839-1914) – something akin to the approach discussed by Vallega 2007. Two demarcations should be made initially:

- It is not argued here that the landscape can readily be read as a text – on the contrary the aim is to dodge what Lefebvre calls 'the illusion of transparency'

(Lefebvre 1991), the tendency for someone to overestimate how well they understand others persons' cognitions. 'A peasant does not perceive 'his' landscape in the same way as a town dweller strolling through it', Lefebvre notes (1991: 113-114), and the same could be said for planners, scientists and social engineers. Unchecked, 'the illusion of transparency' might distort landscape policy and cause excesses of authority.

- There will be analogies between a theory of space and a theory of signs, but it is not argued here that a theory of space is a sub domain in a theory of signs. Landscape, as it is discussed here, is about discourses *about* space and *in* space, not *of* space.

A semiotic approach to landscape theory is an interest shared with other quarters of landscape researchers, e.g. Pitte (1983), for whom the landscape is a sign indicating the whole spectrum of human needs: the production of food, the consumption of other goods and services. The question raised in this chapter is: given a semiotic approach, what methodological principles or leitmotifs follow in the study of landscapes?

Mitchell (1994) discusses the assertion that landscape is solely a European or western phenomenon. He contrasts two rather typical Western approaches, and develops a third that transcends this dualism. The first typical approach is the representation of landscapes in drawings, paintings, maps etc., where the landscape is the content within the medium of (mainly) visual representation. The second is an interpretive approach, where the landscape becomes an allegory of the (ambiguous) social and even psychological that is open to decoding. In this mode, the landscape is the medium itself from which messages can be drawn, so to speak. Meinig (1979) identifies ten message approaches to this discipline, among which landscape as history, as wealth, as problem, as place and as nature.

To the extent 'landscape' is a Western phenomenon, it belongs, according to Mitchell, to these two traditions. Mitchell refutes the idea that landscape must be apprehended this way, and develops an approach that asks not what a landscape 'is' but rather what it 'does' and how it works as social practice. As we shall see, this is a vital point in a semiotic approach too – it is not so much about seeing as doing and conceivable practical effects. And because it is about conceivable practical effects for some (and by logic not for others), landscape is ultimately of importance in issues of power and empowerment, as I have argued in an earlier paper (Arnesen 1998).

Semiosis is any form of activity, conduct, or process that involves signs, including the production of meaning. Semiosis is, according to Peirce, 'an action […] which is […] a cooperation of a sign, its object and its interpretant' (CP 5.484). Briefly said, semiosis is the sign process.

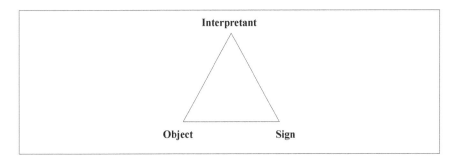

Figure 24.1 Sign triangle

The term was introduced by Peirce to describe a process whereby a sign user is referring signs to their objects. Semiosis is irreducibly triadic; it is by its logic a threesome. Peircean semiotics operates with three elements in semiosis – the object, the sign and the interpretant – the latter being the sign user (community) for whom the sign represents the object in a certain respect or capacity.

A sign user community is not congruent with the group using a natural language. *Prima facie*, the assertion uttered 'this landscape is valuable' means different things (i.e. what values, and how they relate to social practice) to different interpretant communities within an ordinary language, ref. Lefebvre's comment on peasant and city dwellers above. The decisive point is that interpretant communities differ in the respect or capacity in which they relate the sign to the object. This dependence on respect or capacity is not superficial, but constitutive to the meaning and is embedded in a life form. The utterance within an agrarian community has a meaning that is different from the utterance in an urban setting, a cultural heritage setting, a mining community setting or a scientific setting in ways that in principle is relevant in policy discourse issues.

The relation between sign, object and sign user (community) or interpretant is in Peircian semiotics irreducibly triadic – the relation cannot be reduced to any combination of dyadic relations without loss of what constitutes meaning. This again relates to a basic argument in pragmatism as it is found with Peirce and later Ludwig Wittgenstein (1889-1951): the relation between sign and object cannot be understood in abstraction from the sign user community.

In this triadic relation between object, interpretant and sign, what is landscape? Landscape is the object as represented to an interpretant (a sign user community) in a certain respect or capacity. So, in this triad landscape is a sign and as such an aspect of an area. So the assertions made by semiotics and applied to landscape studies are: all cognition is embedded in interpretation, all interpretation is by sign and sign users. All landscapes are interpretations of an area by a sign user community in some respect or capacity. In this semiotic formula, 'landscape' is a sign. As such, landscape is a sign *about* space in a discourse *in* space.

The meaning of any sign cannot be conceived in abstraction from the interpretant – the sign user community. A sign has no universal valid meaning, it is only charged with meaning as a representation of an object in a given sign user situation.

'A sign', Peirce says, 'is a sign *in actio*, by virtue of it receiving an interpretation'. (CP 5.448).

The meaning of a sign is inherently tied to an act, an act whereby signs are used by a sign user community ultimately as an integrated function in a life form – thus pragmatism. In semiosis, a landscape is a (meaningful) representation to an interpretant (community) of an object that is a physical unit – a piece of land, definite or indefinite, with inventory typically at a scale of square kilometres.

Now, whatever a 'landscape' is, it cannot be conceived as contingent on the interpretant alone nor the object only, but they relate to produce or constitute 'landscape' through language use. The relation between the interpretant and physical object is not un-mediated; it is not to be thought of as that of a mirror to its mirrored object. Quite the contrary, it is mediated and thus interpreted.

The landscape is the result of the interpretant (community) act to cognize the object as something in some respect or capacity. In this operation the interpretant is guided by or contingent on intention, purpose, project, enterprise etc. in conceiving the landscape, and not disconnect-able from intention, purpose, project, enterprise etc. The landscape is a sign representing an area as it is conceived by the interpretant as something in some respect or capacity.

I interpret Olwig's (1996) *Recovering the Substantive Nature of Landscape* as advancing basically by performing an similar type of pragmatic analysis: 'It is not enough to study landscape as a scenic text. A more substantive understanding of landscape derives … from the historical study of out changing conceptions and uses of land/landscape'.

I interpret Krogh's concept of 'landscaping' in a similar fashion when he defines 'landscaping' as 'man's process of creating meaning in interaction with his environment' (1995) – the 'outcome' of which are landscapes. Through landscaping we make new relations to an area. It is less obvious that landscapes should basically be understood in the same way. This is Krogh's focus exactly, that we cannot access a landscape but through landscaping.

Methodological consequences

Now, if 'landscape' is a sign in a semiosis, then some methodological consequences can be deduced from that approach. Four of these are discussed below: (1) accessing a landscape; (2) subsumption or multiple landscapes; (3) landscape change; (4) abstraction and landscapes.

(1) Accessing a landscape

What does it mean to enter, be in and share a landscape – in short accessing a landscape? I am not referring to the legal issue of open, restricted or no access, or to physical access. What I have in mind is access in the sense we depend on when doing and communicating a landscape analysis.

Applying the non-reductionist approach in semiotics: accessing a landscape cannot be reduced to being in or travelling through the object (here area, a parcel of land – subjected to interpretation). Any two people may be co-located in the same physical unit, same object, without being in the same landscape.

Since a landscape is contingent on intention, purpose, project, enterprise etc., any two people must master – at some level – the same set of intention, purpose, project, enterprise etc. that is relevant to a given object to be in the same landscape. To have access to a landscape is not the same as to access an *area* where this landscape exits. Any two people can be on the same spot, in the same area, but in different landscapes.

From a pragmatic point of view, to access a landscape means to somehow take part in the life-form/life-style/or simply activity and way-of-doing-things that makes up the core of the interpretants' projects or program. Two people doing the same thing in the same area are more likely to be in the same landscape.

Norman Henderson (1994) doing 'Replicating dog travois travel on the Northern Plains' in Canada in order to understand the area/technology interaction and essential features in the native landscape, is a very good example of a methodology that is in line with pragmatic access as defined above. Another example is Henderson (1996) discussing 'The canoe as failure on the Canadian Planes' in the face of how 'Canadians mythologize the aesthetics and importance of canoe travel in their nations development' – replicating travel being the means for accessing and assessing the landscape.

(2) Subsumption

It also follows that any one object – in this case an area as a physical unit – can be related to any number of interpretation communities or interpretants.

This has some odd implications: the one and the same (even congruent, if you will) piece of land that we wish to subsume under the concept of 'landscape' may or may not contain many co-existent – even congruent – landscapes, related or unrelated.

This situation where one object – in this case an area – is interpreted by several related or unrelated signs/sign systems by different interpretant communities could be labelled the physical mode of subsumption.

Subsumption construed this way is not dependent on different interpretants conceiving mutually inconsistent facts about what the inventory or status in a landscape. They may operate with inconsistent assertions about facts and is such cases at least one part will be wrong. But disputes over facts as such are not the point here. The point relates to relevance. The numbers of facts that can be ascribed to a piece of land are limitless. A selection will be made, and made by the various interpretant communities as part of their sign use – and this selection of relevance is related to the respect or capacity in which the sign is interpreted.

This again relates to the community lifestyle the landscape in question is supporting or embedded in. What are considered relevant facts legitimately and understandably vary across different user groups. These differences may in a developed discourse in, for instance, a landscape analysis process turn up as a use of different data sets, different observation scales, different observation interests, etc. (Sevenant and Antrop 2007).

As is shown in theories of discourse analysis (Brown and Yule 1983), these differences may run deep and are complex to resolve as part of a development process in which a landscape analysis may be one element.

Furthermore, any one sign may or may not relate to any number of physical units – see point (4) Abstraction for more discussion on this. This means that an interpretant (community) may subsume many objects – here, pieces of land – spatially detached – under the same 'landscape'. This could be labelled the mental mode of subsuming and it is, in this approach, a basis for landscape typologies.

(3) Landscape emergence, change and loss

Landscapes change and landscape change is an important part of most landscape analysis, policy making or review. Landscapes even emerge beyond the mere geophysical processes that change our natural surroundings.

Given that 'landscape' is a sign, then in what ways can a landscape emerge, change and even disappear? What modes of change can a landscape undergo according to this approach? A landscape can come into being, change or disappear either by changes in the mental unit – the interpretant (community), in the physical unit – the area as a physical object, or in both.

Material changes (by nature or human made) in an area may or may not be considered a change in, or a threat to, a given landscape. Material change does not, according to this approach, necessarily imply a landscape change – it is a change to the extent that it is relevant to the respect or capacity a sign user (community) relates the sign to the object in an act of representation. Logging in forestry does not change the landscape for loggers as forest landscape, but it may radically change the landscape for an environmentalists or a hiker.

So, the object of a sign is one thing, its meaning another. Whether or not material change is understood as landscape change is dependent on the way the material change affects the landscape, which reflects the effects on an actual change ('conceivable practical effect') for a given interpretant.

This is the essence of the Peircean type of 'pragmatic meaning theory' (Apel 1973). If the change has no conceivable practical effect within an interpretive tradition, or if the practical effects are considered irrelevant or unimportant by that interpretive tradition, material change may not alter their landscape.

The reference to practical effects connects to technology, since possible action and possible practical effects normally refer to or involve technologies within which we generate or discard ideas of how lives could be lived. In this way, landscape is something construed within a technological context.

This is visible as an effect of transportation technology. Travelling has greatly changed our views on landscape (Urry 1990). Often landscapes are to us more about travelling through than being in (de Botton 2002). We even tend to make a distinction in daily speech: 'Have you been there?' 'No, I have only travelled through'. Since we travel through so many more landscapes than we are in, travel-through-landscapes (tt-landscapes) are important to study, not just being-in-landscapes (bi-landscapes). Urry (1990) explores this distinction in tourism:

> The tourist gaze is directed to features of landscape [...] which separate them off from everyday experience. The viewing of such tourist sights often involves different forms of social patterning, with a much greater sensitivity to visual elements of landscape [...] than normally found in everyday life. (Urry 1990: 3)

An interaction between tt-landscape tourism concerns and bi-landscape local concerns plays out in many places in the world, as tourism continues to grow and spread (Hall and Müller 2004). It is often assumed that travelling is a good cure against prejudices to other cultures – and it might indeed be so – but if over time tourism causes landscape policy to maximize tt-landscape values, local values may well be at a disadvantage. Landscape as a sign may be a construct or a model, a perspective, filter or a cognitive aspect which we focus on and fix certain stable features in a continuously changing object, insofar as it represents a special value or attracts the attention of an interpretant. But the construct or model in its turn delivers premises for policy making, resulting in controversies between the tt-landscape perspective and the bi-landscape perspective, not uncommon in many landscapes along tourist trails.

Landscapes as signs may be lost. In the extreme case of the interpretant community disappearing or being unable to relate to the physical unit as before, the landscape disappears. Stonehenge may figure as an example; we still do not understand the doing of the landscape. Blain and Wallis (2007) have pointed to the huge variety of views which show the continued and growing importance of Stonehenge today, but also the loss of what landscape it once was and for whom.

In these situations – oddly enough – the physical unit may *in principle* be the same and undergo no relevant change even though the landscape changes. This represents what we could call the mental mode of landscape emergence, development, change and possible disappearance.

If the physical unit changes in relevant ways to the interpretant community, or disappears, their landscape changes or disappears. This represents what we could call the physical mode of landscape change and development.

Intermezzo: EU and the landscape convention

The first three points discussed are consistent with, but transcend, the definition of landscape as it is found in the European Union Landscape Convention – also signed by Norway.

The definition in the convention is:

> "Landscape" means an area, as perceived by people, whose character is the result of the action and interaction of natural and/or human factors. (ETS No. 176)

The key point here is that landscape is defined as contingent on 'perceived by people'. It follows by logic that if a certain perception disappears, the landscape disappears. If a certain perception changes, the landscape changes accordingly. There is no requisite in the definition for any material changes for the landscape to change – thus landscapes may emerge, change and disappear without any material changes in the area perceived. The only requisite is that the perception changes. And perception might obviously change for many other reasons than a material change in the area being perceived as landscape. New knowledge, new interests and other impulses in the one who perceives might well provide them with a new approach. A good example here would be farmers and other landowners in rural areas who realize that land of marginal interest from a farming point of view might still contain recreational amenities that can be turned into a commodity in a given market e.g. as sale of parcels for second homes. And voilá, a significant change has affected how an area is perceived without any material change taking place. The general point here is that technological, social and cultural development will change landscapes – but not necessarily by causing material change.

And, likewise, a material change in the area being the object of perception might be irrelevant to the area as perceived by people. Again, the object is one thing and the perception of it another – and the object is not the landscape.

And again likewise, different groups of people might – and most probably do – perceive one and same area differently, even at the same time. And thus we may – and probably do – have different landscapes coexisting in the exact, or approximately same, area.

Is the definition offered by the convention sufficient? I think not – the definition should make a better effort at bringing home one vital point: landscape is a representation of an object; and the crucial question in landscape management and policy – which is the main concern with the convention – is not perceptions *per se*, but how these relates to people's lifestyles or life forms, and in the next step how we can gain access to those perceptions.

The phrase 'perceived by people' has a certain ease or lightness attached to it. You might be mistaken to think that pointing at a picture gives you access to its meaning. It does not. A picture does not amount to much.

A picture *per se* is in itself almost void of meaning – as is so brilliant demonstrated by Rene Magritte and his 'The Treachery of Images' painting. The representational use of objects as other than what they seem is typified in this painting which shows a pipe that looks as though it were a model for a tobacco store advertisement. Below the pipe Magritte painted 'This is not a pipe' (*Ceci n'est pas une pipe*), which seems a contradiction, but is actually true: the painting is not a pipe, it is an image of a pipe. It smells of oil paint. You cannot feel its texture or weight and you cannot smoke it. To understand what you are looking at you are dependent on being initiated in a culture, its technology, enterprises and values. None of these are mediated in the picture as such.

To coin a sign to refer to an object to convey meaning in use, a community of sign users must agree on a simple meaning within their sign system. But a sign can transmit that meaning only within the structures and codes of a given sign system. Codes also represent the values of the sign user community culture, and are able to add new shades of connotation to every aspect of life.

From a methodological point of view, little is gained by pictures as the tool to gain access to the landscape various interpretant (communities) may have. Sadly, this is an all too common method used in landscape policy-making processes. From a semiotic point of view, the result returned by this method probably amounts to no more than reproducing the sign repertoire (or even ignorance) of the observer.

Figure 24.2 Memorandum: Whatever is depicted,
it does not short-circuit interpretation
Source: After Magritte's work.

Magritte's work may serve as a methodological reminder, as a memorandum.

(4) Abstraction and meaning

As a sign, a landscape is an abstraction relating to the interpretant community, and its object is the various ways discussed in sign theory. Various signs lend themselves to abstractions in various ways. To discuss how, we have to look briefly at various types of signs.

Any one sign may, or may not, relate to any number of discrete physical units that have common qualities on some respect or capacity. This means that an interpretant (community) may subsume a non-specific objects with general qualities – e.g. splendid crag landscapes for climbing – under the same 'landscape'. This allows for development of a landscape typology; that is signs relating to indefinitely many discrete areas with specific qualities with conceivable practical effects or values in a certain respect or capacity.

A sign is a representamen with a mental interpretant. Peirce operates with an advanced system of classification of signs. If we focus on the relation between the sign and the object, Peirce operates with three qualities of signs which all are present in all signs, but more or less pronounced and of more or less pronounced in their function as signs. There are

- icons: they fulfil their function as signs due to a likeness between the sign and the object it signifies. The map or the picture are good examples;
- indexes: they fulfil their function due to the fact that they are in physical contact with the object, or that the sign is really affected by or dependent on the object, they are in a direct time- and space relation to their object. The thermometer, the flag, 'my hometown', the road sign, these are all examples of indexes, or signs where the indexical quality is important;
- symbols: these are signs who is given a function pr convention (they are also time- and space-independent).

If we focus on the relation between the sign and the interpretant, Peirce operates with several categories, among those 'Qualisign'. Qualisigns are signs where 'quality' (as opposed to convention, law, causality and so on) is an important part of the sign function.

An example is feelings, and how feelings in a specific sign-use-situation are a sign for us. This is important in landscape discussions. In Peirce's classification, the landscape in many practical situations functions as an indexical qualisign. For some, being 'there' and being overwhelmed by the feeling it arouses, is the sign quality which is *sine qua non* – it is identity experienced – it is a practical effect indeed.

Peirce offered a maxim for determining the meaning of a sign:

> Consider what effects, that might conceivably have practical bearings, we conceive the object of our conception to have. Then our conception of these effects is the whole of our conception of the object. (CP 5.402)

This is a meaning principle, and not a demarcation criterion separating the meaningful from the meaningless. It is a method for ascertaining meaning. The maxim reaches beyond the actual practical situation, and allows for the flight of imagination provided this ultimately alights upon possible practical effects (CP 5.196). The important thing here is the internal relation between sign use, cognition and action.

Because landscapes are a product of cognition specific to a sign user community, they become a most general and publicly accessible and shareable aides-memoire of a culture's knowledge and understanding of its past and future (Küchler 1993). In this sense landscape enters politics. As such it is the indexical and qualisign aspects of the landscape as a sign which is important. Because of this, landscapes, and memory of landscapes lost, may serve as an important focus for political organization around the issue of territory (and not around the space-independent class interests for example), and this has been witnessed many times in the course of history. A landscape is not something external to groups in society. It is part of society, of the matrix of action and possible action in social groups, and part of identity.

Conclusion

In the semiotic approach 'landscape' is a sign for an object (a piece of land) for some interpretant (community) in some respect or capacity. Four methodological guiding principles in the study of landscape discourses are argued for:

1. About grasping the essence of a landscape: to grasp the essence of a landscape means to somehow take part in the life-form/life-style/or simply activity and way-of-doing-things that makes up the core of the interpretants' projects or program. Landscapes are something doable.
2. About multiple landscapes in the same area: the one and the same (even congruent if you will) piece of land may contain many co-existent – even congruent – landscapes, related or unrelated.
3. About change in landscapes: the object of a sign is one thing, its meaning another. Whether or not material change is understood as landscape change is dependent on how the material change affects the landscape, which reflects the effects on an actual change ('conceivable practical effect') on a given interpretant (community). In the extreme case of the interpretant community disappearing or being unable to relate to the physical unit as before, the landscape disappears – even without any material change in the area. If the interpretant community changes, the landscape changes.
4. As a sign, a landscape is an abstraction relating to the interpretant community and its object. This allows for development of a landscape typology: as a sign, landscapes have abstract qualities which makes a basis for a landscape typology; signs that relates to indefinitely many discrete areas.

References

Apel, K.-O. 1973. *Das Apriori der Kommunikationsgemeinschaft und die Grundlagen der Ethik*. Berlin: Suhrkamp.

Arnesen, T. 1998. Landscapes Lost. *Landscape Research*, 23(1), 39-50

Blain, J. and Wallis, R.J. 2007. *Sacred Sites, Contested Rites/Rights*. Sussex: Academic Press.

Brown, G. and Yule, G. 1983. *Discourse Analysis.* Cambridge: Cambridge University Press.

CP x.y = *Collected Papers of Charles Sanders Peirce*, volume x, paragraph y. Peirce, C.S., *Collected Papers of Charles Sanders Peirce*, Vols. 1-6, 1931-1935, edited by Charles Hartshorne and Paul Weiss; Vols. 7-8, 1958, edited by Arthur W. Burks. Cambridge, MA: Harvard University Press.

de Botton, A. 2002. *The art of travel*. London: Penguin Books.

ETS no. 176 – European Landscape Convention was adopted by the Committee of Ministers in July 2000 and came into force 1 March 2004.

Hall, C.M. and Müller, D.K. (eds) 2004. *Tourism, Mobility and Second Homes. Between Elite Landscapes and Common Ground.* Toronto: Channel View Publications.

Hartshorne, R. 1939. *The Nature of Geography. A critical survey of current thought in light of the past.* Lancaster, PA: AAG.

Henderson, N. 1994. Replicative Dog Travois Travel on the Northern Plains. *Plains Anthropologist*, 39-148, 145-159.

Henderson, N. 1996. The Canoe as Failure on the Canadian Plains. *Great Plains Research*, 6, 3-23.

Krogh, E. 1995. *The phenomenology of landscape*. Aas: Norges Landbrukshoegskole, Norway.

Kuchler, S. 1993. Landscape as Memory: The Mapping of Process and its Representation in a Melanesian Society, in *Landscape, Politics and Perspectives*, edited by B. Bender. Oxford: Berg Publishers, 85-105.

Lefebvre, H. 1991. *The production of Space*. Cambridge, MA: Blackwell.

Meinig, D.W. 1979. The Beholding Eye: Ten Versions of the Same Scene, in *The Interpretation of Ordinary Landscapes: Geographical Essays*, edited by D.W. Meinig. New York: Oxford University Press, 33-48.

Mitchell, W.J.T. (ed.) 1994. *Landscape and Power*. Chicago: The University of Chicago Press.

Olwig, K.R. 1969. Recovering the Substantive Nature of Landscape. *Annals of the Association of American Geographers*, 86(4), 630-653.

Pitte, J.-R. 1983. *Histoire du paysage français.* Paris: Tallandier.

Sevenant, M. and Antrop, M. 2007. Assessing land use intensity and dynamics in a rural Mediterranen landscape: Lassithi plateau in Crete, in *European Landscapes and Lifestyles. The Mediterran and Beyond*, edited by Z. Roca, T. Spek, T. Terkenli, T. Plieninger and F. Höchli. Lisboa: Edições Universitárias Lusofonas, 175-197.

Syse, K.V.L. 2007. Moving about: An ethnographic approach to landscape research, in *European Landscapes and Lifestyles. The Mediterran and Beyond*, edited by Z. Roca, T. Spek, T. Terkenli, T. Plieninger and F. Höchli. Lisboa: Edições Universitárias Lusofonas, 323-330.

Urry J. 1990. *The tourist gaze: Leisure and travel in contemporary societies.* London: Sage Publications.

Vallega, A. 2007. The landscape: Places and Cultures, in *European Landscapes and Lifestyles. The Mediterranean and Beyond*, edited by Z. Roca, T. Spek, T. Terkenli, T. Plieninger and F. Höchli. Lisboa: Edições Universitárias Lusofonas, 57-70.

Chapter 25

'Participation' between a Must and Negligence: Paradigmatic Resistance to Landscape Assessment of Wind Industry in Gaspésie (Québec) and Finistère (France)

Marie-José Fortin, Sophie Le Floch and Anne-Sophie Devanne

Applying the European Landscape Convention: Methodological or paradigmatic problem?

The European Landscape Convention is ambitious. First of all, it subscribes to a cultural conception of landscape. The adopted definition indeed sets landscape as 'an area, as perceived by people'. It is important to note that this approach does not deny the material aspects of landscape – as indicated in the continuation of the definition[1] – but it implies that interventions in landscapes are considered as being influenced by actors' perceptions and representations. According to this approach, the perceptions related to a same landscape can vary from a culture to another, from a social group to another, from an individual to another. The Convention thus emphasizes the need to recognize this variety in the characterization and assessment of landscapes and in doing so it calls for reviewing the usual methods and tools which mainly rely on the knowledge and the perceptions of experts. In other words, interventions in landscapes must consider other social groups' knowledge and perceptions. This recognition of public actors is normally understood as requiring the implication of the local populations in so-called 'participatory' planning processes.

Under these two aspects, landscape as a socio-cultural construction and as a place of a large social participation in planning processes, the Convention underlies radical changes in the practices (especially of evaluation) that have been traditionally associated with landscape change. To implement these changes, the Convention addresses great challenges, as underlined by Jones (2007) and Olwig (2007).

The difficulties are often considered to ensue from problems of methods, weaknesses in processes and existing institutions, lack of experience and so on. In

---

1 '[…] whose character is the result of action and interaction of natural and/or human factors […]' (CEP 2000).

our opinion, however, the current problems do not rest so much on methodological, but rather on paradigmatic terms. In other words, they touch values and beliefs which structure thoughts, practices and methods in landscape change. Moreover, the Convention fails to integrate a fundamental, yet neglected, dimension of landscape: its political dimension. Landscape is at the centre of contemporary social relations and, as such, is an object of struggle and power, as expressed by Mitchell (2003) or Howitt and Suchett-Pearson (2003), among others. In our opinion, there lies the biggest change brought about by the Convention, the one that creates so many tensions in its implementation and that invites us to qualify it as a paradigmatic shift.[2] This is, in fact, why it is quite difficult to implement participatory planning.

This idea is discussed in this chapter on the basis of two case-studies, one in the Gaspésie region, located in the eastern part of the Canadian province of Québec,[3] and the other in the Finistère area, in the French region of Bretagne. In Québec, practices and original mechanisms of participation were developed for the last 30 years within the framework of large infrastructures and development projects, such as the *Bureau d'Audiences Publiques sur l'Environnement* (BAPE) (Office of Public Hearings on the Environment). However, public participation is still applied in a limited way to the question of landscape change. In France, by contrast, public participation emerged only in the 1990s and gave rise to a vast array of experiences that are still at experimental stages (Mermet and Berlan-Darqué 2009). The current debates around projects of wind farms could create a favourable context to rethink these practices.

Wind energy public policies: Regional development and energy

Rationales to develop wind energy are different in the two case studies, as well as in their effects on wind farms' locations. For instance, an original aspect of wind energy development in Québec is that it meets national objectives of power production with those of regional economic development of the Gaspésie region. These objectives are set in several public policies, such as the 2006 National Energy Strategy, the 2002 Accord agreement adopted in Québec recognizing the wind energy sector as an excellent niche for Gaspésie, and, especially, the Hydro-Québec's tendering system. This system was to favour the construction of large wind parks located in the Gaspésie area by grouping 30, 70, even 150 machines. The idea was that such a concentration of wind farms within a delimited territory would create a market which would be so attractive to foreign manufacturers of wind equipments that they would build factories in the region. The latter would

---

2   After territorial and cultural paradigms, it has to do with the political one. For a detailed description of this emergent paradigm, see Fortin (2005: Chapter 2).

3   The large majority of the 7.5 million residents settled in the southern part of a territory of more than 1.5 million square kilometres.

set up the basis of a new industrial fabric in Gaspésie, thus contributing to the diversification of its fragile economy, and also to the development of a new Québec expertise in that domain anchored in the region. In France, the national policy also combines energy and industry objectives but the latter are less prominent and have no geographical requirements. This energy policy was later reinforced by the liberalization of the electricity market. Numerous small wind farms (groups of 3-8 machines in Finistère) are scattered all over the country.

In the two countries, a tendering system was conceived at first. In Québec, in 2003 and again in 2005, promoters were invited to submit projects adding up 1000 MW and 2000 MW respectively. In addition to the price proposed to Hydro-Québec for purchasing the electricity produced, they had to meet certain requirements such as assuring minimum economic benefits to the Gaspésie area, 'social acceptability' through favourable recommendation from locally elected representatives, and the respect for existing planning rules. In France, the 'Eole 2005 programme' was launched in 1996 with the target of 250 to 500 MW to be installed in 2005. The criteria were the price of electricity, respect for the environment, the economic and industrial importance of the project, the opinion of the locally elected and geographical location. The last element was of great concern because dispersed wind farms appeared to facilitate the integration of the produced electricity in the existing network and allowed to avoid the construction of alternative energy plants in areas of high wind production which is, by nature, inadmissible. In 2000, the policy was revised in favour of a fixed price system in the context of the liberalization of the energy sector encouraged by the European Commission. The production then rose from 11.4 MW at the end of 1999 to 3500 MW in 2009. The wind energy development has been almost entirely entrusted to private developers in a decentralized perspective which breaks away from the dominant traditional perspective. The increase in the number of windmills preceded the elaboration of specific regulation tools, especially concerning land management. This resulted in a spatially scattered great number of small wind farms.

The rise of critics and social disputes

In both countries, the wind energy sector is developed in a similar manner. The settlement of wind farms is delegated to private operators, with specific criteria, which is quite a new situation in both cases. They differ though on two fundamental aspects: the scope of the wind farms planned and the economic spill-off, which is a main concern in Québec. Despite these differences, social contestations occured in both regions during the 2000s, as windmills were installed. The landscape effects became a matter of great local concern, together with other issues such as economic and governance concerns.

In eastern Québec, the concentration effect intended for economic development objectives was seen by some as creating other cumulative impacts considered negative for the quality of life. An important question was raised: what will be the effects on the quality of the landscapes and the residents' living environment, on the regional identity, on the tourists' experience and the territory's attractiveness? Criticism was particularly strong about projects planned in highly valued landscapes, such as those as along the Saint-Lawrence river. In the Finistère area, a key argument involved the impacts of scattering windmills on landscapes (called '*mitage du paysage*'). This scattering was considered particularly negative in those local landscapes that are already characterized by a diversity of components and by their narrow scale. As in Québec, this argument relates to the issue of the quality of life and was often associated with a claim for public participation.

Even if the valorization of wind power in Québec differs from that in France, wind farms projects have been similarly criticized for landscape matters. The development of the wind energy sector is definitely not simple. The link between energy, economy and territory presents important challenges. The 'social acceptability' and participation of the communities are increasingly seen as 'conditions' for the development of wind farms projects and the feasibility of this renewable energy sector.

New participative planning process proposed but not applied

Confronted with the situation, regional development stakeholders rallied in their organizations in Québec and in France. They grabbed the landscape issue in order to design some decision-making tools: guidelines, management plans, planning processes, etc. A common ground in these tools is that they assert the need to broaden social participation.

In Québec, five processes were proposed that insist in various degrees on the necessary consideration of landscape quality in wind farms planning. They come from two ministries and a regional environmental organization (MRNF 2004, 2005; MAMR 2007a, 2007b; CRE 2006). The documents review has shown considerable differences between preferred landscape approaches. Although most refer to a 'cultural' concept of the landscape (namely, values, meanings), it is rarely translated in the subsequent methods that are always dominated by expert knowledge. One initiative only, that of the ministry responsible for territorial planning (MAMR 2007a), is breaking away from traditional practices and could be associated with the cultural and political approach adopted in the European Landscape Convention. Its main objective is to provide a 'method' which allows planners and decision-makers to determine the degree of 'compatibility' between landscapes and wind farms. To do so, the participation of a wider number of groups of actors other than experts constitutes a key principle in the planning process. Several instruments are suggested as well.

The planning process was announced in the spring 2007, during hot debates in Eastern Québec. It was considered a means to 'manage the social stake' resulting from landscape change and to 'ensure the social acceptability of the projects of wind farms' (interview, 30 November 2008). Nevertheless, this planning process has not been applied since. How to explain such little interest in the proposed process? We could obviously evoke the relative 'novelty' of the document, or the fast pace of wind energy development, as well as the too advanced stage of the existing projects, the lack of human or financial resources and the lack of expertise and 'participation' experience in landscape-related concerns. However, as certain facts suggest, there is a more classic explanation. As a matter of fact, several actions were taken to meet the social demand for landscape quality, to influence the projects and the course of the events. Some actions required important human and financial resources for environmental protection,[4] but none implied any form of participation of the local populations, or of the affected groups. These actions always involved only restricted groups in semi-private discussions and strongly relied on classic methods and experts' knowledge.

In Finistère, the importance of taking into account the 'social representations of landscapes' was acknowledged by public actors right from the beginning of wind energy development. Many of them, along with landscape architects, local representatives, and so on, had consulted each other in the process of elaborating a local wind farms plan. The '*Charte départementale des éoliennes du Finistère*' was the first tool of this kind adopted in France (Préfecture du Finistère 2002) and the Regional wind energy plan was adopted at the end of 2006 (Conseil régional de Bretagne 2006). A new category of landscape has emerged during the process, that of 'emblematic landscapes'. These were defined as landscapes playing a major role in the Finistère identity: beside preserved landscapes, there are landscapes depicted by artists, or valued by tourists. The concept of 'social demands towards landscapes' is thus assimilated to 'benchmark' specific landscapes. This new category serves as basis for the elaboration of a zoning planned to protect these highly valued landscapes. However, as soon as it was defined, it started to be highly contested and, in practice, some windmill plans were installed in protected zones, labelled 'emblematic landscapes'.

Some actors, among whom those who actively worked on designing the category of 'emblematic landscapes', underlined its limits: social demand was insufficiently taken into account. Beyond the 'emblematic landscapes', there are many other landscapes that remain unqualified in the decision-making tools, such as the landscapes of everyday life for vast parts of the population. Finally, it is the category of 'ordinary landscapes', which implicitly lies in the planning documents, which appeared to be important. However, defining it was quite a

---

4   For example, a major action was the characterization and assessment study of regional landscapes assigned to *Ruralys* by the elected representatives of the Lower St. Lawrence region. In order to conduct this ambitious study within 12 months, *Ruralys* managed to collect a quarter of a million Canadian dollars (about 155,000 Euros), which, to our knowledge, represents the highest amount ever dedicated to this type of studies in Québec. This mobilization of financial resources is all the more significant in the context of the economic fragility of this region.

challenge. First, it required considering visual, economic, moral and utilitarian issues altogether; second, it forced planners to go beyond the zoning intentions. As a local representative put it, 'emblematic landscapes' are endued with exclusion, 'ordinary landscapes' with inclusion (interview, 27 June 2006).

If there is a consensus in Finistère about the need to develop public participation, as in Gaspésie, there has been little implementation of any large-scale participation. Some of the first wind farm plans involved local inhabitants, but these were restricted groups, at very local scale, led by professional consultants who share the promoters' interests. The public inquiry device, which is quite less extensive and lasting than the public hearings conducted by the BAPE in Québec, has often been considered as little credible and no one perceives any consequences on the land management process (interviews, 3 November 2006, 29 May 2008). Of course, it has become common to organize public meetings anywhere a wind farm plan arises, but, despite the repeated claim expressed by stakeholders and public actors to explain that it is necessary to organize them upstream from the plans themselves, public meetings do not represent a determining factor for public participation for promoters, nor for public actors. As a matter of fact, we noted in the interviews (18 September 2006, 22 June 2006) that they mainly conceived public meetings as mere information/communication operations.

Thus, it is necessary to look for other types of reasons to explain hesitations to invest in participative planning process regarding landscape change. In our opinion, these reasons have to do more with values and beliefs that structure practices in landscape and development.

Area of cognitive resistance:
Example from Gaspésie with two representational systems of landscape

The introduction of the so-called 'participatory' methods aims at documenting different perceptions and, more importantly, stimulating social dialogue. Yet, talking about landscape is not a trivial matter. The discussion goes beyond the acknowledgement of 'beautiful' or 'ugly'. The exercise basically aims at trying to understand *why* it is perceived as beautiful, *why* it is perceived as ugly and for *whom*. In other words, beyond the explicit criteria from the experts ('absorption capacity' integration, etc.) and the discourse on shapes and colours, the question is to know on what experiences, rationalities, beliefs and representations are the appreciations of landscape based. Emphasizing the existing representational systems inevitably calls upon updating the differences that can become sources of tension, even conflicts between individuals or groups that support them. In this context, ways of doing things may be questioned, which may undermine the favourable conditions for certain social groups more than for others. In short, participatory approaches in landscape-related concerns require talking about social interactions and power. This idea can be illustrated with the Québec case study.

The difficulties encountered in the implementation of participatory procedures are particularly striking in this province, which has been experiencing them for several decades through the BAPE. From 1995 to 2007, eight public hearings were held in Gaspésie as part of the environmental assessment procedure. Such hearings allowed local groups to become aware of the large number of proposed wind farms and of their major size: over a period of 10 years, more than 21 projects could be built in the targeted territory. If all were realized, the majority of the thousand wind turbines – or 1,250 – would be located along the Saint-Lawrence river, a highly valued landscape where most people have their home (BAPE 2005, 2007).

According to our analyses, representations associated with the future development of the territory in question are a strong theme for judging the impacts on the landscape and the 'social acceptability' of wind farms projects. Two main representations of the territory appear to be at the heart of the conflict and we named them *productive landscape* and *inhabited landscape* (Fortin 2008). In the former, regions contain 'resources', especially natural ones. Their exploitation will boost regional economy and, in the broader sense, ensure their sustainable development. In the case of wind energy, it will therefore be important to find ways to use the wind resource. The wind industry is not seen as a new source of energy, but rather as a new industrial activity. In the latter representation, territory is primarily a living environment. Its quality is based on a set of considerations, including economic ones. For example, the desire to improve employment opportunities and living conditions, whether for oneself or for other community members, is present in virtually all critical discourses. We even notice that there is no opposition to the idea of developing a new industrial sector to stimulate the region's fragile economy. However, what the groups contesting wind farms stress is that the economic activity must not threaten the other attributes of the living environment, notably the social and cultural factors that affect lifestyles, practices related to the territory, social relations, identity and the sense of belonging, among others. Actually, that is precisely what was feared from the current projects. The protests focus on the implementation of the wind industry, i.e. introducing large wind farms (a fundamental element of the underlying tendering system equation), without taking into account the impacts caused on living environments.

Thus, two representational systems of the territory are at the heart of the conflict. Let us emphasize however that we cannot attribute these two social representations to specific social groups who would oppose, for example, local populations to governmental institutions. Our analyses actually show that if some elected representatives support the representation of the production territory, others favour the inhabited territory. Even the 'local people' do not form a homogeneous group. Some citizens feel marginalized in relation to other members of their community. That is why there is so much tension within communities.

What matters beyond this explanation is the fact that some representations are more supported by some governance instruments that can influence the tangible future of the territories. In Québec, several regulatory frameworks follow one another during the process of setting up wind farms. The one related to the tendering system firmly orients well upstream the design of major projects. This framework is largely based on economic criteria for judging the suitability of projects submitted to Hydro-Québec, such as the buyback rate of electricity, their profitability and compliance with regional spill off.[5] In doing so, it is more consistent with the representation of the productive landscape. The groups supporting the representation of the inhabited landscape must then enforce their concerns through other regulatory frameworks located further downstream, such as at public hearings scheduled by the environmental assessment procedure and using local planning tools, or during private negotiations with promoters. Some of their requests may then find an answer. For example, a wind turbine can be moved a few hundred meters into the forest to avoid the hiking trail and to preserve this valued practice. However, the exercise is more difficult when, for example, it has to do with ensuring that residents can always identify with their local landscape. Moving a hundred wind turbines 300 or 700 meters, as high as a 30-storey building, will change very little the presence of the mega farms in a landscape history. An alternative that would have more impact, that of reducing the number of wind turbines, is more difficult to maintain because, according to promoters, it could jeopardize the financial viability of their project established upstream during the call tenders.

In sum, the conflict about wind farms is partly structured around two representational systems of landscape: as an area of production and as a living environment. They affect the way the future of a territory is secured, as well as the place given to the economy in development strategies. In the former case, economic activity is the purpose, whereas in the latter it is compared to other social needs, without dominating them. The importance given to the landscape is also questionable in these representations: is landscape quality treated as a secondary need, even as a luxury item, in contemporary societies compared to emergencies affecting the economy? Or, on the contrary, is it considered vital in ensuring the sustainability of communities, because it meets important non-commercial needs,[6] like belonging and a sense of identity, which are just as important? In the latter case, landscape actually becomes a tangible expression of the relationships built with the territory of belonging, as well as with other community members (Le Floch and Fortin 2008).

---

5   Other social criteria were added in the second call for tenders in order to meet objections to the farm projects. Three points out of 100 have thus been attributed to 'social acceptability' in the project evaluation grid of Hydro-Québec. However, more than 84 points have had an economic underpinning.

6   It is worth noting that the landscape may be perceived as being in danger even within a market approach. This is particularly true for some tourism stakeholders and citizens, who, at public hearings, have expressed concerns about the impacts on landscape quality, the main 'resource' of the tourism industry, and on local economy.

Conclusion: Toward a relational aesthetic?

The European Landscape Convention aims at promoting new approaches in landscape analysis that involve broader social participation, such as that of local populations. These approaches seem difficult to implement. In our opinion, these difficulties are not manifested in methodological but, rather, in paradigmatic terms, since such approaches include the political aspect of a landscape-related dimension that is currently under (re)definition.

More specifically, participatory approaches strengthen social dialogue. They bring out different perceptions of the landscape that can become an area of confrontation among visions and different expectations related to the same territory and, implicitly, between social groups. Moreover, they reveal power relations. In fact, if the results of landscape analyses must be used to guide decisions regarding the future of a territory in its materiality, the question then remains of detecting which vision will be maintained at the expense of which other one. In other words, which actor will succeed in asserting its physical and symbolic aspirations on a given territory? Participatory approaches promote definition of social representations that guide the actors' behaviour and, most importantly, identify those representations that are materialized in development projects causing landscape change.

Furthermore, the landscape-related discourse spills over into other serious issues, which not only fundamentally affect the relationships between communities and their territory, but also the relationships amongst community members. This issue was particularly evident in conflicts around the wind farms studied in Québec and Bretagne. The discourses of opponents were marked by the feeling of being excluded: excluded from the governance process and even from their community for proposing different visions of their future – at least different from those expressed in the regulatory frameworks recorded upstream in the project implementation process. In other words, landscape quality and change reveal not only the relationships between communities and their territory, but also among members of a community, both the harmonious and the conflicting ones, including exclusion.

In this perspective, the question of landscape aesthetics does not focus so much on the register of shapes and visibility as on the complex issue of social relations. The introduction of such a 'relation' aesthetic (Le Floch and Fortin 2008) constitutes, in our opinion, a paradigmatic turn with regard to traditional practices because it recognizes the political dimension of landscape.[7] It is no wonder then that participatory approaches driven by the European Landscape Convention encounter difficulties.

Acknowledgements

This text is based on studies carried out in the framework of a research project funded by the research program 'Landscape and Sustainable Development' of MEEDDAT (French Ministry in charge of Energy, Environment, Sustainable development and Land Management), as well as by a 2008-2011 programme funded by the Social Sciences and Humanities Research Council of Canada (SSHRC). We thank the MEEDDAT and the SSHRC for their support, as well as Professor Danielle Lafontaine for her sharp comments on the first version of this text.

References

Bureau d'audiences publiques sur l'environnement (BAPE) 2005. *Projets de parcs éoliens à Baie-des-Sables et à L'Anse-à-Valleau. Rapport d'enquête et d'audience publique no. 217.* Québec: Bureau d'audiences publiques sur l'environnement.

Bureau d'audiences publiques sur l'environnement (BAPE) 2007. *Projet d'aménagement d'un parc éolien à Carleton-sur-Mer. Rapport d'enquête et d'audience publique no. 238.* Québec: Bureau d'audiences publiques sur l'environnement.

Conseil de l'Europe 2000. *Convention Européenne du Paysage* 20/04/2001, 2001. [Online] (http://www.nature.coe.int/french/main/paysage/conv.htm).

Conseil régional de Bretagne 2006. *Guide de l'éolien en Bretagne.* [Online] Available at http://www.bretagne.fr/internet/upload/docs/application/pdf/2009-02/int-guide-eolien.pdf [accessed 15 February 2010].

---

7   We discussed this issue elsewhere: 'Aesthetic quality seems less and less like an attribute of an object – the wind turbine – and more and more like a characteristic of social relationships, with the understanding that they are also spatial. It not only occurs at the level of intending to shape the landscape material, but also at the level of intending to build social relationships from which will emerge new spatial shapes. There is a shift from the idea that the future materiality can be built according to a project and that social relationships are based on physical relationships toward the idea that the future continuously emerges, shaped by the interaction between work of the imagination and the shaping of material; the creation of material objects and the building designs of these objects are processes that generate themselves (Healey 2004)' (Le Floch and Fortin 2008: 230, own translation).

Conseil régional de l'Environnement du Bas-Saint-Laurent 2007. *La filière éolienne au Bas-Saint-Laurent: Un outil d'aide à la prise de décision dans le contexte municipal*. Rimouski: CRE du Bas-Saint-Laurent.

Fortin, M.-J. 2005. *Paysage industriel: Lieu de médiation sociale et enjeu de développement durable et de justice environnementale. Les cas des complexes d'Alcan (Alma, Québec) et de Péchiney (Dunkerque, France)*. Doctoral thesis in Regional development, UQAC. [Online] Available at http://theses.uqac.ca/these_24605668.html].

Fortin, M.-J. 2008. Landscape, an interpretative framework for a reflexive society, in *Landscapes: From knowledge to action*, edited by M. Berlan-Darque, Y. Luginbühl and D. Terrasson. Versailles: Quae, 17-27.

Fortin, M.-J. 2008. Paysage et développement: Du territoire de production au territoire habité, in *Sciences des territoires: Perspectives québécoises*, edited by G. Massicotte. Québec: Presses de l'Université du Québec, 55-76.

Howitt, R. and Suchet-Pearson, S. 2003. Ontological Pluralism in Contested Cultural Landscapes, in *Handbook of cultural geography*, edited by K. Anderson, M. Domosh, S. Pile and N. Thrift. London/Thousand Oaks: Sage, 557-569.

Jones, M. 2007. The European Landscape Convention and the question of public participation. *Landscape Research*, 32(5), 613-633.

Le Floch, S. and Fortin, M.-J. 2008. Paysage, co-visibilité et esthétique autour de l'implantation d'éoliennes, in *L'après développement durable. Espaces, Nature, Culture et Qualité*, edited by A. Da Lage et al. Paris: Ellipses, 223-231.

Mermet, L. and Berlan-Darqué, M. (eds) 2009. *Environnement: Décider autrement. Nouvelles pratiques et nouveaux enjeux de la concertation*. Paris: L'Harmattan.

Ministère des Affaires municipales et des régions (MAMR) 2007a. *Guide d'intégration des éoliennes au territoire, vers de nouveaux paysages*. Québec: Gouvernement du Québec [http://www.mamr.gouv.qc.ca].

Ministère des Affaires municipales et des régions (MAMR) 2007b. *Les orientations du gouvernement en matière d'aménagement. Pour un développement durable de l'énergie éolienne*. Québec: Gouvernement du Québec [http://www.mamr.gouv.qc.ca].

Ministère des Ressources naturelles et de la Faune (MRNF) 2004. *Plan régional de développement du territoire public- Volet éolien: Gaspésie et MRC de Matane*. Québec: MRNF, Direction générale de la gestion du territoire public du Bas-Saint-Laurent-Gaspésie-Îles-de-la-Madeleine.

Ministère des Ressources naturelles et de la Faune (MRNF) 2005. *Guide pour la réalisation d'une étude d'intégration et d'harmonisation paysagères. Projet d'implantation de parc éolien sur le territoire public*. Québec: Gouvernement du Québec [http://www.mrnf.gouv.qc.ca].

Mitchell, D. 2003. Dead Labor and the Political Economy of Landscape – California Living, California Dying, in *Handbook of Cultural geography*, edited by K. Anderson, M. Domosh, S. Pile and N. Thrift. London: Sage, 233-248.

Olwig, K. 2007. The practice of landscape 'Conventions' and the just landscape: The case of the European Landscape Convention. *Landscape Research*, 32(5), 579-594.

Préfecture du Finistère. 2002. *Charte départementale des éoliennes du Finistère.*

Chapter 26

New Landscape Concerns in the Development of Renewable Energy Projects in South-West Spain

Marina Frolova and Belén Pérez Pérez

Introduction

The landscapes 'emerging' from the development of renewable energy have become an important point of discussion in many European countries, since the European Union initiated a process of energy policy reform in order to reduce greenhouse gas emissions within the context of climate change. Renewable energy should account for 20 per cent of global electricity generation of the EU in 2020.

Due to its ambitious policies, Spain has achieved a very successful implementation of renewable power. The Spanish Renewable Energies Plan 2005-2010 seeks to increase electricity generation based on renewables by 102,259 GWh in 2010, by which time they will supply over 21 per cent of the national electricity production. Wind power installed potential reached 19,149 MW at the end of 2009. This sector has developed very quickly in Spain. Installed capacity of wind power in Spain doubled from 2003 to 2006, and it is expected to double again by 2010, with a total installed potential of 20,155 MW (AEE 2009).

The environmental impacts of wind power infrastructures are far less dramatic than those produced by other power infrastructures and are mostly limited to the perceived impact on landscape (Pasqualetti et al. 2002; Burrall 2004) and land use conflicts, as well as the problems of noise pollution and hazards to birds (Wolsink 2000). However, wind energy plants have brought some new aspects to land use policy, in relation to the smaller scale and greater dispersion of renewable energy plants, which increase the number of siting decisions that need to be taken. Generally the visual impact of wind generation tends to be higher because the best sites tend to be in scenic mountain and seaside areas. In addition, as renewable energy conversion is characterized by lower energy densities, the relative visual impact (per MWh of output) is usually higher. Another important issue is the tendency of renewable energy conversion to happen closer to where the energy consumer lives, which makes it more visible and brings environmental impact closer to residents (Wolsink 2007; Wüstenhagen et al. 2007).

Therefore, as the use of wind turbines is spreading in Spain, cultural preferences for country landscape and seascape preservation have become an increasingly important factor in the perception of wind farms, which have even been described by some opponents as 'single-crop wind farming' or 'wind-farm monoculture'. Wind power has become an essential factor shaping present-day Spanish landscape, especially in Galicia, Castilla and Léon, Valencia, the Ebro valley and Cadiz, where the proliferation of wind farms has been greatest. Therefore the awareness of the landscape impact of wind power and the growing public participation of local stakeholders in the decision-making processes on energy infrastructures have in some cases become powerful barriers to the achievement of renewable energy targets. In this context, more democratic forms of land-use policies emerged during recent decades which clashed with energy planning policy in Spain, which until quite recently was hierarchical, authoritarian and functional. The UNECE Convention on Access to Information, Public Participation in Decision-making and Access to Justice in Environmental Matters (Aarhus Convention) (1998)[1] linked environmental rights and human rights and established that sustainable development can be achieved only through the involvement of all stakeholders. At the same time, the European Landscape Convention (ELC) (2000)[2] and some Spanish regional landscape laws derived from it[3] highlighted the importance of the citizen as an active player in an increasingly complex environment. In fact the new landscape legislation tries to encourage local and regional groups to take an interest in landscape questions in land-planning projects. At the same time, the laws urge the different stakeholders and users to create these projects together through participation and consensus, so as to be able to achieve a collective appropriation of a culture of projects by and for local people and stakeholders. From this moment on, a landscape which had hitherto earned only a marginal position in the documents relating to the implementation of renewables in Spain is a subject of increasing interest.

This chapter explores the institutional and social processes through which renewable landscapes have emerged to form part of Spanish energy policy. We examine these processes on different political and geographical scales, and compare the different courses taken at national and regional levels. Up until very recently this landscape reflected the interests of central state authority, civil servants and electric companies rather than the interests of the population as a whole or of the local people directly affected by landscape changes. The divergence between landscape policy and the policy of renewable power development in Spain is analysed by comparing the implementation of renewable energies in the national, regional and local contexts. The pilot study focuses on

---

1   The Aarhus Convention was ratified by Spain in 2004 and came into force in 2005.

2   Spain signed the ELC in 2000, ratified it in 2007 and it came into force in 2008.

3   Law for Land-use Planning and Landscape Protection of the Valencia Autonomous Region (4/2004), Law for Landscape Protection, Management and Planning of the Catalonia Autonomous Region (8/2005), Decree for the Landscape Regulation of the Valencia Autonomous Region (120/2006) and Law for Landscape Protection of the Galicia Autonomous Region (7/2008).

the development of wind power in the Cadiz area of the Autonomous Region of Andalusia (south-west Spain).

The research presented in this chapter is exploratory in nature and based on textual analyses of a range of sources at different geographical levels: energy and environmental policy documents for the whole of Spain, for the Autonomous Region of Andalusia, special local plans for wind resources management and projects of renewable energy development in the province of Cadiz, and discourses of nature protection organizations, anti-windpower groups and academic groups.

The implementation context

The Spanish energy policy clearly defined the objectives, conditions and instruments of financial procurement which contributed to the successful deployment of wind power. The rapid expansion of renewables in Spain owes a great deal to a series of laws and national decrees, which provided a favourable framework for renewables implementation. However, the spatial deployment of renewable energy policies depends not only on its contents but also on the social contexts in which it is developed: the degree of stakeholder participation, the presence or absence of debates or conflicts, institutional and legal conditions governing the decision-making process and its monitoring etc. Therefore the successful renewable implementation policies and the public perception of wind farms are related to a range of complex cultural, contextual, socio-economic, political and physical factors (Ellis et al. 2007). Of these the most important are considered to be the geographical potential, the planning regime, the financial support systems which vary in their effectiveness over time, values attached to landscape quality and preservation, and the degree of public participation and of local ownership of schemes to build wind farms (Toke et al. 2008; Wolsink 2007).

Geographical potential

Spain's particular geographical conditions make it an ideal place for the development of renewables. There are numerous suitable sites for wind power generation in the country with extensive mountain and seaside areas and relatively low population density (compared with the European average). The average population density of about 90 people per square kilometre is far less than those of other European countries like the Netherlands (about 400/km^2), Belgium (about 350/km^2) or the United Kingdom (about 250/km^2) (2009),[4] where the renewable energy infrastructures situated too close to towns and villages generate important conflicts on land use.

---

4   See United Nation World Population Prospect: The 2008 Revision Population Database, available at http://esa.un.org/unppl [accessed 8 May 2009].

The recent evolution of Spain's planning regime

The application of wind energy in Spain is a central government policy, but changing a zoning scheme is a regional political decision. Until quite recently the prevailing model of spatial planning in relation to renewables development was hierarchical, authoritarian and functional, as for example in the case of the traditional policy of hydropower deployment. This policy was based on the technocratic management of natural resources irrespective of any physical constraints, in line with the prevailing general economic interests and the specific interests of certain industrial sectors (building companies, dam and irrigation equipment suppliers, electricity sector, etc.) and responded to the corporate pressure brought to bear (Frolova 2010).

Another area that affected the development of renewables in Spain was land use policy. Before 1978 land use policy in Spain was centralized. After the dictator Franco died in 1975, the country underwent a dual economic, political and social transition to democracy and decentralization. The Spanish Constitution of 1978 established a system of decentralized government and shared power between the central and regional governments or autonomous regions. According to the Constitution of 1978, powers which are not considered to be the exclusive domain of the central government can be assumed by regional autonomous governments, if they are established in their Statute of Autonomy. Among these powers, the Constitution names: land-use planning and related policies as well as environmental management, although the establishment of basic environmental protection laws is the State power. Regarding the installation of electricity production, distribution and transport systems, provided that the electricity is not transported outside the region and its use does not affect other provinces or autonomous regions, it can be the domain of the regional government if it is previously established by its Statute of Autonomy. Some of Spain's Autonomous Regions like Catalonia, the Basque Country, Galicia and Andalusia were granted greater powers than other regions due to their special status as 'historical nationalities'. Between 1983 and 2001 all the autonomous regions passed their own laws on land use.

The financial support systems

The development of renewables in Spain has also been accelerated by a strong financial support system, as happened in Denmark and Germany (Toke et al. 2008). The payments for electricity generated by wind farms in Spain are based on a feed-in scheme (IEA 2008). As early as 1997, a payment and support mechanism was enshrined by the parliament through the Law 54/97 on energy policy. According to this law, producers of renewable energy were entitled to receive remuneration for transferring the power to the system through the distribution or transmission grid. Royal Decree 436/2004 has been especially beneficial for wind energy development in Spain, as it guaranteed feed-in tariffs. Still, as a reaction to the increasing price of electricity in Spain, Spanish authorities passed Royal Decree

661/2007 on renewable energy sources, and wind energy, in particular, in order to regulate the price received by wind-farm operators.

Public participation

Increased public collaboration and participation in planning processes is a characteristic feature of many European countries (Wolsink 2007; Healey 2006; Holden 1998). Although this process has also occurred to some extent in Spain, due to the influence of general European trends, and in particular through the adoption of the Aarhus Convention (1998) and the ELC (2000), the tendency towards a top-down, technocratic, hierarchical way of thinking about how the planning system has to be shaped, inherited from the pre-1978 period of centralized policies persists in Spain, as in a number of other countries (Wolsink 2003; Cowell and Owens 2006; Wüstenhagen et al. 2007). Therefore national policy has been keeping grassroots initiatives at a distance in the formal decision-making process on Spanish renewable planning. In addition, the institutionalized power of energy companies up until quite recently has not allowed these groups to influence the spread of wind power (Frolova and Pérez 2008). The siting of wind power schemes in Spain has not been decided at a local level, which creates an important contradiction between the intentions of Spanish landscape policy and the actual mechanics of decision-making processes. Finally, the generalized practice all over Europe to see public engagement as a one-way process the end results of which are determined in advance (Ellis et al. 2007) has also contributed to the very limited role given to the opinion of local stakeholders and of nature protection organizations in the formal decision-making process. This tendency in renewable deployment policy has been changing during the last decade, especially in the case of hydropower, due to the growing influence of public participation (Frolova 2010).

Perceptions of renewable energy

A body of work (Ellis et al. 2007; Devine-Wright and Devine-Wight 2006; Haggett and Toke 2005; Haggett and Smith 2004; Woods 2003) shows how positions of support and objection are not constructed just from a lack of awareness of the benefits of wind power, scepticism of the technology or the location of specific proposals, but also reflect deeper values, wider cultural and institutional contexts and claims over objectivity and truth.

General public support for wind energy technologies is high in Spain: about 67 per cent are in favour of the use of wind energy, with opponents amounting to only 2 per cent of the people questioned (Eurobarometer 2007). The supply of acceptable spaces for wind energy infrastructures is generally thought to be plentiful. As in other countries, public opinion perceives renewable energy technology as benign and beneficial in the context of global warming (Cowell 2007).

Although general attitudes towards wind power are very positive, attitudes toward wind farms can be completely different (Wolsink 2007; Bell et al. 2005). Resistance to wind farms is growing with particular intensity in the areas where they have been installed in high density, such as Galicia, Castilla and Léon, the Ebro valley and Cadiz.

The Spanish anti-wind power initiatives came to the fore at the end of the 1990s when 'The first national meeting to protect landscape from wind energy infrastructures' (1999) took place.[5] At the same time, several regional organizations in favour of the 'rational implantation of wind farms' were organized all over Spain.

The opposition to these forms of energy is made up of a wide variety of social groups: some ecological organizations and NGOs; the people and towns affected that do not benefit from renewable energy installations; some scientists; urban dwellers who enjoy escaping to the countryside; people who run rural tourism businesses, etc. Their arguments vary a great deal and often go beyond aesthetic considerations, although at a local level countryside landscape or seascape values are sometimes a determining factor in public attitudes. Existing research confirms that the strongest impacts on attitudes to wind farm proposals is the projected aesthetic value of turbines and their perceived impact on landscape or negative effects on community identity (Wolsink 2007; Burrall 2004; Woods 2003, Mercer 2003; Pasqualetti et al. 2002), followed by concerns on noise pollution and hazards to birds (Wolsink 2000). Another important objection to wind power in Spain lies in the fact that it has very little effect on local economy, employment or demographics, or on the standard of living of local residents, as the projects require large tracts of land and normally negotiations are entered into with single owners in order to make it easier to buy the land.

Landscape values and landscape policy

It is commonly accepted that the most salient public concerns in considering the costs and benefits of a wind power scheme involve landscape value (Wolsink 2007). Strong and effective opposition to wind developments is always primarily rooted in landscape values (Toke et al. 2008). Is this thesis applicable to Spain?

Could we relate the raising of Spanish landscape awareness with the development of landscape policies? Somehow the absence of specific legislation on landscape management and conservation of landscape policy in Spain until the 2000s and of powerful landscape protection organizations rooted in socio-cultural traditions explain the fact that landscape concerns, in a social and cultural sense, have appeared late in Spain's discourses on renewable power landscapes. This factor influences the lack of strong and effective opposition to wind developments even in the autonomous regions which have landscape protection laws. Frequently, the most effective opposition to wind power in Spain occurs in areas where there

---

5   See http://waste.ideal.es/eolica.htm.

are important conflicts on land use or in relation to the incompatibility of wind-power with local business activities (tourism, fishing, etc.), and the attitudes of residents are not directly related to landscape values.

During the 1990s these new landscape concerns were sparked by a crisis of identity in rural Spain arising from the transformation, and often the complete disappearance, of many traditional ways of life. Various processes were active in bringing about this crisis including urban growth in many areas of Spain, and particularly on the coast, water policy, reforestation and intensive agriculture, which led to the deterritorialization and the destruction and uniformization of traditional rural landscapes. This crisis, which is both ecological and cultural, has contributed to the progressive raising of Spanish society's awareness on landscape questions (Frolova 2009).

New concerns for landscapes have entered energy policies in Spain also as a result of developments at the European level. The application of the ELC encouraged several autonomous regions to incorporate landscape as an important issue in land use regulation.

Emergence of landscape assessment in Spain's energy policy

In spite of essential changes in landscape policies, they are still out of step with the development of renewable energy policies. It has still not been possible to introduce landscape as a transversal element in Spanish energy policy either at national or regional level.

In Spain there is no specific regulation of landscape questions in legislation relating to industrial installations in general and power plant installations in particular, though landscape assessment has been gradually integrated into the Environmental Impact Assessment of industrial projects.

The 2006 Law on Strategic Environmental Assessment, which applied the Directive 2001/42/EC in Spain, established that Reports on Environmental Sustainability should contain possible landscape effects of plans.

Royal Legislative Decree 1/2008 approved the amended text of the Law on the Environmental Impact Assessment of projects and for the first time treated landscape as an important element to consider. As a consequence of this law, landscape should be evaluated systematically in Spain's land-use planning at project scale.

The case study: Wind farm development and regional responses in the Province of Cadiz (Andalusia, South-West Spain)

Infrastructures linked with renewables have become an essential element in shaping the landscapes of some regions and provinces of Spain. Such is the case of the province of Cadiz in the Autonomous Region of Andalusia.

Wind energy development in Andalusia and Cadiz

In 2007 Andalusia became the region in which the greatest capacity (in absolute terms) was installed in Spain, with 853 MW. Later in 2008 Andalusia underwent the largest growth (in percentages) of installed capacity in Spain of 140 per cent and in January of 2009 reached a total capacity of 1,794 MW (AEE 2009). Since 2007 Andalusia has implemented European and national renewable power targets through Law 2/2007 for the promotion of electricity produced from renewable energies and energy saving and efficiency, as well as through the Andalusia Plan for Energy Sustainability of 2007-2013 (PASENER). This Plan forecasts the installation of 4,800 MW by 2013, which will make the region one of the leaders in wind power deployment in Spain.

Inclusion of landscape issues in Andalusia's land use policy

Andalusia took an important step forward in the international development of landscape policies when it signed the Mediterranean Landscape Carte (the precursor of the ELC) in 1993 together with the regions of Toscana (Italy) and Languedoc-Rousillon (France). However, it was not until 2006 that the Andalusia Land-Use Plan included references to landscape assessment. In 2007 Andalusia's Statute of Autonomy recognized the right of Andalusians to enjoy their landscapes and their duty to use them responsibly. In the same year, Law 7/2007 on Integrated Management of Environmental Quality in Andalusia applied new norms for landscape assessment, established by Directive 2001/42/EC.

There are few local instances in which local landscape value considerations have been integrated into wind power planning. One of these examples is the Special Plan for wind resources management in the La Janda area of the Province of Cadiz.

Pioneer experiences in the Province of Cadiz

The case of the Province of Cadiz, strongly affected by renewable power infrastructures, is an interesting example for studying the conditions from which landscape concerns in relation to the development of renewable energy emerge.

Cadiz is one of the main provinces in implementing Andalusia's plan for wind power. Due to its almost constant exposure to winds from the Atlantic, Cadiz is extremely well suited to wind energy development. Cadiz became the leader of wind energy production in Andalusia, with a total of 63 wind farms (in 2010), which generate about 43 per cent of Andalusia's installed capacity. There has therefore been a leap from the 776 MW of power installed in 2007 to the 1,238 MW completed on 1 January 2010 (OAER 2010).

However, relatively well conserved natural landscapes especially abundant in the La Janda area of Cadiz and important wildlife resources, favoured by intensive bird migration through the Strait of Gibraltar, and a variety of sea species,[6] are being harmed by the proliferation of new wind power projects. In addition, there is a problem of lack of coherent planning of both on-shore and off-shore wind farms, which aggravates conflicts with regard to land use and to the fact that various business activities of the local population are incompatible with on-shore and off-shore installations (fishing, tourism).

Therefore in the Cadiz area there are places in which there is a strong current of opposition to the installation of new wind-energy projects. This opposition culminated in the creation of the Forum on off-shore wind energy and sustainable development in 2004, in which groups representing all the stakeholders in the conflict took part: ecological associations, town councils, development companies, scientists, fishermen etc.

All this has led Cadiz to become the first province in Andalusia in which town councils have raised the need for planning instruments to be established for wind-power resources that take environmental and landscape issues into account through wind energy plans at both municipal and higher levels (Diputación de Cádiz 2004). The Councils want to be able to participate in the processing of applications for wind-energy plants alongside the Regional Government of Andalusia, so that they can play an active role in the development and planning of wind energy resources in their municipal areas.

The Special Plan for wind resources management in the La Janda area (Diputación de Cádiz 2004) was one of the first plans of this type drawn up in Spain. La Janda is an area covering 3,560 km^2 on the Atlantic Coast of Andalusia between the Bay of Cadiz and the Bay of Algeciras, standing opposite the coast of Morocco. The gentle climate and almost unspoilt natural and historical landscape of its coastline has made this region very attractive for tourism.

Within the context of the lack of specific regulation for the spatial planning of wind power, the proliferation of projects for the installation of wind turbines and the growing social rejection of wind farms, the Special Plan for Wind Resources Management was drawn up in La Janda by eight town councils and the Cadiz Provincial Council. For the first time, this Plan has cited landscape criteria as determining factors in land-use planning. It included a map with a zoning system (Exclusion Areas, Areas with Limiting Factors, Areas without Specific Limiting Factors) based on their compatibility in environmental and landscape terms with wind power projects, which enable them to guarantee the conservation of pre-existing values and create a stable framework for decision-making.

However, in spite of the progress achieved through dialogue between the different stakeholders with regard to social, environmental and landscape impacts of some projects and the achievements made in the spatial planning of wind power in certain towns in Cadiz, its rapid deployment has led to even stronger social opposition. For instance, the plan to install twelve off-shore wind-parks in Cadiz by the year 2012 produced a wave of protests on the beaches near Cape

---

6   The Atlantic coast of La Janda is the largest fishing ground in the Iberian Peninsula.

Trafalgar. Many town councils and other bodies oppose the development of more projects even though these have been given favourable Environmental Impact Assessments.

The rapid development of land-based wind-parks in Cadiz has now been combined with the uncertainty surrounding possible off-shore wind-power projects. The Marine Wind-Power Map has recently been approved. This will serve as basis for the development of this industry in Spain, and will protect a strip of sea covering the first eight kilometres out from the coast, with the rest of the coastline divided according to environmental criteria into exclusion zones, zones with limitations and zones considered suitable for wind-power mills.

This map, which was published in April 2009, was drawn up by compiling a number of existing studies that had been presented up to December 2008. It contains a number of sectors or areas in which applications can be made for the development of wind-power projects. These projects must be accompanied by technical studies and an Environmental Impact Assessment and their acceptance or otherwise is in the hands of a committee made up of representatives of five ministries and a spokesperson for the particular region affected. This procedure does not have a planning instrument in which the cumulative and synergic effects of projects located in the same area can be taken into account, nor does it permit the participation of local or subregional stakeholders, and nor have any landscape criteria been taken into account in the drafting of the map as no studies of this kind existed. All of this is creating a strong social backlash against these kinds of project by different local groups and stakeholders who fear their interests may be harmed. They believe that the vibrations and the noise made by the turbines will affect the fishing grounds and the migration of birds, whales and red tuna; they claim that these parks are not compatible with underwater archaeological heritage, and warn that the turbines will be visible from the beach, will have negative effects on tourism by changing local identity, will alter the clarity of the water and the coastal dynamics, etc.

In this first stage of our research, the Cadiz wind power planning case demonstrates that the centralized government is left with important powers in the final decision and could impose planning views on local wind power development as occurs in a number of cases in countries with similar planning traditions (Nadaï 2007). The environmental issues at stake, combined with the perception of wind power infrastructures as being imposed by the central government and of little benefit to the local population of Cadiz, created a feeling of local identity, as has happened in other environmental disputes (Wolsink 2006; Dalby and Mackenzie 1997).

Conclusions

The institutional emergence of a 'renewable energy landscape' in Spain is underway in a context in which the European Union provides a framework for the development of landscape legislation and the implementation of alternative energies in different European regions. Spain's renewable power policy has been developing very rapidly and the implementation of renewable power has been carried out in favourable political conditions. The evolution of landscape policy however has been rather slow.

While in many other European countries a new sensitivity towards landscape questions has become one of the main barriers to the development of renewable energies (Wüstenhagen et al. 2007), in Spain relatively limited resistance to a massive expansion in renewable power made it largely unnecessary for government to direct local siting processes (Cowell 2007; Devine-Wright 2005). It could be explained by a specific context of spatial planning in this country, where the supply of acceptable spaces for certain category of renewable energy infrastructures is generally thought to be plentiful, where there is a strong financial support regime for wind power, the model of spatial planning related is rather authoritarian and functional, and grassroots initiatives in the formal decision-making process are kept at a distance. The late appearance of landscape policies in Spain also explains why landscape concerns, in a social and cultural sense, have emerged late in Spain's discourses on renewable power landscapes.

So far national policy objectives in wind power have been easily translated into regionally and locally accepted policies, but this has not always occurred with hydropower development due to social resistance. An important policy tendency towards top-down planning of large-scale development, which was the dominant trend in Spain until recently, has already become an important obstacle to the successful implementation of hydropower development in Spain (Frolova 2010). Taking into account the growing opposition to wind power projects, as has happened in the province of Cadiz, it is foreseeable that this trend could affect the implementation of wind power targets in Spain.

Acknowledgements

This chapter is part of the research project 'Landscape Policies in Spain: Landscapes, Water Problems and Sustainable Development', funded by the Ministry of Science and Technology of Spain (Program Ramon and Cajal). We are grateful to Juan Requejo Liberal for his valuable comments.

References

AEE, Asociación Empresarial Eólica 2009. *Observatorio eólico. Potencia Instalada.* [Online] Available at www.aeeolica.es/observatorio_potencia.php [accessed 25 May 2009].

Bell, D., Gray, T. and Haggett, C. 2005. The 'social' gap in wind farm siting decisions: Explanations and policy responses. *Environmental Politics*, 14, 460-477.

Burrall, P. 2004. Putting wind farms in their place. *Town and Country Planning*, 73(2), 60-63.

Cowell, R. 2007. Wind power and 'the planning problem': The experience of Wales. *European Environment*, 17(5), 291-306.

Cowell, R. and Owens, S. 2006. Governing space: Planning reform and politics of sustainability. *Environment and Planning A*, 24(3), 403-421.

Dalby, S. and Mackenzie, F. 1997. Reconceptualising local community: Environment, identity, and threat. *Area*, 29, 99-108.

Devine-Wright, P. 2005. Local Aspects of UK renewable energy development: Exploring public belief and policy implications. *Local Environment*, 10(1), 57-69.

Devine-Wright, P. and Devine-Wright, H. 2006. 'Social representations of intermittency and the shaping of public support for wind energy in the UK'. *International Journal of Global Energy Issues*, 25(3/4), 243-256.

Diputación de Cádiz 2004. *Plan Especial Supramunicipal de Ordenación de Infraestructuras de los Recursos Eólicos en la Comarca de La Janda (Cádiz)*. [Online] Available at http://www.fmedioambienteyenergia.es/apec/index.php?option=com_remository&Itemid=36&func=fileinfo&id=57 [accessed 15 February 2010].

Diputación de Cádiz 2005. *Documento de síntesis de las conclusiones y trabajos realizados por el Foro de la energía eólica marina y desarrollo sostenible de la Diputación Provincial de Cádiz*. [Online] Available at www.foroeolica.dipucadiz.org [accessed 24 May 2009].

Ellis, G., Barry, J.M. and Robinson, C. 2007. *Many ways to say 'no' – Different ways to say 'yes'. Applying Q-methodology to understand public acceptance of wind farm proposal*. [Online] Available at www.qub.ac.uk/research-centres/REDOWelcome/filestore/Filetoupload,40560,en.pdf [accessed 15 February 2010].

Eurobarometer 2007. *Energy Technologies: Knowledge, Perception, Measures*. A report produced by the European Opinion Research Group for the Directorate-General for Research, Luxembourg. [Online] www.managenergy.net/products/R1597.htm [accessed 15 February 2010].

Frolova, M. 2009. La evolución reciente de las políticas de paisaje en España y el Convenio Europeo del paisaje. *Proyección*, 6. [Online] Available at http://www.cifot.com.ar/proyeccion/admin/app/webroot/index.php?frontend/fichaRevista/38 [accessed 15 February 2010].

Frolova, M. 2010. Landscapes, Water Policy and the Evolution of Discourses on Hydropower in Spain. *Landscape Research*, 35(2), 235-257.

Frolova, M. and Pérez. B. 2008. El desarrollo de las energías renovables y el paisaje: Algunas bases para la implementación de la Convención Europea del Paisaje en la política energética española. *Cuadernos Geográficos de la Universidad de Granada*, 43, 289-309.

Hagget, C. and Toke. D. 2006. Crossing the Great Divide – Using Multi-method Analysis to Understand Opposition to Windfarms. *Public Administration*, 84(1), 103-120.

Healey, P. 2006. *Collaborative Planning. Shaping Places in Fragmented Societies*, 2nd ed. London: Macmillan.

Holden, E. 1998. Planning theory: Democracy or sustainable development? – Both (but don't bother about bread, please). *Scandinavian Housing and Planning Research*, 15, 227-247.

IEA 2008. *Wind Energy 2007. Annual Report*. A report produced by The Implementing Agreement for Co-operation in the Research, Development, and Deployment of Wind Energy Systems of the International Energy Agency. [Online] Available at www.ieawind.org/AnnualReports_PDF/2007/ 2007%20IEA%20Wind%20AR.pdf [accessed 24 May 2009].

Mercer, D. 2003. The great Australian wind rush and the devaluation of landscape amenity. *Australian Geographer*, 34, 91-121.

Nadaï, A. 2007. 'Planning', 'siting' and the local acceptance of wind power: Some lessons from French case. *Energy policy*, 35, 2715-2726.

OAER, Observatorio Andaluz de Energías Renovables 2010. *Informe del febrero de 2010*. [Online] Available at http://www.aprean.com/observatorio/obs_6_ 1.pdf [accessed 8 November 2010].

Pasqualetti, M.J., Gipe, P. and Righter, R.W. 2002. *Wind power in view: Energy landscapes in a crowded world*. San Diego: Academic Press.

Toke, D., Breukers, S. and Wolsink, M. 2008. Wind power deployment outcomes: How can we account for the difference? *Renewable and Sustainable Energy Reviews*, 12, 1129-1147.

Wolsink, M. 2000. Wind power and the NIMBY-myth: Institutional capacity and the limited significant of public support. *Renewable Energy*, 21, 49-64.

Wolsink, M. 2003. Reshaping the Dutch planning system: A learning process? *Environment and Planning A*, 35, 705-723.

Wolsink, M. 2006. River basin approach and integrated water management: Governance pitfalls for the Dutch Space-Water-Adjustment Management Principle. *Geoforum*, 37(4), 473-487.

Wolsink, M. 2007. Planning of renewables schemes: Deliberative and fair decision-making on landscape issues instead of reproachful accusations of non-cooperation. *Energy Policy*, 35, 2692-2704.

Woods, M. 2003. Conflicting environmental visions of the rural; windfarm development in Mid-Wales. *Sociologia Ruralis*, 43(3), 271-288.

Wüstenhagen, R., Wolsink, M. and Bürer, M.J. 2007. Social acceptance of renewable energy innovation: An introduction to the concept. *Energy Policy*, 35, 2683-2691.

Chapter 27

GIS Approach to Historical Landscape Changes: The Case of Modernization at Lake Biwa in Central Japan

Carlos Zeballos, Caroline Borré, Kati Lindström and Junzo Uchiyama

Introduction: The Neomap Project

The *Neolithization and Modernization: Historical Landscape Change on East-Asian Inland Seas* Project (NEOMAP)[1] is a large-scale international and interdisciplinary project that focuses on landscape on two axes: a temporal one, comparing the historical processes during the Neolithization and Modernization periods;[2] and a spatial one, contrasting several experiments in the East Asian inland seas (the Japan Sea and East China Sea).[3] The project is, by definition, a comparison in space and time.

---

1   The NEOMAP Project is being carried out in the framework of the Research Institute for Humanity and Nature (RIHN) in Kyoto, Japan. RIHN is a governmental financed research institute that aims to solve present and future environmental issues through interdisciplinary research uniting both humanities and natural sciences.

2   'Neolithization' and 'Modernization' are not set chronological periods that are universal for all areas but rather a set of historical processes including the following elements. Neolithization is the period of 1) introduction of a sedentary lifestyle, rapid population increase, and appearance of large settlements; 2) introduction of agro-pastoral economy; 3) expansion of trading activities and appearance of inter-societal division of labour; and 4) intensive resource exploitation by newly introduced complex technologies. In spite of different scales, these events show considerable similarities with the ones that take place during the Modernization period. As defined by the project, Modernization is the period characterized by 1) large-scale urbanization; 2) further dependence on domesticated products; 3) globalization of trading activities and expanding inter-regional hierarchies; and 4) further exploitation of resources by newly introduced complex technologies. Both during Neolithization and Modernization, people changed local ecosystems and human-nature relationships to a large extent. Therefore, analysing the processes which occurred during these two periods and comparing them can give us unprecedented insight into the emergence of the present environmental issues and possible future development. The exact times when these periods occurred have to be defined separately for each region. This enables us to compare similar phenomena in different cultures without being confined to standard chronological periodizations.

3   NEOMAP has eight regional work groups at the key spots on the East Asian inland seas' rim: Hokuriku in North-Eastern Japan, Lake Biwa in Central Japan, Kyushu in

Through the research of the relationships between humans and landscape in the past, NEOMAP aims at understanding the environmental issues of the present. In that regard, it is very important for the NEOMAP project, not only to carry out research on historical landscape changes, but also to communicate the outcomes to the public, in a clear and didactic way.

Communicating historical research to the public

While the use of GIS in historical and landscape scholarship is gradually spreading (Knowles 2002, 2008; Chapman 2006), there have been few examples of it transcending from the academia to the general public.

Because of the nature of the NEOMAP project, namely the incorporation of specialized and interdisciplinary studies, it is necessary to develop a method in which the outcomes of the research can be explained to and shared with a general audience. Also, the fact that the NEOMAP project is both interdisciplinary and international and that it includes eight distinct research areas that are studied in a long period, spanning from prehistory to modern times, poses a serious challenge to sharing and managing data and research outcomes. It has been necessary to create a general base of information that might otherwise be inaccessible to the scholars of specialized fields who are incapable of reading historical manuscripts by themselves.

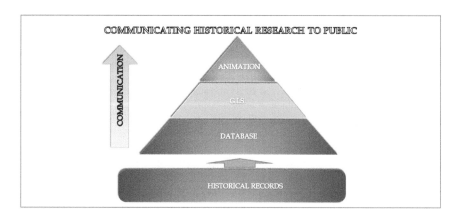

Figure 27.1 The work process from collecting historical material to public outreach

Southern Japan, Hokkaido in Northern Japan, the Ryukyu Islands in the Southern Japan (the latter two are historically non-Japanese cultural areas), Northern Zhejiang in China, the Southern Coast of Korea, and the Primorye region in Russia.

In order to achieve these goals, we propose the following work model (Figure 27.1). It summarizes the consecutive stages, starting with the stage of compiling and organizing the historical records into a database, to the stage of analysing the data using GIS and finally to communicating the results to the general public through the use of animation, video and multimedia.

The database/GIS approach

To meet the challenges posed by the organization, the NEOMAP project has among other measures opted for the Database/GIS approach. The NEOMAP Project is using database and Geographical Information Systems (GIS) in order to collect, manage, analyse and produce information related to historical landscapes. GIS provides the perfect platform for researchers to share interdisciplinary results in different scales (both historical and geographical).

The GIS unit of the NEOMAP project is composed of two sections: Neolithization and Modernization. The objectives of this unit are:

- To collect information on landscape change in both the Neolithization and Modernization periods in the overlapping research areas, in order to provide a comparative vision of the human settlements and their relationship with the landscape throughout time;
- To translate information from ancient sources, such as historical records and maps (only accessible to the professional reader) into ideas that can be comprehended by the general public;
- To use the research results as a tool for education and communication;
- To gain, through the analysis of the past, a better understanding of the perception of landscape and the environmental problems in present times.

Collecting historical data

The project aims at constructing a database with basic information on settlement patterns, production data, etc. for the Neolithization and Modernization periods in seven project research areas, excluding the Northern Zhejiang work group in China from the plan, due to the lack of comparable data source. For the rest of the areas, we can find rather detailed data compiled by the new colonial powers (in Russia, Hokkaido, Ryukyu Islands and Korea) or by the central government that dates from the nineteenth century. Whereas the exact occurrence of Modernization varies from one region to another (and in some regions, possibly coincides with Neolithization), these data can be considered to represent fairly well the state of affairs in the areas at the peak of Modernization processes, just before the big leap into contemporary landscapes.

While the recording of information in Japan during the late Edo Period (first half of the nineteenth century) both in data and maps was not always accurate, that situation changed after the Meiji Restoration (1862-1869). One of the first tasks undertaken by the authorities when rebuilding the country during the Meiji Period (1868-1912), was the meticulous documentation of the imperial dominions. In just a few decades, the detail of population and production records, as well as the accuracy of map survey, was notoriously enhanced by the government technicians in collaboration with the citizens. They produced books, census, records and maps.

However, we still encountered many problems that hindered the extensive use of these data. For one, the language and references used at the time of compilation differ from the current ones, which make these documents difficult to understand, even for scholars. Additionally, the data came from various sources and often made use of different systems of measurement. Also, the accuracy of the late nineteenth century maps obviously does not reach the level of the current maps used by GIS, which use satellite precision. Therefore, a system was needed to arrange the data within a flexible and organized database.

Organizing the database

Any information processing system must make assumptions about the nature of information, and GIS assumes that the information to be captured and stored in its database would be contained in maps (Goodchild 2008). Therefore, the first step to systematize the data was to identify every place with a unique geo-code. This code reflects the political division in Japan during the early Meiji Period: *Ken* 県 (prefecture), *kuni* 国 (historical prefecture),[4] *gun* 郡 (county) and *mura* 村 (village).

For example:

Place ID:
□□ □□ □□ □□□
県   国   郡   村
Shiga-ken 01 00 00 000
Oumi-kuni 01 01 00 000
Kurita-gun 01 01 02 000
Eda-mura 01 01 02 007

Subsequently, a relational database was created, assuming the geo-code as the ID of each record, and attaching several fields to that ID code, such as agricultural production, population, livestock, number of boats, etc. (Figure 27.2). Since various systems of measurement were used in old Japan, the ways to quantify volumes, weights and areas were translated and standardized to a more easily comparable metric system.

---

4 Historically Japan used to be divided into *kuni*, political units created in the Nara Period (710-784). At the end of the Edo Period (1603-1867), there were 60 *kuni*. However, during the Meiji Period (1868-1912) these *kuni* were transformed into *ken* or prefecture.

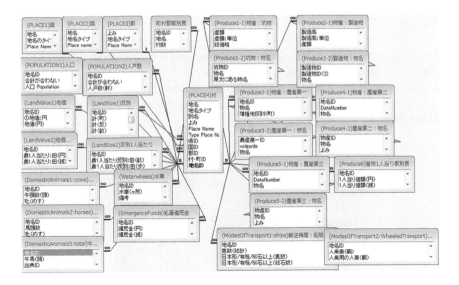

Figure 27.2 Relational database: Several tables inter-related using common ID codes

GIS analysis

GIS links the information stored in the databases to features in georeferenced maps. Anne Kelly Knowles (2002) defines georeferencing as the process of assigning coordinates from a known reference system, such as latitude/longitude, UTM or State Plate, to the page coordinates of an image or a planar map. However, different sorts of maps had to be produced according to the kind of data analysed and the scale in which they needed to be represented.

In this way, several levels of georeferenced maps were produced, each storing different kinds of feature classes:

- National level, including several prefectures represented by polygons;
- Prefectural level, including several counties also represented by polygons;
- County level, that is a polygon including several points or villages; and
- Village level, including several types of data represented by polygons, lines or points).

While the national and prefectural level were basically used for general demographic and economic analysis, the county and village levels were deemed more appropriate for landscape analysis.

The case of Lake Biwa

To illustrate the methodology that we are proposing for the analysis of landscape during the Modernization period in Japan, we will use the area of Lake Biwa, which is one of the 8 study areas within the project (Figure 27.3). We will also attempt to summarize some of the preliminary results of the research performed in this area.

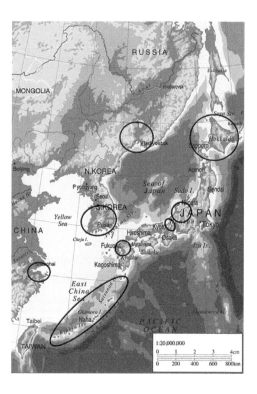

**Figure 27.3 Eight project areas at East Asian Inland Seas
with Lake Biwa in the middle**

Firstly, it is worth noting that the territory around Lake Biwa has been a place of human settlement from as early as the Neolithic times and its historical importance therefore goes without saying. Since the Lake Biwa area is easily accessible from both the Pacific Ocean and the Japan Sea, it acquired strategic importance in history as well, as it was considered a main target in warfare for controlling land and trade routes (both marine and continental).

Lake Biwa can be considered one of the most important areas in Japan not only from a historical viewpoint, but also from an environmental perspective. Lake

Biwa is the largest freshwater lake on the Japanese archipelago and situated in Shiga Prefecture, Central Japan. It lies on the borderline between broad-leaved and ever-green broad-leaved forests. Within this rich natural environment, the area itself forms a basin, of which the plains are seasonally flooded by lake waters and thus structure a vivid area of dynamic landscape. The varied terrain combined with traditional rice paddy agriculture utilizing the seasonal flooding and fish spooning movement along the creeks, has contributed to the rich biodiversity of the area.

In fact, Lake Biwa is one of the oldest lakes in the world, dating back about four million years. This long history allowed for the development of a rich ecosystem containing more than 1,000 different plant and animal species inside the lake itself (Mori and Miura 1990). Lake Biwa was registered with the Ramsar Convention on Wetlands in 1993 as a wetland of international importance.

County level

At the beginning of the Meiji Period, several maps showing the location of towns and villages were produced. One hundred and thirty years later, however, many of these towns had merged with other (bigger) towns, were renamed, or had simply disappeared. Moreover, since the difference in accuracy between the old map and the current map was substantial, the Meiji Period map could not be used directly as a cartographic base for locating and georeferencing the villages.

This problem is not uncommon in the use of GIS for digitizing historical maps, for obvious reasons. When the old maps present a uniform distortion in comparison to an accurate geographical position, a georectification process can be applied. Knowles (2008: 12-13) defines georectification as the process that assigns geographical coordinates to the map's digital image so it can be aligned with other data in GIS. However, since the county maps from the Meiji Period had an uneven distortion, they had to be translated village by village, which were thus located and placed in a georeferenced map and identified with a unique geo-code (Figure 27.4). It was through this code, the same one used in the database, that we connected the map to the stored data.

Subsequently, using current elevation data, a topographical map was produced in order to have a better understanding of the spatial relationship of the villages and their surrounding physical landscape.

The following are some of the primary results we have achieved so far. In the test analysis, we focused on the case of Shiga County, which is located in the southwest of Shiga prefecture and has Lake Biwa as a natural border on its eastern side. The methodology used for the analysis will be replicated in the other counties of Shiga Prefecture and other project research areas.

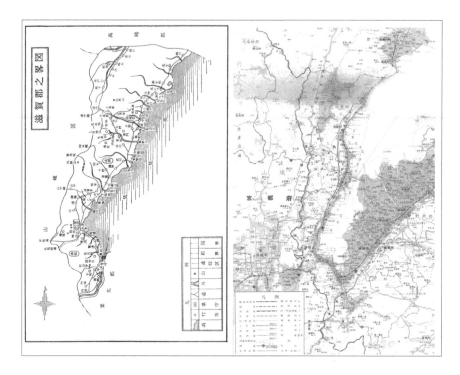

Figure 27.4 Early Meiji Period map of Shiga county
Note: Left = 1870; Right = Current map, 2007.

Population

GIS analysis helps to show not only the location of settlements in the topography but also the size of their population. (Figure 27.5a) Superimposing the Modernization map of settlements with that of Neolithization, we will be able to see the dynamics of settlement patterns and the tendencies in the change of population size.

Most of the larger towns were located in the flat lands and on the shores of the lake. The only exception is Sakamoto, a town on the foot of Mt. Hiei, which served as a living quarters for the servants of the very important temple Enryakuji on the top of Mt. Hiei.

There are more waterwheels in the small villages than in the big ones. This is interesting, because it contradicts the image that most Japanese have of rural villages from that time, namely that of large villages with many waterwheels and smaller and therefore less developed villages with few waterwheels. However, the data show the opposite is true.

This might be explained by the fact that small villages, in comparison to the bigger ones, were relatively new and had a high number of samurai population (Figure 27.5b). At the beginning of the Meiji Period, Japan abandoned the feudal

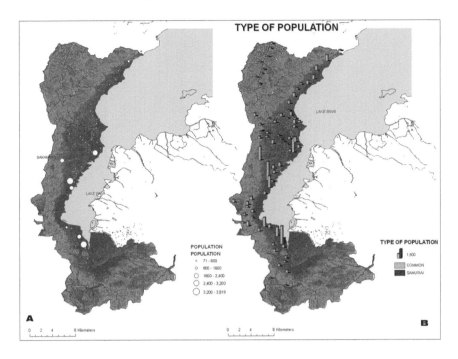

Figure 27.5a Location and population of the urban settlement
Figure 27.5b Composition of the population

class system and many ex-samurais found themselves without jobs. They were forced to turn to agriculture to make a living and they settled in new small villages, setting up irrigation systems with many waterwheels to facilitate rice cultivation.

Water transportation

Based on the number of boats per village, the distribution confirms Katata as the central node in the water transportation, outnumbering the ships of the Otsu area, which is considered central today (Figure 27.6). Located at the narrowest point between the Southern and the Northern part of the Lake and at the foot of the mountains where there is a passage to Kyoto and Mt. Hiei, Katata had an obvious advantage for becoming a harbor town.

Rice production

Two types of rice were produced: glutinous and non glutinous rice, of which the latter was the predominant one.

    Rice was mainly cultivated near the shore of Lake Biwa and along the waterways towards inland (Figure 27.7a). This coincides with the fact that rice

Figure 27.6 Number of ships per village

paddies need a lot of water, especially during planting season (rainy season), and depended heavily on the flooding of the lake and its rivers. Rice paddies were also used for fish cultivation and the paddies along with their irrigation canals were an important spooning area.

An exception was the mountainous area of Ogi, where rice terraces were used. The data clearly show that most rice fields in Shiga county were near waterways, and that rice terraces like the ones in Ogi were rather exceptional at the time. However, most Japanese people's mental image of rice fields is one of rice terraces like the one in Ogi. This mental image of the landscape is different from the actual one.

Other products

There are detailed accounts of several products that were cultivated in the area, like sweet potatoes, wheat, barley, millet, tobacco, tea, etc. (Figure 27.7b). It was not just rice that was cultivated in large quantities. This, again, contradicts the eternal image of the Japanese being solely dependent on rice cultivation. Historically, the complex agricultural system of rice paddies implied also harvesting other crops in the levees of the rice paddies and the backyards not suited for rice.

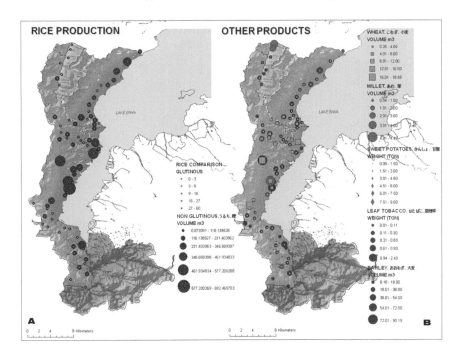

Figure 27.7a Rice production
Figure 27.7b Other agricultural products

Productivity

By comparing land size (converted into hectares; Figure 27.8a) and the amount of production (converted into cubic meters) of every place, the productivity of every place could be calculated (Figure 27.8b). Interestingly, some areas were, despite their smaller sizes, more productive than bigger ones.

Village Level

As we mentioned above, at the beginning of the Meiji Period a lot of detailed maps were produced in order to have a more accurate record of the Japanese Empire. In order to tie the places on these maps to a real location on the surface of the earth, we had to apply a georectification process to them, stretching some areas and shrinking others in the process.

While the information on these maps was rich, it was also quite difficult to understand and analyse properly. Consequently, datasets were produced from the maps' information in order to facilitate their interpretation. Information was

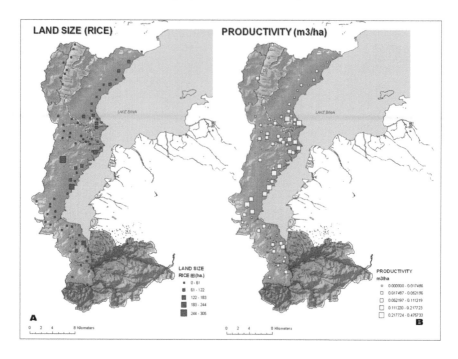

Figure 27.8a Land size in hectares
Figure 27.8b Productivity m³/ha

organized in several layers, such as vegetation, urban areas, constructed areas, ponds, roads, paths, bridges, shrines, temples, schools, waterwheels, graveyards, train stations and railroads (Figure 27.9a). GIS analysis of these historical maps has shown proximity relationships between temples and graveyards, as well as tea fields, used mainly for ceremonies or meditation.

Even if there is some distortion, these layers could be superimposed onto other kinds of geographical representation, including topographical maps, satellite imagery or aerophotography, in order to analyse changes in the landscape trough time. In our case, we applied the dataset layers to Google Earth (Figure 27.9b), where 3D visualization and navigation was displayed in real time.

Visibility analyses and sacred land

Another interesting possibility for applying GIS to landscape research is visibility analysis. Visibility analyses open a possibility to see how more perceptional and individual features like visibility contribute to processes in a landscape.

Michinori Hashimoto has carried out research trying to clarify the true reality of the medieval ban on fishing and hunting in the vicinity of temples, focusing on the range of vision from them (Hashimoto 2009a, 2009b). According to ancient

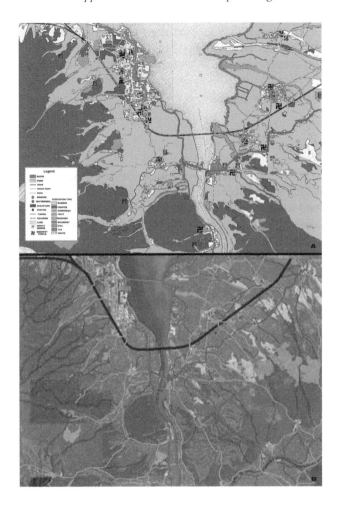

Figure 27.9a The map reworked into dataset layers
Figure 27.9b The map superimposed with the image from Google Earth

laws, hunting and fishing was prohibited in the distance of 2 *lis* (approximately 1,300 meters) from any temple (calculated from the position of the Buddha seated in the Main Hall), but it appears to have been defined differently at the medieval Choumei and Ishiyama temples at Lake Biwa.

Due to the difference in vegetation and changes in the coastline as a result of land reclamation, it is not possible today to retrace the exact boundaries of the visibility range from the temples which defined the 'sacred land' at the time, nor can we completely know the impact it had on urban and agricultural development. Thus, GIS visibility analysis was carried out using a Digital Elevation Map of the area. Visible and non-visible areas showed that, in the case of the Ishiyama

Temple, the views were mainly focused along the river (Figure 27.10a), whereas in the case of Choumei Temple, the view shed was very broad (Figure 27.10b). These visibility areas correspond to the areas of restriction on fishing and hunting as they can be retraced to medieval documents. Visibility-based restriction areas were maintained until the pre-modern era.

Animation

While GIS is a useful tool for analysing historical and geographical information and for organizing it and publishing it in atlases and other printed media, it can be effectively complemented with 3D modelling and animation. Multimedia software, such as Autodesk 3DStudio, has proven to be a very useful tool for communicating results to the general public.

Before the 3D modelling and the rendering of the animation takes place, scholars have to synthesize their research and imagine a simple way 'to tell the story' in an attractive visual way. Because 3D modelling, especially if it is very detailed, can be an extremely time-consuming process, clear outlines or storyboards can effectively facilitate the animation process.

The experimental animations that the NEOMAP project has produced so far have received enthusiastic reactions, both from scholars and non-scholars. In future, the virtual reconstructions of landscape features and human settlements according to historical documents could add an interesting new dimension, offering the possibility of 'taking a walk' in historical environments.

Conclusion

The mutual use of databases and GIS is an effective method to compile, standardize, organize, display and analyse historical records from various sources, and thus it facilitates collaborative interdisciplinary research on landscape changes and evolution. Technical difficulties behind the use of such data and the confusion that may arise from the translation and standardization of measurement systems and historical terms into modern language, as well as the georectification of historical maps, might render the outcome dubious for some scholars who would need very detailed and specialized original data, but makes all the historical information easily accessible and comprehensible for people who are not trained in reading historical documents. The combined use of GIS with 3D animation and other multimedia applications helps further to convey the idea of landscape as a whole to the general public.

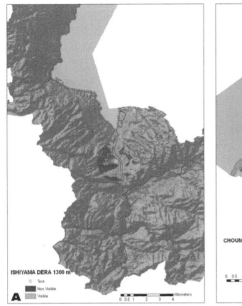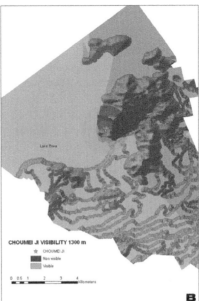

Figure 27.10a Visibility analysis in Ishiyamadera
Figure 27.10b Visibility analysis in Choumei ji

Acknowledgements

We are deeply indebted to Dr. Michinori Hashimoto, Senior Curator of Lake Biwa Museum, Shiga, Japan. His ideas are a constant inspiration and his research findings have been very useful during the process of GIS analysis shown in this article.

Special thanks to Prof. J. Christopher Gillam of the University of South Carolina for his continuous assistance and insightful ideas.

We would also like to thank the students and researchers that have collaborated on the construction of the database and GIS maps for the Modernization section: Tomohiko Matsumori, Akinori Jikihara, Machika Yamamoto, Maki Sato, Hiroaki Morita.

References

Chapman, H. 2006. *Landscape Archeology and GIS*. Stroud, Gloucestershire, Great Britain: Tempus Publishing Limited.
ESRI 2006. *Using ArcGIS Desktop*. Redlands, California: ESRI Press.

Goodchild, M. 2008. Combining Space and Time: New Potential for Temporal GIS, in *Placing History*, edited by A. Knowles. Redlands, California: ESRI Press, 179-197.

Hashimoto, M. 2009a. On the Ban on Killing in the Vicinities of Lake Biwa's Temples – Towards a Theory of Modernisation, in *NEOMAP Interim Report 2008. Neolithisation and Modernisation on East Asian Inland Seas*, edited by J. Uchiyama, K. Lindström, C. Zeballos and O. Nakamura. Kyoto: Research Institute for Humanity and Nature, pp. 163-169 (in Japanese).

Hashimoto, M. 2009b (forthcoming). Understanding the Ban on Killing in Temple Vicinities – Landscape Created by Religious Taboos, in *Keikan kara mieru atarashii rekishi to mirai. Keikanshi to kankyou*, edited by J. Uchiyama and K. Lindström. Kyoto: Showado.

Knowles, A. 2008. GIS and History, in *Placing History*, edited by A. Knowles. Redlands, California: ESRI Press, 2-25.

Knowles, A. 2002. Glossary of GIS terms, in *Past Time, Past Place. GIS for History*, edited by A. Knowles. Redlands, California: ESRI Press, 2-25.

Mori, S. and Miura, T. 1990. List of Plant and Animal Species Living in Lake Biwa (Corrected 3rd Edition). *Memoirs of the Faculty of Science, Kyoto University* (Biology series), 14 (1-2), 13-32.

Chapter 28

Beyond Regional Landscape Typologies: A Multi-Scaled and Trans-Regional Landscape Characterization for the Federal State of Belgium

Veerle Van Eetvelde and Marc Antrop

Introduction

Rapid changes in the landscapes have occurred very frequently in the European continent in the last decades. These changes, characterized as a 'crisis of the landscape' by Lemaire (2002) and described in the Dobříš Assessment (Stanners and Bourdeau 1995) caused loss of regional diversity, landscape character and identity. Many new European initiatives were set up, indicating a general renewed interest in landscape research and policy. The most important is the Council of Europe's European Landscape Convention. Its aim is to promote integrated landscape planning. One of the specific measures is to identify and describe the landscapes covering the entire territory, by making an inventory of the significant features that characterize them (Council of Europe 2000). Even before it entered into force on 1. March 2004, many initiatives started in different countries to set up new landscape inventories. However, these different landscape typologies do not fit at the national administrative borders and are not comparable because of different approaches, data sources, and methods (Wascher 2005). Trans-border problems in landscape classifications also exist at regional level within states. Examples are differences between England, Scotland, and Wales, in the UK, between Flanders, Wallonia, and Brussels Capital region, in Belgium, between the regional atlases in France (Lunginbühl 1994), in the biotope mapping of the federal state of Germany (Jessel 2006). Even between smaller administration units (counties, provinces) the classifications may not match at the borders (Stiles 2005). In addition, landscape inventories of different subject (forest areas, build up areas, heritage values, …) can vary between regions (Antrop and Van Eetvelde 2000). These differences are caused by the use of different source materials (scale, period, data quality), different methods and scale levels and also different aims established for the inventories or classifications, which makes comparison of data difficult (Antrop and Van Eetvelde 2000; Jessel 2006).

The European Landscape Convention was also signed by Belgium, having entered into force on 1 February 2005. In the federal state, the three regions (Flanders, Wallonia, and Brussels Capital region) are autonomous in policy and legislation for territory-related issues such as spatial planning, environment, agriculture, and landscapes. On the contrary, the communities (Flemish, French- and German-speaking communities) have authority over cultural matters, including heritage and landscape. As a result of these autonomies, there are different conceptions and approaches to landscape, even within Belgium. One of the results of the differences in landscape management and protection in the Flemish and Walloon regions is the different landscape classifications (Antrop et al. 2004; Van Eetvelde and Antrop 2009a), namely the classification of Flanders's traditional landscapes (scale of 1:200,000) (Antrop 1997; Antrop et al. 2002) and a physiographic classification of contemporary landscapes in Wallonia (scale 1:50,000) (Droeven et al. 2004). These differences derive from different goals in making landscape inventories, often linked to a specific scale, and different datasets used. The comparison of the two regional landscape classifications made clear that no integration is possible. As a consequence, it is necessary to compile a new landscape typology of contemporary landscapes at the Belgian scale level to meet the European Landscape Convention's measures (Van Eetvelde et al. 2006).

In this contribution, a new method for a multi-scaled approach in landscape typology will be presented, which forms the basis for trans-border integration. In the first section, the theoretical framework and methods of landscape classification, typology and characterization will be discussed. Second, the integrative method, combining different approaches and resulting in a multi-scale classification and GIS-database, will be described in detail for the case of Belgium, including a description of available source materials. The case of Belgium is used to demonstrate the integration of different trans-border regional classifications in a landscape characterization at a higher scale level and the applications of a landscape database. In the discussion, the integration of historical data sources into the GIS-database will be considered.

Landscape classification and characterization: Theoretical framework and approaches

Generally, landscapes can be classified in a typology or a chorology. Landscape typologies are systematic classifications of (abstract) landscape types which are based on attributes describing landscape properties of interest (i.e. land use, scenic properties, or cultural characteristics or history). Landscape types are defined by unique relations between natural components (such as geology, soil, morphology, land cover) and human components (such as settlement and field patterns, land use, building and farming styles). They often reflect a specific landscape history or are formed by specific processes like open field and enclosed (bocage) landscapes, pastoral landscapes, polderland, heathland, and Mediterranean polyculture (Lebeau

1972; Meeus 1995; Pinto-Correia and Vos 2004). This means that landscape types are generic, i.e. they may occur in different areas and in different geographical contexts (Mabbutt 1968). When one knows the characteristics of a landscape type, it is possible to identify it everywhere.

On the other hand, landscape chorologies take into account the spatial patterns of landscape types to form unique spatial arrangements with a distinct identity. So the horizontal relations and spatial distributions of landscapes are also considered; the area, situation and spatial associations with neighbouring areas are important. The chorological approach determines regions (choros = region), as they are often unique, which is reflected by a proper name given to the area. Generally, landscape chorology is a hierarchical spatial classification at different scale levels. In classical land evaluation, landscape typology and chorology are often combined (Mitchell 1973; Webster and Beckett 1970; Zonneveld 1994).

The holistic and parametric methods are two methodological approaches for compiling landscape classifications. Figure 28.1 illustrates the principle of the holistic method, an approach strongly developed with the rise of aerial photographs, introduced in landscape ecological studies by Troll (1939). He stated that aerial photographs were the optimal source for studying the relations between different landscape elements, as they give a synoptic view of the landscape from a bird's eye perspective. Aerial photographs represent the holistic landscape; so, based on the Gestalt-abilities of our perception, one is able to interpret complex patterns. The holistic classification procedure is very similar to analogue photo interpretation. The synoptic documents are stepwise interpreted to delineate hierarchical spatial landscape units. The first level is based on the most important features, which can be easily identified. The resulting units are subdivided into smaller units based on more detailed characteristics. The holistic method is typically a process of zooming in on the landscape, starting from total overview going further into detail (Figure 28.1). The parametric method (Figure 28.2) means that different properties of the landscape are mapped in separate thematic maps (i.e. soil characteristics, slope, land use). The set of thematic maps are then combined in an overlay. The resulting composite map uses many polygons to define the landscape units and describe the landscape types (Mitchell 1973). This method can be compiled almost automatically in GIS, by which slivers can be created.

In general, older landscape typologies are classifications of geographical regions, often holistic and generic in nature. Examples are the traditional landscapes of Flanders (Antrop 1997; Antrop et al. 2002), the geographical regions of Belgium (Brulard et al. 1970), and the landscape regions of Belgium (Christians and Daels 1988). One of the oldest known landscape classifications using the parametric approach is the landscape regions of Estonia by Granö (1929). He used topographical maps to draw up thematic maps (which he named 'Stoffen'); combined, those maps using overlay and introduced transitional zones that appear on the analytical maps. More recent typologies use GIS-overlay of digital thematic maps, combined with spatial and statistical analysis to define landscape types (some examples: Kilchenmann 1973; Lioubimtseva and Defourny 1999; Mücher et al. 2003).

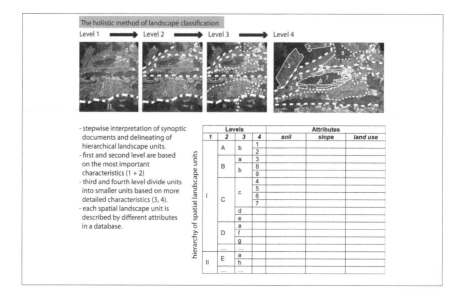

Figure 28.1 The holist method of landscape classification

Source: After Antrop 2007 and Van Eetvelde and Antrop 2009a.

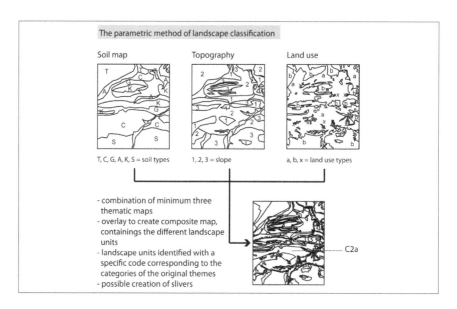

Figure 28.2 The parametric method of landscape classification

Source: After Antrop 2007 and Van Eetvelde and Antrop 2009a.

The choice between the two methods will mainly depend on the nature of available data sources, scale level (extent), available time and the researcher's expertise. The holistic method is very useful for landscapes with clear spatial associations between components (such as land use, soil conditions, and geomorphology) and complex patterns (such as field systems), which are clearly revealed on the image or maps. The method can be finished fast, allows deficiencies of the maps and creates the possibility of building up an open framework to be completed consecutively. The parametric approach is efficient when high quality digital maps are available. It is a semi-automatic technique in GIS, where the outcome is highly dependent on map properties, such as legend, scale, and quality. Often, digital maps with sufficient detail are commonly available only for elevation (digital terrain models) and land cover. Thus, other significant properties may be excluded from the classification, so the result of the classification will depend on the selected themes. Using a large number of themes makes statistical data analysis and clustering necessary.

Since the 1990s, and especially since the implementation of the European Landscape Convention, numerous landscape classifications have been produced at different scale levels. In the United Kingdom, the concept of landscape character was introduced in Landscape Character Assessment, an approach containing methods, tools and procedures to identify and assess landscape character. Landscape is seen as in integrated concept, since character is defined as a distinct, recognisable and consistent pattern of elements in the landscape that makes one landscape different from another, rather than better or worse (Swanwick 2002). For the benefit of archaeological and historical landscape research, and due to the lack of historical and cultural dimensions in Landscape Character Assessment, a similar approach was developed such as the Historical Landscape Characterisation (HLC) (Fairclough et al. 2002; Fairclough and Macinnes 2003; Clark et al. 2004). Groom et al. (2006) reviewed several landscape classifications in Europe, made in the mainstream of landscape character assessment. They stated that the majority of the classifications at national level were based on expert knowledge in a top-down approach. The distinction between landscape character types and landscape character areas is similar to the differences between landscape typology and chorology used in geography and land evaluation (Antrop 2000).

A trans-border characterization of Belgium: Method and results

Figure 28.3 gives an overview of the new method that was set up based upon the existing geographical approaches to landscape classification like those described above. The whole procedure consists of five steps and is a combination of the parametric and holistic methods. The result is an identification of contemporary landscapes at national level at two scale levels. Spatial units are defined and landscape types are assigned to each spatial unit at both scale levels. The first scale level uses a tessellation of regular grid cells to assign landscape types, using a parametric approach. The second level defines landscape units by aggregating the grid cells (holistic approach). The two levels are not hierarchical, in the sense of the holistic approach and Landscape Character Assessment (i.e. nested levels), but they represent two different landscape typologies at different scale levels.

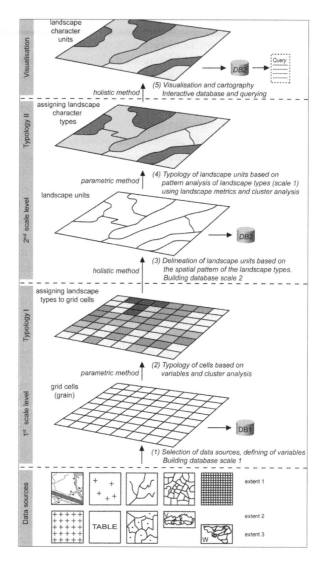

Figure 28.3 The 5 steps in the method model and the method used for the typology and characterization of Belgium's contemporary landscapes

Source: After Van Eetvelde and Antrop 2009a.

STEP 1: Selection of data sources, defining variables,
geocoding grid cells, and building the database

All relevant data sources covering the whole of Belgium were collected, representing in a balanced way both natural and social/cultural components of the landscape (Van Eetvelde and Antrop 2007). For the Belgian case, most data are nowadays collected and structured differently in the Regions, due to their political autonomy, and data covering the whole of Belgium have consistently become rare. Only four basic data sources are available and are significant for contemporary landscapes. The CORINE Land Cover map (1990) describes the main land cover types classified based on Landsat 5 TM satellite imagery. The soil association map published in the National Atlas of Belgium (Maréchal and Tavernier 1974) is a synthesis of Belgian soil maps, indicates the main physical and geological characteristics of the landscape and is useful to define the hydrographical network. The digital terrain model indicates relief and slopes. A satellite image Landsat 5 TM (1989-1990) gives a more synoptic, holistic view of the area. Socio-demographical census and economic data are available at federal level, but these consist of aggregated data generalized by administrative units (municipalities) and are thus of little use in the method of landscape characterization (Van Eetvelde and Antrop 2007).

At the first scale level of the typology, the tessellation of regular grid cells of 1 km² used as spatial unit define the grain of the classification (Figure 28.3). Each grid cell is a polygon in the GIS, which makes it possible to compile a GIS-database and to integrate other data sources more easily. Based on the four selected data sources, 18 differentiating variables were defined, shown in Table 28.1. The variables derived from each map were integrated to the grid cells by GIS overlay of the data sets with the grid cells (parametric method). To reduce the maximum possible combination, the thematic maps were reclassified. The CORINE Land Cover map was reclassified into six main land use categories, setting the distinction between main perceptive characteristics like urban and non-urban areas, open spaces and forested land. The proportion of each of these categories was calculated per grid cell (in percentages) and then used as variables (1-6 in Table 28.1). The soil association map was reclassified into nine categories; their proportion covers were used as variables 7-15 (Table 28.1). Based on the digital elevation model the topographic variation by grid cell was calculated, which was used to define flat and undulating areas. The average elevation per grid cell was grouped into classes, characterizing the characteristic of the main natural regions. The satellite Landsat 5 TM image (1989-1990) has a ground resolution of 30 m which was resampled to 25 m by Eurosense. For each grid cell, the Shannon information entropy of the image reflectance classes (0-255) was calculated, indicating the landscape heterogeneity based on the main field pattern (Forman and Godron 1986).

Table 28.1 Variables derived from the four basic data sources in the case of Belgium

Land Cover
(derived from CORINE Land Cover 1990, NGI)

1.	% per km² of urban fabric, artificial non-agricultural vegetated areas
2.	% per km² of industrial, commercial and transportation infrastructure, mining areas and waste dumps and construction sites
3.	% per km² of arable land, permanent crops, mixed crops agricultural areas
4.	% per km² of pastures, natural grassland
5.	% per km² of forest and semi-natural areas; inland wetlands
6.	% per km² of wetlands, water bodies, open spaces with little or no vegetation

Soil Association
(derived from the Soil Association map of Belgium (Maréchal and Tavernier 1974)

7.	% per km² of polders soils
8.	% per km² of alluvial soils
9.	% per km² of sandy soils
10.	% per km² of loamy soils
11.	% per km² of inceptisols
12.	% per km² of soils on steep slopes
13.	% per km² of sandy and clayish soils
14.	% per km² of peat soils
15.	% per km² of un-mapped or disturbed soils

Landform and topography
(derived from DTM Belgium, National Geographic Institute)

16.	Average elevation in meter per km²
17.	Standard deviation of elevation per km²

Landscape heterogeneity
(based on Landsat TM imagery 1989-1990)

18.	Entropy of image reflectance per km²

STEP 2: Defining landscape types by grid cell at the first scale level

The national grid of 1 km², according to the Lambert 72/50 reference system, was used as grid cells for the characterization at first scale level. The different variables were integrated into the 31,473 grid cells covering Belgium. This integration was used to define homogeneous groups, considered as possible landscape types (Figure 28.3). Multivariate analyses are frequently used in landscape analysis as non-spatial operations to group similar objects and to obtain landscape types (e.g. Kristensen 2003; Bryan 2006; Owen et al. 2006). Due to the high number of cells, in relation to the extent of Belgium and the grain of the cells, it was necessary to perform a *k*-means clustering, using SPSS 12.0 (Norušis 2006). The number of clusters, similar to the landscape types, must be defined in advance. No clear solution could be found in the literature to determine the number of clusters, so the final number is based on the possible combinations of the nominal variables (land cover and soil association), possible correlations between these variables (tested with Pearson correlation). The different possible outcomes were tested using information theory (Kilchenmann 1973; Van Eetvelde 2007). Finally, 6 variables of land cover, 8 of soil association, 2 derived from the elevation model and the landscape heterogeneity were selected to perform the cluster analysis (Table 28.1). The category 'un-mapped or disturbed soils' of soil association (variable 15) was not included due to the high correlations with land cover (variable 1, 2 and 6). Thus, clustering resulted in 48 landscape types characterized by land cover, soil association, mean elevation, relief variation, and landscape heterogeneity. The landscape types were described with a name and code, assigned to each cell and added as an attribute in the database. The visualization of each type in the GIS was based upon the characteristics of the landscape type. The choice of the colour is based upon the most dominant, characteristic properties and on colour schemes commonly used in cartography, for example red for urban fabric (Brewer 1994). To improve the readability of the map, the relief variation is represented by colour saturation. Figure 28.4 (top) shows the final result of this typology.

STEP 3: Delineation of landscape units at the second scale level

As illustrated in Figure 28.3, the landscape units at the second scale level are formed by unique combinations and distinct spatial pattern of adjacent grid cells with different landscape type at first scale level. The landscape units are delineated manually by holistic interpretation of these patterns, similar to the holistic interpretation of aerial photographs. For Belgium, 222 landscape units were formed by unique combinations of adjacent landscape types of the first scale level. This delineation was refined and adjusted manually using the Landsat 5 TM satellite image (1989-1990) to assist, where possible, with the mapping of landscape unit borders associated with distinct image features, for example forest complexes.

STEP 4: Defining landscape types at second scale level

For the typology of the landscape units, the spatial pattern of the grid landscape types within each unit is quantified using landscape metrics (McGarigal et al. 2002). Landscape metrics describe the configuration and composition of the grid patterns of the first scale level within each landscape unit at the second scale level. Configuration metrics describe the spatial configuration of patches referring to spatial arrangement, position, orientation of the patches, while composition metrics take into account the variety and abundance of types. In typology, adjacent grid cells of similar landscape types form homogenous units and define patches (Forman and Godron 1986). The first selection of the landscape metrics is based on the significance of the metrics for a landscape typology. In Fragstats (freeware program to analyse spatial patterns of categorical maps), six groups of metrics can be calculated, giving an indication for (1) the shape of landscape units; (2) the spatial pattern of the patches in landscape units; (3) the edge metrics of the patches in landscape units; (4) aggregation of the patches in landscape units; (5) landscape types within landscape units; and (6) diversity of landscape units (Farina 1998; McGarigal et al. 2002; O'Neill et al. 1988; Turner et al. 2001). The final set of landscape metrics is selected after correlating and analysing principal component between a large series metrics (Cain et al. 1997; Haines-Young and Chopping 1996; Herzog et al. 2001; Lausch and Herzog 2002; Li and Wu 2007; Riitters et al. 1995). For Belgium, 25 landscape metrics were calculated, describing both composition and configuration, and the Pearson correlation coefficients and principal component analysis (Varimax rotation method with Kaiser Normalization) were performed in SPSS. The final four selected landscape metrics, number of patches (NP), patch richness density (PRD), Shannon diversity (SDHI), and contagion index (CONTAG), were combined with the proportion of 48 landscape types, to define groups in landscape units (Van Eetvelde and Antrop 2009a). They are considered variables in a hierarchical clustering of the spatial units to define 45 landscape character types at the second scale level. The names of the landscape character types give an indication of the land cover, soil properties, relief and heterogeneity of the landscapes.

STEP 5: Visualization of landscape character areas

To visualize the result of the second scale level of the typology, adjacent landscape units with similar types were joined into 197 landscape character areas (Figure 28.4, bottom). The 54 landscape character types are represented in nine families. Three types are named as urban landscapes (type 1-3), showing urban agglomerations and highly urbanized cities, and occur in 29 areas. Suburban landscapes (29 landscape character areas) are characterized by four landscape character types (4-7), composed of a matrix of build up area and more rural areas. Two types (8-9) represent industrial and harbour landscapes, situated in five landscape character areas. The urbanized coast and dunes form one landscape type (10), fragmented into three areas along the Belgian coast line and fragmented by an urban and harbour landscape. The family of the Polderland (indicated by A on the top map) consists of two distinct landscape character areas with different landscape character types: the Coastal Polderland (type 11) and polderland of the Scheldt River (type 12). Most of central Belgium consists of rural landscapes, with a mixture of build up land, which are dominated by arable land forming 74 landscape character areas, characterized by 18 landscape character types (13-30). The types vary in combinations of agricultural land and build up land, pasture land, and forests. Ten areas, situated in the Kempenland (D), the Fagne (I), eastern part of the Ardennes (L), and in the Lorraine (N) in the south are dominated by pasture land (7 types, 31-36). The landscape character types of forest landscapes consist of 14 different types (37-50), including landscapes dominated by forest,

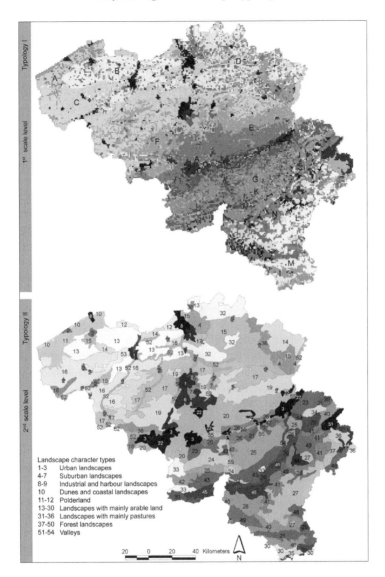

Figure 28.4 Landscape characterization of contemporary Belgian landscapes

Note: Top = classification of square kilometre cells. The main geographical regions are indicated with letters; Bottom = classification of landscape character areas, defining 54 landscape character types based on the first scale level's pattern. For a coloured version see www.geoweb.ugent.be/lr/.

Source: CORINE Land Cover 1990, Soil Association Map of Belgium, Digital Elevation Model (NGI), Landsat TM 1989-1990 (Eurosense).

as well as combinations of forest with pastures, arable land, and build up land. These forest landscapes are scattered over 35 landscape character areas in the Kempenland (D), Fagne-Famenne (I-J) and the Ardennes (L), and on the *cuestas* of Lorraine (N). The last group of landscape character types are the valleys (four types, 51-54), with heterogeneous land use, fragmented in 19 areas.

The characterization of the grid cells and its cartographic representation (Figure 28.4, top) allowed the identification of some typical patterns, synthesized in the landscape character areas and illustrating the holistic interpretation of the types' spatial pattern. For example, the Polderland (A) is characterized by a pattern of only two landscape types and forms landscape character area 11. On the flat sandy soils of northern Flanders (B), the enclosed landscapes show a different spatial pattern from the ones on hilly terrain on loamy soils (C), so they form (at the second scale level) two distinct landscape character areas (13 and 16). The Kempenland in the north (D) is characterized by a coarse grained mosaic of open agricultural land, woodland, and urbanized areas, forming a heterogeneous and well structured landscape (landscape character area 15). On the loamy plateaus, the landscape character areas 19 and 20 consist of open fields that are highly homogeneous in the east (E) and more dissected and heterogeneous in the west (F). The landscape character area 25 of the hilly land with large farms of the Condroz (G) is clearly distinct from the industrial belt of the Sambre-Meuse rivers (H, landscape character area 26) north of it. It is also distinct from the hilly karstic wooded land of the Calestienne (K) in the south. The region of Ardennes consists of steep forested slopes (M, landscape character area 43) and the upland is characterized by a mixture of grasslands and forest (L, forming landscape character area 27).

Applications of the landscape database:
The integration of cultural and historical data sources into the GIS-database

Besides the two resulting maps of landscape characterization, there is added value in the landscape database. It consists of the GIS-layers at the two scale levels, created by the thematic variables. Not only the landscape units but also the grid cells are represented as polygons. This allows us to easily link the spatial units and grid cells to an open descriptive database for more detailed characterization, extendable to add new information. Although the landscape typology maps at both scale levels offer a synoptic and coherent picture, the analogue version is difficult to interpret because of the large number of categories. Thus, since the map is linked to a GIS database, it allows easy and interactive querying of the spatial units and access to their characteristics. Furthermore, the database allows integration of the different scale levels, for example, with existing landscape classifications and inventories, different regional classification in particular. In the Belgian case, at the first scale level, the landscape database contains 18 differentiating variables, to which 104 descriptive variables were added. The Belgian landscape characters areas are mainly based on natural features, and cultural and aesthetic themes are not included to define landscape types but are added as descriptive variables in the database. Also, perceptive characteristics are used in the cartographic visualization and presentation. Furthermore, links are made to the corresponding categories in the existing regional classifications, to the Landscape Atlas of Flanders, to forest inventories. Additional variables describe the cultural, historical and visual characteristics of the landscape. Examples are significant place names, architectural properties, information from other inventories and local projects. This open database structure allows features to be added that were not used in the classification. For example, information related to cultural, social and archaeological features could be added, like settlement type, cultural heritage, architectural properties, place names and even photographs and descriptions of local projects.

One of the applications of the landscape database is monitoring historical and current landscape changes. Based on inventories of historical maps, the evolution of forest and build up area between the eighteenth and the twentieth century, for instance, can be analysed per km². Inventories are integrated per grid cell, so it is possible to compile change indicators per km² or the time depth and trajectories of forest and build up patches. Figure 28.5 illustrates the evolution of forest between the eighteenth and the twentieth century in the area of the Brugse Veldzone (Van Eetvelde and Antrop 2009b). Generally, it shows a decrease per km² between 1775 and 1900 and an increase between 1900 and 1990. However, this result may be influenced by the data sources. The data quality, expressed by the scale of the data sources, completeness (commission and omission), positional and temporal accuracy and thematic consistency must be taken into account when analysing landscape changes.

The example of the integration of data derived from historical maps shows the lack of suitable historical and cultural variables in defining landscape character types and delineating character areas. This is explained by the restrictions in data describing these themes, as they are highly fragmented, often specifically local. Data on cultural, historical, archaeological, architectural and aesthetic features seldom have consistent coverage of the entire country in sufficient detail and quality to integrate small scale landscape assessments and characterizations. In most cases, the historical and cultural qualities are often linked to landscape elements like monuments and sites or they are aggregated by administrative units, e.g. the number of archaeological sites per municipality. Also, the spatial distribution of historical and archaeological sites is often biased as high densities of findings are mainly found in undisturbed areas away from the effects of urbanization, for example (Fairclough et al. 2002, Van Eetvelde and Antrop 2005). When using the holistic approach to analyse aerial photographs, it is easier to include historical and cultural qualities, like the examples of the Historic Landscape Characterisation. Although the proposed method does not follow the HLC-approach, it offers the possibility of adding such information as descriptive variables in the database. The landscape characterization maps on both scale levels can be used as entries for querying these data as well.

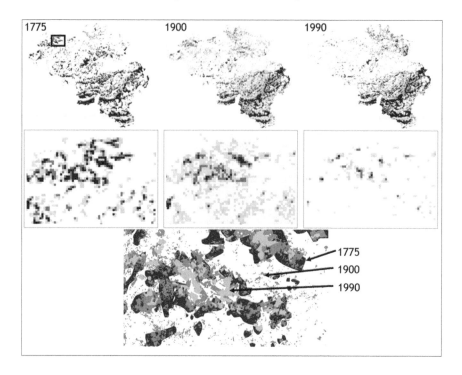

Figure 28.5 Application of the landscape database

Note: Top = evolution of forest in Belgium between the eighteenth and the twentieth century; Middle = detail of Brugse Veldzone; Bottom = original data sources used to calculate the proportion of forest for every grid cell.

References

Antrop, M. 1997. The concept of traditional landscapes as a base for landscape evaluation and planning. The example of Flanders Region. *Landscape and Urban Planning*, 38, 105-117.

Antrop, M. 2000. Geography and landscape science. *Belgeo*, 1(4), 9-35.

Antrop, M. 2007. *Perspectieven op het landschap. Achtergronden om landschappen te lezen en te begrijpen*. Gent: Academia Press.

Antrop, M., Belayew, D., Droeven, E., Feltz, C., Kummert, M. and Van Eetvelde, V. 2004. Landscape research in Belgium. *Belgeo*, 2, 205-218.

Antrop, M. and Van Eetvelde, V. 2000. Holistic aspects of suburban landscapes: Visual image interpretation and landscape metrics. *Landscape and Urban Planning*, 50, 43-58.

Antrop, M., Van Eetvelde, V., Janssens, J., Martens, I., and Van Damme, S. 2002. Traditionele landschappen van het Vlaamse Gewest. Ghent University, Department of Geography. [Online] Available at http://www.geoweb.ugent.be/

landschapskunde/projecten/traditionele-landschappen-vlaanderen [accessed: 15 February 2010].

Brewer, C.A. 1994. Color use guidelines for mapping and visualization, in *Visualization in modern cartography*, edited by A.M. MacEachren and D.R.F. Taylor, Oxford: Pergamon, 123-147.

Brulard, T., Dussart, F., Nicolai, H. and Snacken, F. 1970. *Regionale Indelingen. Blad 50a, Atlas van België*. Brussel: Nationaal Comité voor Geografie. Commissie voor de Nationale Atlas. Koninklijke Belgische Akademie.

Bryan, B.A. 2006. Synergistic techniques for better understanding and classifying the environmental structure of landscapes. *Environmental Management*, 37, 126-140.

Cain, D.H., Riitters, K.H. and Orvis, K. 1997. A multi-scale analysis of landscape statistics. *Landscape Ecology*, 12, 199-212.

Christians, C. and Daels, L. 1988. *Belgium. A Geographical Introduction to its Regional Diversity and its Human Richness*. Liège: Université de Liège.

Clark, J., Darlington, J. and Fairclough, G. 2004. *Using Historic Landscape Characterisation*. English Heritage and Lancashire County Council.

Council of Europe 2000. *European Landscape Convention and Explanatory Report*. Florence: Council of Europe, Document by the Secretary General established by the General Directorate of Education, Culture, Sport and Youth, and Environment.

Droeven, E., Feltz, C. and Kummert, M. 2004. *Les territoires paysagers de Wallonie*. Namur: Ministère de la Région wallonne.

Fairclough, G., Lambrick, G. and Hopkins, D. 2002. Historical landscape characterisation in England and a Hampshire case study, in *Europe's Cultural Landscape: Archaeologists and the management of change*, edited by G. Fairclough, S. Rippon and D. Bull. Brussels: Europae Archaeologiae Consilium, 69-80.

Fairclough, G. and Macinnes, L. 2003. *Landscape Character Assessment. Guidance for England and Scotland. Topic Paper 5 – Understanding Historic Landscape Character*. The Countryside Agency, Scottish Natural Heritage. [Online] Available at http://www.landscapecharacter.org.uk/files/pdfs/LCA-Topic-Paper-5.pdf [accessed 15 February 2010].

Fairclough, G., Rippon, S. and Bull, D. (eds) 2002. *Europe's Cultural Landscape: Archaeologists and the Management of Change*. Brussels: Europae Archaeologiae Consilium.

Farina, A. 1998. *Principles and Methods in Landscape Ecology*. London: Chapman and Hall.

Forman, R.T.T. and Godron, M. 1986. *Landscape Ecology*. New York: John Wiley.

Granö, J.G. 1929. *Reine Geographie. Eine Methodologische Studie Beleuchtet miet Beispielen aus Finnland und Estland*. Helsinki: Helsingfors.

Groom, G., Wascher, D., Potschin, M. and Haines-Young, R. 2006. Landscape Character Assessments and fellow travellers across Europe, in *Landscape*

Ecology in the Mediterranean: Inside and outside approaches; Proceedings of the European IALE Conference, edited by R.G.H. Bunce and R.H.G. Jongman. IALE Publication Series 3, 221-231.

Haines-Young, R. and Chopping, M. 1996. Quantifying landscape structure: A review of landscape indices and their application to forested landscapes. *Progress in Physical Geography*, 20, 418-445.

Herzog, F., Lausch, A., Müller, E., Thulke, H.H., Steinhardt, U. and Lehmann, S. 2001. Landscape metrics for assessment of landscape destruction and rehabilitation. *Environmental Management*, 27, 91-107.

Jessel, B. 2006. Elements, characteristics and character – Information functions of landscapes in terms of indicators. *Ecological Indicators*, 6, 153-167.

Kilchenmann, A. 1973. Die Merkmanalyse für Nominaldaten – eine Methode zur Analyse von qualitativen geographischen Daten. *Geoforum*, 15, 33-45.

Kristensen, S.P. 2003. Multivariate analysis of landscape changes and farm characteristics in a study area in central Jutland, Denmark. *Ecological Modelling*, 168, 303-318.

Lausch, A. and Herzog, F. 2002. Applicability of landscape metrics for the monitoring of landscape change: Issues of scale, resolution and interpretability. *Ecological Indicators*, 2, 3-15.

Lebeau, R. 1972. *Les grands types de structures agraires dans le monde*. Paris: Masson.

Lemaire, T. 2002. *Met open zinnen. Natuur, landschap, aarde*. Amsterdam: Ambo.

Li, H. and Wu, J. 2007. Landscape pattern analysis: Key issues and challenges, in *Key Topics in Landscape Ecology*, edited by J. Wu, J. and R. Hobbs, R. Cambridge: Cambridge University Press, 39-61.

Lioubimtseva, E. and Defourny, P. 1999. GIS-based landscape classification and mapping of European Russia. *Landscape and Urban Planning*, 44, 63-75.

Lunginbühl, Y. 1994. *Méthode pour des atlas de paysages – Identification et qualification*. Strates/CNRS, Ministère de l'Aménagement du Territoire, de l'Equipement et des Transports/Direction de l'Architecture et de l'Urbanisme.

Mabbutt, J. 1968. Review of Concepts of Land Classification, in *Land Evaluation*, edited by A. Stewart. Melbourne: Macmilan of Australia, 11-28.

Maréchal, R. and Tavernier, R. 1974. *Pédologie. Comments to the sheets 11A and 11B of the National Atlas of Belgium*. Brussels: Nationaal Comité voor Geografie.

McGarigal, K., Cushman, S.A., Neel, M.C. and Ene, E. 2002. *FRAGSTATS: Spatial pattern analysis program for categorical maps*. University of Massachusetts, Amherst. [Online] Available at http://www.umass.edu/landeco/research/fragstats/fragstats.html [accessed: 15 February 2010].

Meeus, J.H.A. 1995. Pan-European landscapes. *Landscape and Urban Planning*, 31, 57-79.

Mitchell, C. 1973. *Terrain Evaluation*. London: Longman.

Mücher, C.A., Bunce, R.G.H., Jongman, R.H.G., Klijn, J.A., Koomen, A.J.M., Metzger, M.J. and Washer, D.M. 2003. *Identification and Characterisation of Environments and Landscapes in Europe.* Wageningen: Alterra, rapport 832.

Norušis, M.J. 2006. Cluster Analysis, in Norušis, M.J., *SPSS 14.0 Statistical Procedures Companion.* Englewood Cliffs: Prentice Hall.

O'Neill, R.V., Krummel, J.R., Gardner, R.H., Sugihara, G., Jackson, B., DeAngelis, D.L., Milne, B.T., Turner, M.G., Zygmunt, B., Christensen, S.W., Dale, V.H. and Graham, R.L. 1988. Indices of landscape pattern. *Landscape Ecology*, 1, 153-162.

Owen, S.M., MacKenzie, A.R., Bunce, R.G.H., Stewart, H.E., Donovan, R.G., Stark, G. and Hewitt, C.N. 2006. Urban land classification and its uncertainties using principal component and cluster analysis: A case study for the UK West Midlands. *Landscape and Urban Planning*, 78, 311-321.

Pinto-Correia, T. and Vos, W. 2004. Multifunctionality in Mediterranean landscapes – past and future, in *The New Dimensions of the European Landscape*, edited by R.H.G. Jongman. Dordrecht: Springer, 135-164.

Riitters, K.H., O'Neill, R.V., Hunsaker, C.T., Wickham, J.D., Yankee, D.H., Timmins, S.P., Jones, K.B. and Jackson, B.L. 1995. A factor analysis of landscape pattern and structure metrics. *Landscape Ecology*, 10, 23-39.

Stanners, D. and Bourdeau, P. (eds) 1995. *Europe's Environment. The Dobris Assessment.* Copenhagen: European Environment Agency, EC DG XI and Phare.

Stiles, R. 2005. Landscape change: Can we find a theoretical basis to integrate temporal, spatial and perceptual aspects? in *Landscape Change. Conference Proceedings ECLAS*, edited by E. Göker, D. Oguz, N. Karadeniz, I. Çakci. Ankara, Department of Ankara University, 14-18 September 2005, 23-33.

Swanwick, C. 2002. *Landscape Character Assessment. Guidance for England and Scotland.* The Countryside Agency, Scottish Natural Heritage. [Online] Available at http://www.landscapecharacter.org.uk/files/pdfs/LCA-Guidance. pdf [accessed: 15 February 2010].

Troll, C. 1939. *Luftbildforschung und Landeskundige Forschung.* Wiesbaden: Erdkundliches Wissen. Schriftenreihe für Forschung und Praxis, Heft 12, F. Steiner Verlag.

Turner, M.G., Gardner, R.H. and O'Neill, R.V. 2001. *Landscape Ecology in Theory and Practice. Pattern and Process.* New York: Springer-Verlag.

Van Eetvelde, V. and Antrop, M. 2005. The significance of landscape relic zones in relation of soil conditions, settlement pattern and territories in Flanders. *Landscape and Urban Planning*, 70, 127-141.

Van Eetvelde, V. and Antrop, M. 2007. Integrating cultural themes in landscape typologies, in *European Landscapes and Lifestyles: The Mediterranean and Beyond. Proceedings of the 21st PECSRL Conference 'One Region, Many Stories: Mediterranean Landscapes in a Changing Europe'*, Limnos and Lesvos, 2004, edited by Z. Roca, T. Spek, T. Terkenli and F. Höchtl, F. Lisboa: Edições Universitárias Lusófonas, 399-411.

Van Eetvelde, V. and Antrop, M. 2009a. A Stepwise Multi-Scaled Landscape Typology and Characterisation for Trans-Regional Integration, Applied on The Federal State of Belgium. *Landscape and Urban Planning*, 91, 160-170.

Van Eetvelde, V. and Antrop, M. 2009b. Indicators for assessing changing landscape character of cultural landscapes in Flanders (Belgium). *Land Use Policy*, 27, 901-910.

Van Eetvelde, V., Sevenant, M. and Antrop, M. 2006. Trans-regional landscape characterization: The example of Belgium, in *Landscape Ecology in the Mediterranean: Inside and outside approaches. Proceedings of the European IALE Conference, 29 March-2 April 2005, Faro, Portugal*, edited by R.G.H. Bunce and R.H.G. Jongman. IALE Publication Series 3, 199-212.

Wascher, D. (ed.) 2005. *European Landscape Character Areas. Typologies, cartography and indicators for the assessment of sustainable landscapes.* Final report as deliverable from the EU's accompanying measure project European Landscape Character Assessment Initiative (ELCAI), funded under the 5th Framework Programme on Energy, Environment and Sustainable Development (4.2.2).

Webster, R. and Beckett, P. 1970. Terrain classification and evaluation using air photography: A review of recent work at Oxford. *Photogrammetria*, 26, 51-75.

Zonneveld, I.S. 1994. Basic principles of classification, in *Ecosystem classification for environmental management*, edited by F. Klijn Dordrecht, Boston. London: Kluwer Academic Publishers, 23-47.

Chapter 29

Back to the Future:
Landscape and Planning History as a Means
to Understand Modern Landscapes

Mattias Qviström

Introduction

One of the major challenges within contemporary planning is to find ways to foster sustainable development and multifunctional use of peri-urban landscapes. However, planning is powerless without a thorough understanding of the processes shaping the land. Furthermore, there is a need for methods for landscape analysis, providing an interface between landscape studies and planning. In a peri-urban context, limited understanding of the processes shaping the land and a shortage of methods for peri-urban landscape analysis are major problems (Qviström et al. 2007). Within the discourse concerning the European Landscape Convention, the importance of integrative landscape analyses that incorporate different aspects of landscape values and potentials has been emphasized.

Contemporary humanistic landscape research provides opportunities to vitalize landscape analysis as a tool to understand landscapes in a planning situation – or to understand the planning situation through the study of a landscape. However, contemporary education and practice in landscape analysis are still based on methodologies developed in the 1960s and 1970s, with few exceptions (Qviström et al. 2007; Vouligny et al. 2009). These methodologies are constrained by divisions taken for granted between rural-urban and nature-culture and do not acknowledge the importance of landscape/planning inertia, which precludes an analysis of the peri-urban condition. Furthermore, they lack a nuanced discussion of the power relations shaping the land. Such methods will fail in an analysis of the peri-urban context, a landscape that is *characterized* by contested land uses and rural/urban hybrids (c.f. Murdoch and Lowe 2003; Gallent et al. 2006).

In order to bridge the gap between humanistic landscape research and methodological discussions on landscape analysis, there is a need for devising a method for peri-urban landscape analysis beyond the rural-urban divide that would acknowledge the contested character of modern landscapes and the appearance of rural-urban hybrids, but also the importance of former discourses and plans for the future of the landscape. Such a method could be considered a 'shadow analysis' by its reference to 'the shadow of the city' (a synonym for peri-urban areas) and also to the imminent threat (or promise) of plans and dreams not yet fulfilled. In a landscape which has been dealt with within planning for decades, such threats and promises of future development are part of the everyday landscape. In Sweden, plans, legislation and the public discourse have dealt with peri-urban development from the 1930s, so in order to guide future countryside development in peri-urban areas, we need to be aware of this legacy. With a case study of the urban fringe of Lund, this chapter illustrates the need for a shadow analysis.[1]

Case study

> According to a common view, nature starts where the city ends. Here, on the edge of the city, lies the boundary between nature and culture, between *red* and *green*, that is: between the built environment and untouched landscape. [...] the city is the enemy of nature and the front-line is the edge of the city. (Tjallingii 2000: 105)

The urban fringe is characterized by land use conflicts, conflicts between rural and urban activities, between the interests of international companies with formal plans on the one hand and local interests and informal land use on the other, and between ingrained ideas and ideals on Nature and Culture, Rural and Urban. In other words, the urban fringe crystallizes the conflicts characterizing modern landscapes. Hence, case studies at the urban fringe can offer knowledge on the interaction between spatial planning and landscape change in general. Such case studies can be especially informative concerning conflicts characterizing peri-urban landscapes. When searching for new ways to understand peri-urban landscapes, the fringe is an excellent point of departure.

The divide between rural and urban areas is one of the reasons why the city edge has been submerged in spatial plans and public policy considerations for a long time (Allen 2003; Qviström 2007). As in many other countries, the rural-urban divide described in Tjallingii (2000) is prominent within Swedish planning.

---

1   This case study was carried out as part of a 2008-2010 research project, entitled 'Shadow analysis: Examining a method for peri-urban landscape analysis beyond the rural-urban divide', implemented at the Department of Landscape Architecture of the Swedish University of Agricultural Sciences at Alnarp. The project is primarily focused on green structure planning in peri-urban areas, as well as the actual development of the green structure during the last decades. This should facilitate an analysis of the embedded ideals of Rural and Urban within spatial planning, since green planning and nature protection have often been regarded as the antithesis of urbanization. However, to reveal land use conflicts as well as the possibilities and constraints for multifunctional land-use, other fields of interests are also studied, in particular farmland preservation and transportation planning. The case studies focus on the period between the 1930s and today. In the 1930s, parallel discourses on the urban fringe were developed in North America and Europe, including Sweden (cf. Matless 1998; Bruegmann 2005).

In general, only densely built areas can be incorporated into legally binding and detailed plans, leaving the countryside and the urban fringe with strategic and advisory plans without any legal power to affect land use changes. In spite of the last decade of criticism, the rural-urban divide is mirrored in the administrative structure at local, regional and national level, the decisions of which materialize in the landscape. The divide is not only promoted by administrative structures, but is also sanctioned by disciplinary interests (Tjallingii 2000). For instance, there are plenty of academic journals on urban studies *or* rural studies, but only a few that pay any attention to rural-urban interfaces. According to Allen (2003: 135), the rural-urban divide is one of the main reasons for the difficulties handling the urban fringe: 'a distinction that (mis)informs not only the setting up of institutional arrangements but also, and more broadly, the deployment of planning approaches and tools'. To accomplish sustainable development or indeed a comprehensive understanding of the environmental impacts of urban growth, we need to overcome this divide. First, however, we need to acknowledge how deeply embedded this dichotomy is in our modern society. We will not be able to go beyond the rural-urban divide unless we acknowledge the inertia of former decisions and the way this divide has been made manifest in the landscape. Thus, we need to highlight the inertia of planning decisions based on a rural-urban divide and the manifestations of this ideology.

The neglect of the city edge is illustrated in a minor case study of land use transitions at the urban fringe of Lund in southern Sweden (Figure 29.1). The study was based on field studies in 1997, 2007 and 2009, combined with an analysis of local spatial plans (from the 1960s onwards) and of cartographic material revealing land use changes. The study employs landscape history in order to understand how planning affects the landscape, and how to improve the planning of tomorrow. Seamless overlapping between historical studies up to the present time, field studies and discussions about the landscape of tomorrow are crucial components of the analysis.

Surrounded by flat agricultural land, Lund is dominated by a densely built environment and, compared with other towns and cities in Sweden, comprises extremely few natural features that support green structure (Kristensson 2002). Furthermore, the value of the agricultural land surrounding the city has nurtured an explicit and long-lasting strategy within local and regional planning to maintain a dense city. As a consequence, Lund demonstrates an extremely pronounced border between city and country compared with other cities and towns in Sweden. A number of investigations of the planning history and urban history of Lund provided a solid background and an excellent analysis of the planning context for the detailed studies (e.g. Kristensson 2002). However, the urban fringe and the interactions between landscape change and planning history were not discussed in these studies.

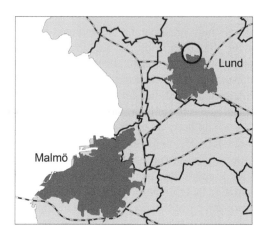

Figure 29.1 Lund and the south-western part of Scania in southern Sweden
Note: Major highways and municipal borders are marked on the map. The case study area
is indicated with a circle.
Source: Sverigekartan@Lantmäteriverket, Gävle 2010. Medgivande MEDGIV-2010-2169.

Landscape change at the urban fringe of Lund

In the spring 1997, I witnessed the demolition of a derelict house and abandoned
garden on the northern city edge of Lund (Figure 29.2). Beside the dirt road next
to the house, vast arable fields surrounded the garden and every hundred metres
or so another garden was scattered on the vast plain. A beautiful pear tree was just
about to be crushed when an old couple who lived in the neighbourhood paused to
watch the demolition. They told me they frequently passed the house on bicycle
and they mentioned how much they had enjoyed the flowering pear tree over the
years. In a landscape dominated by large arable fields, a lush garden means a
great deal, not only to humans. There is a deficit of green areas in the vicinity of
Lund, a city located in a region comprising some of the most productive arable
fields in Scandinavia and with no hills, rivers or other obstacles to the large-scale,
monotonous agricultural landscape. A week after the event, the house was gone
and all that was left was the odd mark on the ground, indicating the position of the
water-pipe.

Lund houses one of the largest universities in Scandinavia, as well as rapidly
expanding high technology industries, causing the city to grow out of its former
boundaries. The number of inhabitants of the city has increased steadily since the
1950s, and today the population (~100 000) is more than twice that of 50 years ago
(Lunds kommun 2005). Still, no matter how expansive a city is, there are always
places open to development that are not utilised. Almarcegui (2005) provides a
number of colourful illustrations of such derelict sites in the central parts of Lund,
which are currently being used for informal activities. Furthermore, one part of

Figure 29.2 Demolition of a house and a garden at the urban fringe of Lund in 1997

Note: Twelve years later, the place has still not been utilized for urban development.

a city might be expanding, whereas other parts remain static. The northern edge of Lund is one of these areas. Considering the postponement of plans for urban expansion, the northern fringe of Lund might be compared with the situation of a shrinking city; in the shadow of future developments, the landscape has been paralysed rather than utilised for development.

The northern city edge of Lund had been designated for urban development since the early 1960s, although the highly productive arable land made the city expansion problematic. Although the city did expand to the north in the early 1970s, vast areas allocated for development were not used at that time. One of these areas, between *Norra Fäladen* (the northern part of the city) and *Stångby* (a village two kilometres to the north of Lund), is the topic of this chapter (Figure 29.1). The entire case study area, approximately two square kilometres in size, was the subject of large-scale plans for city expansion in the early 1970s and in the 1980s. For instance, a plan presented to the public in 1986 contained housing for 15,000 inhabitants and 6,000 jobs at the northern fringe. However, the city has mainly expanded in other directions, partly due to conflicts with the interests of agriculture (Qviström, unpublished).

In order to prevent land speculation and to secure possibilities for the city to expand, vast areas at the fringe of the city were bought in the 1960s by the municipality of Lund (as with many other municipalities in Sweden at the time). These areas have been leased ever since (primarily to farmers), while waiting for expansion to happen. In 2005, 59 hectares of land were designated for urban development in the near future, but as a complement, the municipality of Lund owns 1,200 hectares of land called 'land reserve', which is intended to be used for urban development in the future and which is managed accordingly. Agriculture in the land reserve areas is dependent on short-term contracts (usually one-year),

Figure 29.3 The northern fringe of Lund

Note: The northern fringe of Lund is dominated by agriculture, scattered settlements and communication infrastructures. However, it is also characterized by development plans, which have affected the landscape for decades. In a modern landscape, an analysis of unfulfilled plans and strategic documents has to be a part of the analysis of the landscape.

which fosters short-term use of the land. Nevertheless, 80 per cent of the land in the reserve has been designated as such for more than 30 years (Tekniska förvaltningen i Lund 2005).

City expansion to the north has been delayed time after time, although some minor areas have been developed. The rest of the landscape has been affected by 30 years of short-term leasing contracts in the shadow of future city expansions. As far as visual analysis of 12 years of landscape changes can show, short-term agriculture seems to have fostered economically viable activities with disregard for recreational values, biodiversity or the interests of its neighbours in the city. Short-term contracts are likely to endorse short-sighted land use.

In a landscape dominated by arable land, abandoned gardens (as well as small biotopes along roads and railways) are important for biodiversity and recreation. In the comprehensive plan of Lund, the importance of such biotopes is clearly stated and the need to preserve and develop green structures at the urban fringe is mentioned. In particular, the need for clumps of trees in the open landscape and vegetation along roads and boundaries is mentioned in the plan (Stadsarkitektkontoret i Lund 1998). However, as mentioned before, the comprehensive plan is a weak planning instrument and the case study tells a different story as regards the past decade of landscape change. Prior to the visions of city expansion, there were no abandoned gardens in the case study area. In 1970-1997, 13 settlements (primarily small farmsteads) were abandoned in the area, possibly due to the threat of city expansion. Nine of these were still there in 1997 (Qviström, unpublished). Visiting the area in 2007, I find it difficult to locate the site of the bulldozed garden; the land has been ploughed, but is now uncultivated and it is impossible to detect any differences between the arable field and the former garden. Two of the abandoned gardens mentioned above have been bulldozed and replaced with new settlements, a third one has been covered with excavated material and yet another is partly used as a dump for garden waste.

After decades of utopian plans for city expansion, most of the study area was designated a valuable cultural landscape to be protected in the 1998 local comprehensive plan (Stadsarkitektkontoret i Lund 1998). However, recent plans for a golf course and school in the area illustrate how weak this 'protection' is. In reality, urban expansion is still a threat, and unless the municipality decides to act differently, the landscape will probably be even more monotonous and dominated by large-scale agriculture in a few years' time.

Discussion

In current planning, ideas of multifunctionality and the urban fringe as an asset for recreation have started to affect strategic plans. However, the former ideals prevail, not only in the administrative and legal structures, but also due to the impacts of former planning documents and leasing contracts on the landscape. As illustrated in the case study, past decisions within planning materialize in the landscape long after they have been approved. A landscape analysis of peri-urban landscapes needs to include an analysis of former plans and decisions that could affect the landscape. A limited understanding of the planning incentives that created the present day situation makes it difficult to manage this landscape within planning. We need to explore the interactions between previous planning and landscape changes, as a basis for understanding present day landscape.

Landscape changes taking place at the fringe while waiting for the city to expand are far from easy to understand, but an analysis of the interplay between spatial plans and everyday actions in these landscapes can reveal some of the processes shaping the land. Qviström and Saltzman (2006) studied this interplay in an area at the city edge of Malmö (Sweden), illustrating the development of a green area in the shadow of city expansion. The new settlements were postponed time after time, and finally the area was rediscovered by planners and protected as a nature park. Conversely, the plans for future development can decrease variation in the landscape at the fringe, as illustrated in this chapter. In contrast to Malmö, the case study area in Lund was dominated by short-term land use contracts. The importance of such contracts for the development of the landscape at the urban fringe has so far not been studied thoroughly, at least not in Sweden. There is a need for further landscape studies exploring the gap between utopian dreams of city expansion and the present day situation at the city edge, if we are to take full advantage of the possibilities for sustainable development of the city.

Acknowledgements

The research was funded by FORMAS, the Swedish Research Council for Environment, Agricultural Sciences, and Spatial Planning. The author wishes to thank Lars G. B. Andersson for GIS support.

References

Allen, A. 2003. Environmental planning and management of the peri-urban interface: Perspectives on an emerging field. *Environment & Urbanization*, 15, 135-147.

Almarcegui, L. 2005. *Guide to the Undefined Places of Lund*. Lund: Lunds konsthall.

Bruegmann, R. 2005. *Sprawl. A Compact History*. Chicago: The University of Chicago Press.

Gallent, N., Andersson, J. and Biaconi, M. 2006. *Planning on the Edge: The Context for Planning at the Rural-urban Fringe*. London and New York: Routledge.

Kristensson, E. 2002. *Relationer mellan natur och kultur i Lunds stadsplanering 1950-2000*. Del 6: *Den kultiverade naturen: Grönskan på Östra Torn*. Lunds: Inst. för arkitektur, Lunds universitet.

Linehan, J. and Gross, M. 1998. Back to the future, back to basics: The social ecology of landscapes and the future of landscape planning. *Landscape and Urban Planning*, 42, 207-223.

Lunds kommun. 2005. *Lunds statistik 2005*. [Online] Available at www.lund.se/templates/Page____1363.aspx [accessed: 1 February 2010].

Matless, D. 1998. *Landscape and Englishness*. London: Reaction Books.

Murdoch, J. and Lowe, P. 2003. The preservationist paradox: Modernism, environmentalism and the politics of spatial division. *Transactions of the Institute of British Geographers*, 28, 318-332.

Qviström, M. 1998 (unpublished). *Om förloppslandskapet*.

Qviström, M. 2007. Landscapes out of order: Studying the inner urban fringe beyond the rural–urban divide. *Geografiska Annaler Series*, 89B, 269-282.

Qviström, M. and Saltzman, K. 2006. Exploring landscape dynamics at the edge of the city: Spatial plans and everyday places at the inner urban fringe of Malmö, Sweden. *Landscape Research*, 31, 21-41.

Qviström, M., Larsson, A. and Ode, Å. 2007. A peri-urban landscape analysis as a strategic instrument in environmental planning, in *Landscape Assessment – From Theory to Practice: Applications in Planning and Design*. Belgrade: University of Belgrade.

Stadsarkitektkontoret i Lund 1998. *Översiktsplan för Lunds kommun*.

Stadsbyggnadskontoret i Lund 2004. *Program till detaljplan för ladugårdsmarken 5:9 m fl (Golfbana) och del av 5:8 m fl (Rekreations- och idrottsområde)*.

Tekniska förvaltningen i Lund 2005. *Markpolicy för Lund.* Lund.

Tjallingii, S.P. 2000. Ecology on the edge: Landscape and ecology between town and country. *Landscape and Urban Planning*, 48, 103-119.

Vouligny, É., Domon, G. and Ruiz, J. 2009. An assessment of ordinary landscapes by an expert and its residents: Landscape values in areas of intentive agricultural use. *Land Use Policy*, 26, 890-900.

Chapter 30

A New GIS Procedure for Peri-Urban Landscape Change Detection

Enrico Borgogno Mondino, Barbara Drusi and Roberto Chiabrando

Introduction

This work is aimed at stimulating the adoption of advanced tools of GIS (Burrough 1986) to improve the reading and interpreting of the territory for urban planning purposes. The basic idea is to proceed to define a suitable way to formalize and represent spatial dependent functions for territory (or landscape) qualification by recognizing, from existing maps, those themes describing, directly or indirectly, the conditioning factors. This makes possible a change detection analysis between consequent periods, allowing planners to evaluate and quantify (both in strength and direction) the forces that drive urban growth. According to this purpose the proposed methodology is mainly in charge of generating graphical representations (maps) explicitly showing the dynamics of urban expansion towards rural areas. It is worth stressing that the main goal of this work is to show how GIS approaches operate effectively in this context; it is not the authors' intention to face and solve the problem of which landscape quality function is the most suitable. The one proposed during the tests should be seen as an example aimed at showing how the procedure runs. From this point of view, GIS can represent a useful device in the hands of planners to make their decisions more objective and their plans clearer. This is obtained by mapping space dependent landscape quality functions whose definition is strictly related to information derivable from existing maps. Their numerical nature permits us to consider them as universal descriptors.

Test area

The application of GIS technologies, both as an archive and a dynamic tool aiming at town-planning management, has been tested upon a peri-urban field within the municipal territory of Grugliasco. The investigated area lies within the first urban belt of Torino and is mainly characterized by the permanence of historical, valuable farmhouses surrounded by agricultural facilities which are prevailingly disused at present. The erosion of the traditional rural landscape and the pressure on farmhouses due to urban expansion after the Second World War are especially evident on the north-western fringes of the settlement, near the University campus.

Radical changes have deeply modified the urban frame starting in the second half of the last century, when major engineering industries and car factories – such as Pininfarina, Bertone, Westinghouse, etc. – settled in Grugliasco. Building activity frenetically increased as well: houses, infrastructures, industrial and commercial firms storm or annihilate canals and trails, rural buildings and productive fields. Some historical farmhouses progressively turned to dereliction and the rural landscape became more and more unreadable up to now.

The comparison between historical maps – from the *Napoleonic Cadastre* through the *Rabbini* maps to the *Origin Cadastre*, respectively dating back to the first decades, the second half and the end of the nineteenth century – as well as two topographical surveys dating back to the second half of the eighteenth century allow the recognition of the historical 'permanence' of any landscape structural invariants compared to the 'fragile' elements susceptible to change. In facts, such diversity effectively appears meaningful to the aim of landscape town-planning.

Regarding the historical farmhouses still existing within the investigated area, both the analysis of cartographic sources and the in-field survey support the permanence of a building type built in parallel upon a courtyard with an east-west orientation.

Available data

In order to exemplify the procedure, the following two maps describing the test area were selected:

1. the 1:10000 scale Technical Map of the Piemonte Region (CTRN_10K);
2. the 1:25000 scale map of the Italian Military Geographical Institute (IGM_ 25K).

The CTRN_10K dates from 1992 and was supplied directly in a digital vector format.

The IGM_25k is dated 1968 and was supplied as paper copies of two map sheets named respectively *056 III-SE (Torino)* and *056 III-SO (Rivoli)*. The reference system of both the data is the projected Italian National one named GAUSS-BOAGA.

Data pre-processing

All the described operations were done using the tools available in the commercial software ESRI ArcGIS 9.2.

Hardcopy map scanning

While dealing with old maps, it is common to obtain paper copies that must be preventively transformed in digital data. The first step is certainly their digitizing using appropriate scanners. During this operation the scanning resolution must be defined according to the nominal accuracy of the considered map. As far as this work is concerned, the IGM_25K sheets were scanned with a resolution of 300 DPI, considering as nominal accuracy of the map that declared for the Italian cartographic context: 0.2 mm at the scale of the map, which is 5 meters on the ground.

This means that the physical size of the pixel of the scanned image must be largely lower than 0.2 mm. If a resolution of 300 DPI is selected the resulting pixel size of 25.4/300 = 0.085 mm is appropriate.

Figure 30.1 Left = IGM_25K map (1968); Right = CTRN_10K map (1992)
Source: Piedmont Region and the Military Geographical Institute.

Image georeferencing

In order to proceed to a multi-temporal analysis between maps of different periods, the first mandatory requirement is that they must be co-registered. A vector map is a natively georeferenced data, while a scanned map is just a digital image without any geocoding information. In order to assign it a reference system a georeferencing process must be performed (Anibaldi et al. 2007). If one of the data to be compared is already georeferenced (CTRN_10K) this can be assumed as reference map to collect the Ground Control Points (GCP) required to calibrate the selected geometrical transformation. In this case, assuming that the scanned maps (two sheets) originally represented a flat representation, a simple bi-dimensional correction model was selected. In particular, a second order Polynomial was needed to keep the Root Mean Squared Error (RMSE) low enough. Sixteen well distributed GCPs were collimated for each of the scanned maps. The RMSE was respectively 5.3 meters for the *056 III-SE (Torino)* and 4.6 meters for the *056 III-SO (Rivoli)*. These values were considered satisfactory when compared to the declared accuracy of the maps (5 m). A unique digital map was then obtained by mosaicking the two georeferenced images. These operations were done using the *Georeferencing Tool* available in ESRI ArcGIS 9.2.

Layer extraction

While trying to qualify the territory the first step is to select from maps that information useful to describe it appropriately. The task is not so easy, as it requires experience in map reading and content synthesis. Furthermore, the selected information must necessarily be converted in such a representation suitable to be managed inside GIS. At this stage of the process, landscape experts should converse with surveyors, or map experts, to define the type of information useful for them; the main constraint is that the selected information must be derived from the available maps in order to ensure they can be spatially represented. For this work, to exemplify the procedure, extremely basic and simplified information was extracted from both maps and represented as three vector layers. According to an approach aimed at representing how the urban growth pushed against the rural context in the considered period, two themes ('roads' and 'buildings') were retained as degrading factors, while the third one ('open land') was deemed an improving factor.

The layers were generated for both periods. For the CTRN_10K (1992), as it was supplied in a vector format, the operation was done simply by selecting the already present layers and saving them separately. For the IGM_25K, the operation was done opening an editing session inside ArcGIS and interpreting the symbology of the map itself. Finally the following layers were available:

1. Buildings (*BUI*): polygon layer representing built areas (or single buildings);
2. Roads (*ROAD*): road network represented as a polyline layer;
3. Open land (*LAND*): polygon layer representing rural or natural areas;

The vector format is the most suitable to be managed inside a GIS. The main peculiarity is that a vector layer is always equipped with an attribute table containing information useful to qualify each geometrical feature of the considered layer (Burrough 1986). For this work an attribute was added to the attribute table of all of the layers and filled with a numerical code to be used as a weight, subjectively defined by the expert, addressed to quantify the impact (improving or degrading) of each factor (layer) onto the global quality of the territory.

For example, an industrial area will be described by a higher weight because it is reasonable to suppose that it acts with stronger degrading force over the rural territory than a residential one (lower weight). This attribute, once assigned, enables us to compute space-dependent indices aimed at continuously measuring the quality of the landscape around the area.

This editing step of the process thus concerns both the geometrical and the semantic aspect of the layers. It is quite obvious that, at this stage, there is a considerable element of subjectivity in the interpretation, requiring, as already told, good synergetic work between landscape and map experts.

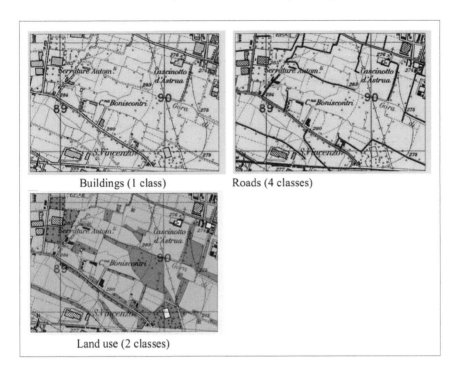

Buildings (1 class)   Roads (4 classes)

Land use (2 classes)

Figure 30.2 New layers containing information useful to qualify the territory were extracted or vectorized by editing from available maps

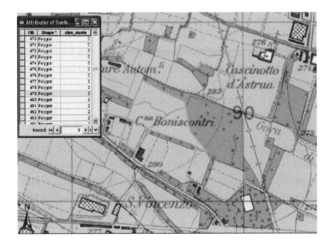

Figure 30.3 Weight assignation to the features of the vector layers

Mapping the landscape quality function

For each time (1992 and 1968), a landscape quality function was defined as a regular sampling of an appropriate numerical index (Wissen et al. 2005). According to this definition, the most suitable format to represent it is the raster one: the quality map must be thought as a matrix having an adequate resolution (pixel size) that describes the spatial variability of an index measuring the local landscape quality (Figure 30.4).

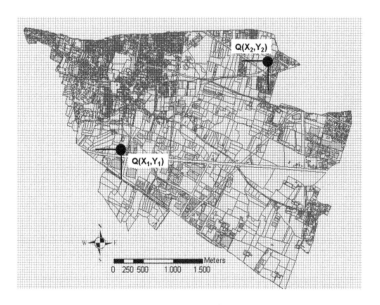

Figure 30.4 The landscape quality function $Q(x,y,t)$ can be considered a regular sampling of a numerical index (spatial and time dependent)

Note: The GRID format is the best one to represent this event.

A huge amount of suggestions is given in literature to define landscape quality indices. It is not our task to state which is the best to be adopted; thus, for this project a very simple definition is given [1]. Any other definition can be adopted; the only constraint is that it must be based on environmental factors that can be mapped.

$$Q(x,y,t) = [a_1 LAND(x,y,t)] - [a_2 ROAD(x,y,t) + a_3 BUI(x,y,t)] \qquad [1]$$

where a_i are the weights assigned to each conditioning factor whose representation is obtained by generating the LAND, ROAD and BUI index matrices. Such elementary definition, based on a sum, allows one to easily separate those factors 'usually' improving (plus sign) the rural landscape (open land) from those 'usually' degrading it (minus sign). Nevertheless, this structure gives the expert the possibility to operate some exceptions during the assignation of the weights while filling the attribute tables of the layers. In fact, sometimes, a feature 'usually' degrading can be retained as improving the landscape (e.g. an historical building). A simple change of the sign (from plus to minus) of the weight in the table produces this effect for the selected feature.

Index maps

During the definition of $Q(x,y,t)$ a fundamental role is played by the $ROAD(x,y,t)$, $BUI(x,y,t)$ and $LAND(x,y,t)$ index maps. During the tests, they were all produced with a cell size of 30 meters.

The *LAND* index map was obtained by direct rasterizing of the correspondent polygon layer, defining as reference field of the attribute table the one containing the previously assigned weight.

ROAD and *BUI* index maps, instead, were generated exploiting the Spatial Analyst tools of ArcGIS; in particular the correspondent *Allocation* and *Distance* matrices were calculated.

The *Allocation* matrix $A_i(x,y)$ contains for each pixel (whose size is defined by the operator) the value of a selected attribute (field) of the feature, of the processed layer, nearest to that position.

The *Distance* matrix $D_i(x,y)$ contains for each pixel the value of the Euclidean horizontal distance that separates its position from the nearest feature of the processed layer.

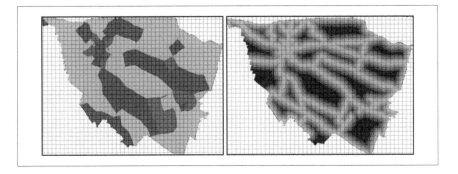

Figure 30.5 *Allocation* (left) and *Distance* (right) obtained from the layer ROAD

Both for the ROAD and the BUI layers the correspondant $D_i(x,y)$ and $A_i(x,y)$ matrices were combined according to [2] under these hypotheses:

1. the contribution of the considered factor decreases while increasing the distance from the nearest feature;
2. its initial (and maximum) value is the one corresponding to the recorded weight in the attribute table of the processed layer for the nearest feature.

The constant values 1000, 10 and the INT() operator were just introduced to exclude numerical problem during index computation.

$$I(x;y) = \text{int}\left[\frac{1000 \cdot A_i(x,y)}{(D_i(x,y)+10)}\right] \qquad [2]$$

As the assigned weights can have different scales for the different layers, before proceeding to combine the index maps along the quality function formula, it is better to normalize them. During this experience the normalization was obtained in a statistical way by calculating the Standard Normal Variable

$$I_z(x,y) = \frac{(I(x,y)-\mu(I))}{\sigma(I)} \qquad [3]$$

where $\mu(I)$ and $\sigma(I)$ are, respectively, the mean and standard deviation values of I.

Landscape quality function computation

The normalized indices, mapped in the format of matrices, were then combined according to the definition of $Q(x,y,t)$ by using the *Raster Calculator Tool* available in ArcGIS. A new raster map of Q was obtained both for the 1992 and 1968 times (Figure 30.7). It can be noticed that a considerable element of subjectivity is still present in the declaration of the a_i coefficients.

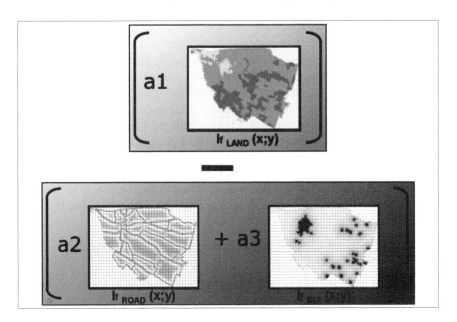

Figure 30.6 Index maps were combined through the raster calculator of ArcGIS according to the formulation of *Q(x,y,t)*

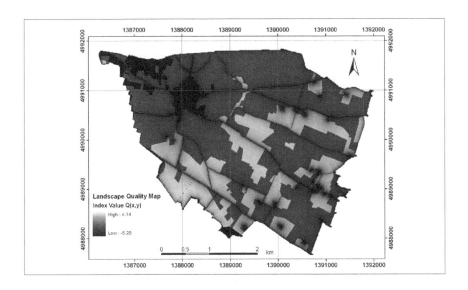

Figure 30.7 Landscape Quality Map *Q(x,y,t)* **for the test area as derived from CTRN_10K (the adopted weights values are** $a_0 = a_1 = a_2 = a_3 = 1$**)**

Landscape change detection

Landscape changes can be represented in different ways. The authors' intention was to produce some valuable representations in which both the technical aspects and the communicative ones can be equally and effectively perceived by planners.

The proposal presented here is a type of representation in which changes are formalized through a vector map where the strength and the direction of the acting degrading forces of the urban growth against the rural areas are shown.

Change means movement and movement means time passing. Thus it is just by comparing landscapes of different times that it is possible to give a reasonable answer to this need. From a practical point of view the solution is to compare the 1992 and 1968 quality maps $Q(x,y,t_1)$ and $Q(x,y,t_2)$ through a simple matrix difference (see [4]).

$$D\ (x,y) = Q(x,y,t_1) - Q(x,y,t_2)$$ [4]

The obtained result is a new map showing the local variations in landscape quality which occurred within the considered time period. This representation permits one to investigate the spatial distribution of the changes, recognizing not only the degraded, or improved areas, but also the direction and strength of the forces that produced those changes.

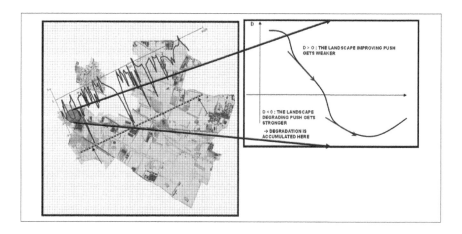

Figure 30.8 (Left) Landscape Quality Difference map *D(x,y)*
in the period 1992-1968

Note: The black straight line shows how, from this representation, it is possible to extract arbitrary profiles (sections) of the D function; (Right) A particular of an extracted example profile is shown. Consider how this type of representation permits to recognize the strength and the direction of the forces that produced the changes along the profile.

Formalization of the urban growth forces

According to the available tools inside a GIS, the visualization of results (landscape change maps) can be performed in different ways that depend on the required application. In this work, we propose a bi-dimensional representation where the Landscape difference map $D(x,y)$ is considered as a field of variation of the investigated force (which is the one causing rural degradation).

The desired map was generated using a vector symbology able to explicitly show the direction and strength of the degrading push. The orientation and size of the vectors showing the local direction and strength of the forces were calculated by considering $D(x,y)$ as a three dimensional surface, that is a sort of Digital Surface Model. According to this approach, by using the *3D Analyst Tools* of ArcGIS, it was possible to calculate for each pixel its *slope* and *aspect*. *Slope* was assumed as indicator of the strength of the local force, while *aspect* as indicator of its direction. Two raster maps showing the local slope and aspect, respectively, where generated with a cell size of 100 meters. From these maps a new vector representation of point type was derived by transferring the information of strength and direction of the changes into two appositely designed fields of the correspondent attribute table. The new point layer was than used to generate the result shown in Figure 30.9.

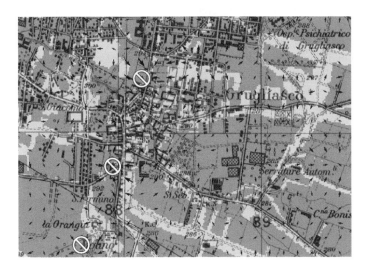

Figure 30.9 Landscape change map showing, through vector symbology, the direction and strength of the degrading forces occurred in the considered period (1968-1992) produced against the rural areas

Note: Colour codification is used to enhance those areas where a landscape improvement (light grey) or degradation (dark grey) was observed. White circles identify those points that suffered from the highest negative changes.

Conclusions

The performed tests demonstrate that GIS can effectively be used as a powerful tool to carry out landscape assessment and especially to represent the obtained results. Specifically, it is worth stressing the importance of the space-dependent index maps, whose correct definition allows one to locally measure the importance of those factors retained as conditioning the landscape. By opportunely combining them it is then possible to map landscape quality function useful to immediately identify at a given time the most afflicted or thriving areas. Furthermore, by comparing quality maps of different periods as we proposed, it is possible to calculate and represent the direction and strength of the acting degrading, or improving, forces at play in the reference period. If the compared situations are referred to the past, the information we gather helps us understand past dynamics. But if the same approach is adopted comparing the present situation with a planned future one, the information we obtain can be used to evaluate the worth or limits of the defined solutions, giving planners a further tool to assess their interventions and eventually re-calibrate them. Nevertheless, some limitations can be easily recognized: firstly, the proposed methodology is based on simplified hypotheses mainly related to the definition of the indices and of the landscape quality function. Anyway, if the definition changed, the methodology would continue to operate appropriately. The only required constraint is that different definitions permit a spatial representation of the function based on factors that somehow depend on the information that available maps can supply.

A second limitation is related to the persistence of a substantial element of subjectivity during the extraction of the starting information (the layers) from the available maps. Bad interpretation can lead to a wrong description of the reality. At this stage, good synergetic work must be done by landscape and map/surveyor experts.

References

Anibaldi Ranco, M., Borgogno Mondino, E. and Garnero, G. 2007. *Mosaicatura e georeferenziazione della carta degli Stati Sardi*, Atti XI Conferenza Nazionale ASITA, Torino.

ArcGIS 9 – Using 3D Analyst (2004), Software User's Manual, ESRI, USA.

ArcGIS 9 – Geocoding in ArcGIS (2004), Software User's Manual, ESRI, USA.

ArcGIS 9 – Using ArcMap (2004), Software User's Manual, ESRI, USA.

Burrough, P.A. 1986. *Principles of geographical information systems for land resource assessment.* Oxford: Clarendon Press.

Chiabrando, R., Drusi, B. and Garnero, G. 2000. GIS as an instrument for historical territory analysis and its sustainable development management. A methodological application in the historical centre of Aosta, in *Histocity Book. The best of 1998-2000 Network Research in the historical cities sustainable development using GIS*, edited by M.A. Esposito. Atti del Convegno Internazionale Histocity 2000. Firenze: Alinea, 173-186.

Drusi, B. 2005. *Permanenze rurali nel paesaggio urbano. Aosta fra XVI e XVIII secolo*. Torino: Celid.

Wissen, U., Schroth, O. and Schmid, W.A. 2005. *Comprehensive Evaluation of Future Landscape Quality by Joining Indicators and 3D Visualisations*, Proceedings of the Conference on 'Visualising and Presenting Indicator Systems', Neuchâtel: Swiss Federal Statistical Office.

In Lieu of Conclusion: Changing Scope and Ambitions in Landscape Concerns

Paul Claval

This chapter brings forward a contextualization of the evolving landscape research, planning and policy concerns, which culminated in The European Landscape Convention, and a review of the contents of this book in terms of contribution to new insights about the landscapes-identities-development interface.

From landscapes as stable structures to landscapes as perceived realities

The initial emphasis on agrarian landscapes

The study of agrarian landscapes appeared as a central theme in human geography or rural history in the fifties. Agrarian landscapes were conceived as structural realities shaped by the logics of agricultural production. They depended on topography, the nature of soils, crop rotation, the combination of cultivation with cattle, sheep or goat breeding, the technology of ploughs and other agricultural tools, the way land plots were delimited and in many cases enclosed, the need to develop sustainable forms of production in societies where there were neither sources of concentrated power usable for farm work nor mineral fertilizers. They reflected the social organization of rural cells.

Because of the lack of communication facilities, the diffusion of innovation was a slow process: it explained partly the homogeneity of most of the agrarian landscapes and the occurrence of the same features over wide areas.

At that time, landscapes were conceived as stable realities or structures. This was true both in the natural field (climax vegetation) and the human (agrarian landscapes).

Modernization and the new dynamics of rural landscapes

Rural landscapes changed dramatically during the last 50 years. The dynamics at work resulted from the universal availability of concentrated forms of energy, which facilitated the mechanization of work, the consolidation of farms, the use of fertilizers and pesticides and a deeper specialization of productions. With higher productivity, the share of agriculture in employment decreased dramatically, which resulted in a rapid migration of rural populations towards urban areas. With higher mobility, competition between agricultural producers became harsher: hence the need to adapt permanently productive activities to changing market conditions, which destructured the former stable agrarian landscapes and generally prevented the formation of new ones.

More recently, and because of higher mobility and new communication media, the composition of rural population changed. Many areas underwent processes of suburbanization or rurbanization, with a lower proportion of farm hands, more industrial workers, clerks and retirees and the growing impact of tourism. Rural landscapes increasingly appear as landscapes of consumption rather than landscapes of production.

The progress of communication media did not transform only local or regional relations. For the first time it allowed contact with faraway correspondents instantaneously and without any filtering process; in this way, it induced glocalization, i.e. the possibility for every locality to access directly to global networks of communication. Hence a deep psychological transformation of the relations between former rural and urban areas.

In our world, rural landscapes are never stable structures. They are undergoing a permanent process of change, because of the new economics of farming or cattle breeding, the progress of suburbanization and rurbanization, the increased dynamism of tourism, and the changing status of those who dwell in rural areas (Renes).

Other orientations in the study of landscapes

There were other ways to develop the study of landscapes. From the end of the nineteenth century, *ecologists* built an interpretation of landscapes focused on living beings and the relations they had with their environment. Because of the role of landscape in painting and the taste for gardens and parks, the *aesthetic dimension* of landscape perception attracted *art historians;* they stressed the symbolic dimension of many man-made environments, and, with the new taste for the sublime, of natural ones. Landscapes bore witness to the deep forces of nature or were associated with the history of human groups, shaped their sensitivity and were marked by the heroes who built them: as a result, they took on a symbolic dimension and appeared as valuable *heritages*, which had to be *preserved*. There were also *planning and development* specialists who knew how to transform a natural area into a successful agricultural tract, how to design gardens or parks, how to develop human settlements, either rural or urban. Planners are increasingly aware of the role of initiatives coming from below in the process of development: this means that local conditions are often strategic variables; proximity allows for easier diffusion of pertinent information; amenities give places their attractiveness.

As a result of the transformation of rural landscapes, it increasingly appeared, in the 1970s and 1980s, that it was impossible to understand how rural areas were shaped without a cooperation between ecologists, geographers, historians, art historians, landscape architects, natural or historical preservation specialists, and planners. At that time PECSRL – Permanent European Conference for the Study of the Rural Landscape, organized by Xavier de Planhol in Nancy in 1957, became a forum where all with an interest in landscape could meet, exchange their views and develop shared projects. It became obvious that the dynamics of rural landscapes shared many facets with that of townscapes. The distinction between rural and urban realities ceased to be as pertinent as in the past.

The time of sustained development

The growing use of fossil energy had deep impact on ecosystems: an increasing part of them is now unable to recycle all the elements which are injected into them. They have ceased to be resilient. It was already true, locally, in the black countries at the time of the industrial revolution; it became a serious problem in all urban areas since the beginning of the twentieth century. There are now dangerous regional imbalances. The whole planet is threatened by the transformation of the atmosphere induced by the increasing use of concentrated forms of energy. Hence the emphasis on what the Brundtland's Commission called sustained development.

This means that the study of landscape as a social construction or as a perceived value cannot be dissociated from the ecological approach: in order to restore global balance on the earth, all ecosystems – and all landscapes – have to be monitored and, in many cases, transformed in order to reduce the production of greenhouse effects gases and increase the production of renewable forms of energy.

The epistemological turn of the last 30 years

During the nineteenth and early twentieth centuries, human and social sciences were generally structured in accordance with the model of physical or natural ones. Major epistemological turns occurred during the last thirty years of the twentieth century: the linguistic turn in history, the cultural turn in geography and the spatial turn in most social sciences. Social scientists discovered that they had no direct access to human reality (they know it through the filters of language) or to environment (they see it through their own perception of the world, or the perception that the human beings they study have of it). Since women and men never have direct access to reality, the way they perceive it depends on the time and place they live or they visit: hence a new cultural, historical and geographical relativism.

The European Landscape Convention

Landscape policies were for long developed for the protection of species and the preservation of the Earth's biodiversity: they were conceived by natural scientists.

Natura 2000, with its emphasis on the preservation of sites, is a good example of such an orientation.

New trends appeared more recently. The European Landscape Convention was issued by the Council of Europe in 2000. Its two first articles involve a deep change in the way landscape policies are thought:

> Landscape means an area, as perceived by people, whose character is the result of the action and interaction of natural and/or human factors. (art.1)

> [...] This Convention applies to the entire territory of the Parties and covers natural, rural, urban and peri-urban areas. It includes land, inland water and marine areas. It concerns landscapes that might be considered outstanding as well as everyday or degraded landscapes. (art. 2)

The first article defines landscapes as perceived realities: they cease to appear as a field dominated by experts, either ecologists, geographers, art historians, landscape architects, preservationists, planners or developers. To know the value of visual environments, the ultimate reference is the population, its sensibility, its preferences, its memory. This means that landscape policies have to be developed on new bases.

The second article enlarges the scope of landscape policy to the whole Earth: the aim is no more to protect areas of outstanding beauty, landscapes of emblematic or symbolic significance or nature reserves. Ordinary landscapes, those of daily life, have to receive as much attention as the most remarkable ones.

New sensitivity to landscapes

To what extent does the identity crisis of modern societies result from the destruction and disruption of traditional landscapes? Is it possible to solve it through the development of new attitudes and new policies concerning landscapes? This book offers responses to such dilemmas and expresses the new sensitivity to landscapes, the new issues they raise – particularly those of heritage and sustained development – and the new way landscape policies are set.

Emerging realities and concepts

In contemporary economies, landscapes do not appear only as the visual translation of economic activities; they are valued for their harmony, their beauty, their quietness; their economic role is increasingly located on the consumption side of the economy: they contribute much to the attractiveness of places for tourism, or new housing – but the transformations they fuel often induce a decay of the landscapes they thrive upon (T. Terkenli). With the global imbalance of the Earth's system, which results from the growing use of fossil fuels, people have increasingly to rely on renewable forms of energy (M.J. Fortin et al.; M. Frolova and B. Pérez Pérez).

This transformation in landscape economy has had consequences on the study of landscape history: it ceased to be mainly focused on the genesis of agrarian landscapes. It stresses more the new perception of landscapes as heritage, and the way new forms of consumption economies are built on it. As we live at the time of the common quest for sustainable development, planning has to reduce some forms of activities and favour new ones. These transformations have deep impact on landscapes: how to reconcile these changes with the preservation of valuable inherited features and the shaping of harmonious new forms?

In order to deal with the new perspectives on landscapes, new conceptual tools have to be developed.

How to grasp, for instance, the subjective nature of landscapes? T. Arnesen suggests relying on Peirce's semiosis when studying the way we perceive the environment: the process involves an object, a sign and an interpreter. The sign reflects both the nature of the object and the intentionality of the participant. It means that the same object may be perceived differently by different participants (on a somewhat similar line, A. Printsmann et al. speak of *ideational landscapes*). In a way, there as many readings of a landscape as there are individuals or groups with different traditions and prospects. The dynamics of subjective landscapes reflect both transformations in the objects and changes in the conceptions and aims of those who observe and use them. A homogeneous environment gives birth to different subjective landscapes, which may evolve because of the transformations of human sensitivity and interests.

Forty years ago, Yi-fu Tuan showed the significance of *topophilia* in the study of human geography. The term is a general one: people like places because they are quiet and their landscapes harmonious, or because they are busy and lively. The taste someone develops for a particular place does not mean that they are ready to participate actively in its development. Hence the need to distinguish, within the wide array of feelings that are summed up as topophilia, the part which is geared to developmental action. Z. Roca et al. propose to call it *terraphilia*. Its assessment is essential for understanding the contribution of local initiatives to sustained development.

Landscapes are generally multifunctional realities which offer goods and services. This allows an understanding of the economics of landscapes (Simoncini et al. 2009): the assessment and financing of landscape goods (agricultural or forestry production) may be provided by market mechanisms. Landscape services are rooted in the (mainly visual) externalities they offer: amenities, tourism, the building of identities, etc. Their creation, management or preservation has a cost, and the services they provide have a value, but the persons who benefit from them have nothing to pay for their use. To be financed, these services have to be considered public goods. The problem in liberal systems is that actors do not care very much about public good and public interest (which landscape definitely is), but more about their own interests.

F. Day and J.W. Vaughan show, for instance, how the incorporation of landscape values in the choice of local residents is responsible for the strong and in many ways irrational segregation between the hill country and the black soil plain to the West and the East of the Austin-San Antonio corridor.

Landscapes are increasingly significant on the consumption side of the economy. Most traditional landscapes date from a time when they reflected production. The farming systems and communities that created them are no longer functional, or have ceased to be viable. How to finance landscape quality? It cannot be guaranteed by preservation only, and has to incorporate change.

Choosing an economic approach shows how important it is to develop research on property rights and property regimes, i.e. the combination of property rights and administrative regulations: it is through their definitions that the economic value of landscapes is integrated into land price mechanisms. As a consequence, property rights could play a central role in building more coherent and efficient landscape policies. A systematic study on the landscape regimes of Europe should certainly be rewarding.

Methodologies, old and new

The main traditional approach to the study of landscapes was functional: it analysed the way each plot of land was used and contributed to the economic, social or cultural life of human groups; it showed how this role shaped its form. Landscape analysis was also archaeological, since some structures were born at a different time, in another social and economic context, and were adapted to new conditions without undergoing overall restructuring. In order to reconstitute past landscapes, scientists had to rely on the ruins or remnants that could be still observed, on the landscape structures which had survived changing conditions, and on historical documents: descriptions, drawings, paintings, old maps and plans, cadastres. These tools have always to be used when dealing with past landscapes: M. Abreu had to work as private eye to build an image of the landscape of sugar plantations around the Bay of Rio in the seventeenth century! Other good examples of this approach are provided by W.E. Doolittle concerning the Spanish aqueducts in Mexico, or by H. Skokanová and Tereza Stránská in the South Moravian region of Czech Republic, where forested areas were stable, but the land use of other areas often changed. G. Isachenko and A.Y. Reznikov rely on the historical documentation on the transformations of the suburban landscapes of Saint Petersburg and on the dynamics of vegetation to feature the future of leisure areas in the vicinity of this great city. The encroachment of farms on peat bogs mirrored a long period of demographic pressure and the arrival of Karelian refugees; their later retreat results from the industrialization and urbanization of Finnish society (M. Tanskanen).

Most historical studies of landscapes now focus on recent evolution. It is certainly not a good thing. J. Renes considers that 'we need more insights into the processes of landscape change and in the resilience of different landscape features'. He pleads for long-term studies, *biographies* of landscape: they rely on the documents explored by historians, archaeologists and geographers. And he goes on: 'It is not the primary task of landscape historians to make landscape museums […]. But our primary task is to help in developing concepts for ways of modernisation that combine a living landscape, a prosperous population, a high biodiversity and an interesting heritage'. Long term historical evidence is needed in order to achieve such a result.

The shift towards perceptual studies of landscapes has had important methodological correlates. In order to understand how landscapes are valued, interview techniques are increasingly used (B. Castiglioni et al., A. Mavridou and T. Kizos, M. Frolova and B. Pérez Pérez). Z. Roca et al. organized workshops and discussions in order to assess the *terraphilia* of the populations they studied. This is a fundamental change when compared with the methods which dominated landscape studies a few decades ago. Narratives on landscapes are explored. Films may offer efficient means to communicate the universal importance of European cultural landscapes, and contribute to an understanding of our common European cultural landscapes.

Online participation is sometimes used for assessing scenic landscape quality. Sophisticated techniques are often mobilized to assess the psychological dimension of what we observe. In some cases, landscape research explores the possibilities offered by the cognitive sciences.

Innovative cartographic techniques and the use of GIS

In order to map present landscapes, studies rely on photos and remote sensing. For past landscapes, studies rely on topographical or cadastral maps and the sketches often prepared by travellers, explorers or topographers (C. Zeballos et al., Skokanová and T. Stránská). They sometimes take advantage of statistical evidence. They turn to paintings, novels or travel accounts when looking for the forms which were then preferred.

In order to identify and describe landscape units, which is important when striving to understand how identities are built, reconstruct the dynamics of landscape or plan their future forms, GIS is often used. It also helps to detect landscape change (E. Borgogno Mondino et al., C. Zeballos et al.); GIS and cluster analysis may be combined, as shown by V. Van Eetvelde and M. Antrop in Belgium. In order to take account of all the information they gather, landscape cartographers are increasingly imaginative, as shown by many of the chapters.

Landscape cartography appears as a fundamental tool for planning, since it offers comparative perspectives on overall patterns, the unequal pressure on environments and the vulnerability of their visual dimension.

Landscapes as repositories of resources

The exploitation of material and intangible landscape resources – amenities, tourism, the building of identities – plays a growing role in contemporary society: the visual consumption of beautiful scenery, the pleasure to live, relax or work in harmonious and pleasant environments, the taste for open air activities have transformed poor farming regions into prosperous tourist areas, rurban settlements for upper classes, or zones of refuge for marginal communities. It is in no way a novelty: Roman senators or Emperors had second homes in the bay of Naples, and hermits congregated in the desert areas of the Middle East. The Grand Tour of British aristocrats signalled the rebirth of this way of consuming landscapes, but the consequences of this evolution changed with mass tourism and suburbanization, and in parallel, with the decline of traditional forms of farming.

Most services provided by landscapes are based on forms generated by traditional land uses. The areas which enjoyed valued environments take advantage of resources which appeared as renewable ones, but are not renewed in present conditions: the social and economic conditions in which they were born and maintained – often for long periods – are over. The new consumers are encroaching on the most genuine areas, either natural or cultural, and destroy their soils, vegetation and structure because of overcrowding (T. Terkenli, A. Printsmann).

Because of the energy crisis, they are valued for the energy resources they offer, which are no longer only those based on photosynthesis: they rely on wind, or photovoltaic and photothermic energies. The problem of these new resources is that they generate visual and sound pollution. What is best: to preserve beautiful landscapes, or to rely on non-exhaustible sources of energy? This new form of development was initially conceived mainly as a top-down process. With the new emphasis on public participation, it has to become a bottom-up process. The topic has been thoroughly explored (M. Frolova and B. Pérez Pérez, M.J. Fortin et al.).

Landscape consciousness and construction of identities

Daily life environments

The landscapes of daily life and familiar environments always played a central role in the construction of local identities, even if, in many cases, the notion of landscape was not made explicit in traditional societies. The environment was perceived as made of fields, meadows, woods, fences, walls, farms, villages: each element was important because of its role in the economy and life of the local cells. As a result, the environment was not perceived as a unified whole, a landscape – A. Berque (1995) and A. Roger (1997) stress the fact that landscape consciousness appeared late in human history (in the forth century AD in China, in the fifteenth century in Europe). In earlier societies, or outside the Far East and Europe, people certainly experienced emotions before landscapes, but they were not subsumed as aesthetic: they were generally associated with religious life.

There were huge tracts of land, in yesterday's Europe, where local societies did not develop landscape consciousness. Even today, many people do not use the word landscape and do not understand its meaning: T. Terkenli, or A. Printsmann et al. signal that the idea of environment is more easily grasped in many parts of Greece or Estonia.

Even if they are not conceived as landscapes, daily life environments help to build local identities. It is through them that the integration of young immigrants often occurs, as shown by B. Castiglioni et al. in Northern Italy. In Nidzica (former Neidenburg), in Poland, the children of the Polish people who settled the city after the expulsion of its former German population do not consider the local landscape as foreign: they have always lived in it, and incorporate it into their memory and identity (B. Borkowska).

We live in urbanized societies where most people ignore how rural landscapes were created and operated, and have no feeling for them. Such a situation means that rural landscapes have to be taught to the groups that had no reasons to know and appreciate them. Landscapes are fluid realities.

Landscape iconographies

The relation between landscapes and the constitution of identities is complex: most identities are not local; the new ease of transport enlarge people's spatial experience; because of migrations, voluntary or forced, many people live in environments which were foreign to them; many tourists discover the sun and coconut trees of tropical beaches, Mediterranean islands, Alpine mountains, and faraway cultures; modern communications allow a wide – sometimes worldwide – diffusion of the iconography of landscapes.

What happens when a society has been conquered by a foreign one? In order to resist the new rulers, people value their traditions (even if they conceive traditional cultures superficially), and the landscapes, mainly rural ones, where they were imbedded. A few years ago, E. Bunkše (1999) stressed the fact that, during the Soviet period, for the majority of Latvians living in new huge urban development, the essence of Latvia was its rural landscapes. A. Printsmann et al. explain in the same way the early creation of Lahemaa National Park on the north coast of Estonia.

The foreign rulers often tried to nationalize the territories they conquered by imposing their own equipments and settlement modes: military camps and huge forms of settlements in Estonia (A. Printsmann et al.), aqueducts in colonial New Spain (W. Doolittle).

For migrants who settled in a foreign region, as was the case of Polish people in Nidzica after 1945, landscapes appear as completely alien. It was against them that identity was built for 45 years. Since 1990, attitudes are changing, and the German heritage is progressively integrated in the local consciousness (B. Borkowska).

In many parts of rural Europe – and increasingly elsewhere in the world – people are leaving the poor rural areas where they were born and raised to settle in cities. The impact of these migrations on landscapes is considerable: many arable land or grassland turn to bush or forest; terraces and irrigation systems decay. R. Beilin et al. and B. Freund analyse such situations in Sweden, Australia and Portugal. Many of those who left Portugal for France, Belgium or Germany 40 or 50 years ago still have a strong sense of belonging to their place of origin, where they frequently build huge houses for their holidays and retirement – but until now they do not display real landscape consciousness. There are also newcomers in these areas of depopulation: their interest in landscape is more based on problem of soil conservation than on aesthetics.

T. Terkenli observes in Greece a recent and deep evolution in the way landscape is associated with identity. Traditionally, and partly for religious reasons, there was no landscape-consciousness in Greece. Today, and thanks to the development of tourism, first foreign and then domestic, attitudes are changing: Greek people have discovered that landscapes are valuable resources – but until now they have been unable to build an image of tourist areas which really fits Greek traditions. They are just adopting the images developed by west European artists, and more recently, travel agencies.

Scale and imagination hierarchies

Feelings of identities are hierarchically structured: personal, local, regional, national or European for instance. At what scale have landscapes been, and are, useful in constructing local, regional, national or European identities?

E. Bunkše's experience of building a second home on the coast of Latvia offered him the opportunity to discover new feelings: he speaks of the ethereal and ineffable in rural landscapes. They first enrich his personal sense of place, but are also worthy for local, national and European identities.

The way landscapes play a role in identity building differs according to scale. It is more direct at the local scale, where people become spontaneously attached to the ordinary landscapes, either rural or urban, in which they live. The link between identity feelings and local landscape is stronger in vernacular societies.

A generation ago, Benedict Anderson attracted historians' attention to imagined communities – those which are so large that nobody is able to develop a direct experience of them, nations mainly. What was the role of landscapes in the building of imagined communities? Great Britain is often taken as a case study: nation-building was contemporary with the industrial revolution; it was paradoxically based on the glorification of the landscapes of traditional rural England as painted by Gainsborough, or on the romantic views of the Peak District or Wales. J. Agnew shows that Italians tried to develop their national identities in the same way, relying first on Florence and Tuscany, and then on Rome – but they did not succeed. This means that the links between imagined communities and landscape are complex. It is important to be aware of that fact at a time when many think that rural landscapes have to be used as a tool for strengthening European consciousness.

Reflecting upon the experience of *Terraforming* in Dubai (the construction of artificial islands in the shallow waters of the Gulf), M. Jackson and V. della Dora analyse an interesting experience in landscape engineering and architecture, and clearly show the limits of the ideological manipulations it involved.

The increasing mobility and the growing efficiency of communication systems have an important impact on territoriality – and landscape consciousness – as showed by R. Haesbaert: after an initial phase of deterritorialization, new forms of territorialization appear. They differ from traditional ones: they have ceased to be based on continuous tracts of land; they take the shape of archipelagos.

Landscape planning today

A new context

Landscape studies are increasingly motivated by the will to develop landscape policies. L. Scazzosi presents a clear picture of the way the European Landscape Convention has been translated into laws and rules in the Italian (and up to a point French) context. The aim is to take into account ecological imperatives, economic necessities, various subjectivities, the preservation of nature and historical heritage. This means that some limits have to be set on the transformation of natural or man-made features of the environment by human action. Landscape development has to be *appropriate*.

In order to develop a useful approach for planners, it is necessary to understand the process of political decision-making in contemporary society. As stressed by D. Terrasson, the distribution of power is changing in contemporary societies; instead of a unique power centre, there is now a plurality of political actors who are considered legitimate. Planners have to present projects so that they can be understood by them all; and have to be prepared for harsh criticism from many of them.

D. Terrasson then shows that landscape policies have ceased to be mainly focused on preservation: they try to conceive forms of change which pay respect to the different uses of landscapes (as a source of aesthetic experience, as inheritance, as a means to achieve economic results, as a negotiation tool, as an alibi since it appears politically correct). Landscape research produces expertise, leads to the elaboration of landscape policies. It helps develop cooperation between academics and decision-makers. It is in this new context that European landscape policies have to be analysed.

A critical view of the way urban plans were conceived and operated in the past is welcome. M. Qviström stresses thus the negative role of unused planning regulations on the evolution of the suburban areas of Lund.

Landscape planning, modernity and postmodernity

In order to understand the problems of landscape research and development planning today, it is well to distinguish two phases in the genesis of contemporary problems: (i) the explosion of modernity and the decline of traditional forms of landscape; (ii) the energy crisis and the search for the sustainable management of emerging types of landscapes – i.e. postmodernity.

Most of the problems of landscapes arose during the first phase, i.e. in the 50s and 60s in western Europe, and since 1989 in eastern Europe: mass tourism along the coasts, especially in the Mediterranean, or in Alpine mountains, the multiplication of second homes in Scandinavia, France and some Mediterranean countries, the development of facilities for proximity tourism in the most populated parts of Europe. Maps of tourist pressures are fascinating in this respect: the impact of the new activities is important all over the margins of Europe, Ireland, Scandinavia, and even more in the Mediterranean countries, but the higher levels are more or less coextensive with the zones of high densities: from south-eastern England to northern Italy through Belgium, the Netherlands, Germany and Switzerland. In this core area, the retreat of pure rural landscapes is often dramatic (Mücher and Wascher 2007).

The problems of landscape planning result mainly, today, from this first phase of rapid change. The new ecological and social philosophies which tend to prevail are responsible for different concerns: how to maintain the links that people had with the land? How to avoid the disappearance of valuable natural or cultural landscapes? How to pay for their maintenance? How to shape the landscapes which will result from the new forms of social life? What about landscape preservation when the main objective is sustainable existence?

A. Printsmann et al. for Estonia and T. Terkenli for Greece present interesting examples of the dilemma of contemporary planning. When established in 1971, the aim of Lahemaa National Park was clear: 'to protect the characteristic north-Estonian landscapes and the national heritage of the area, and to preserve the harmonious relations between man and nature'. It also 'enabled to limit the soviet ruler-led planning and large-scale amelioration, thereby preserving the local peculiarities'. Forty years later, the image is blurred. There was a dual dimension in the initial plan: the conflicting opposition of the park construction versus local identity. 'Due to historical reasons, [Estonians] find themselves in a situation where the identity of Estonia is confusing, also the identity of Lahemaa National Park is contested, landscape and heritage are understood differently'. 'Discrepancy between (Estonian national) ideational and (local) physical, memorized and remembered' landscape is the central problem for planners. This contradiction is present as soon as planning becomes attentive of people's preferences.

In Greece, landscape consciousness developed recently: it mainly resulted from the development of tourism. People know that landscapes are fundamental for the development of Greek economy, but there is no agreement on what a Greek landscape is, and what features have to be preserved or developed. Hence the lack of any coherent landscape policy in this country (T. Terkenli).

The Brioni archipelago, in Croatia, offers another wonderful case study: there are numerous archaeological sites, an old medieval borough, a wealthy late nineteenth century tourist development. Tito decided to expel all the native population to the continent. Hence a fundamental question: for whom to plan (I. Schrunk and V. Begović)?

The question of expertise and the democratization of landscape planning

This general approach to landscapes, with its increased emphasis on their subjective dimensions, according to the European Landscape Convention, results into a dramatic change in planning strategies. Landscape planning ceases to appear as a privilege of learned elites. It has to be based on more democratic principles.

It is not easy to implement this new procedure, as shown by the studies on the new wind power equipments in Southern Spain (M. Frolova and B. Pérez Pérez) and in Québec and Western France (M.J. Fortin et al.). Do windmills have to avoid emblematic landscapes? In fact, their presence in ordinary landscapes appears more disturbing: it is certainly the main conclusion of the study, but it is never clearly expressed by local populations.

The whole process of landscape planning was based on expertise, either in the ecological field or in the human history field. Today, new forms of expertise are required in order to evaluate the value and limits of participation. In a way, the idea of expertise itself is undermined by the participation project.

The landscape planner increasingly appears as an interpreter and a mediator. The idea that landscapes are never stable realities is today widely accepted. The way they are evaluated changes also, and varies according to social groups or their cultural heritage. As a result, expertise has taken a different meaning: experts have ceased to be those who design the best environments according to prevailing values; they are those who facilitate transformations and minimize disruptions in conflictive evolving settings.

Final notes: Landscape studies and newly spatial conditionality

Contemporary landscapes reflect social situations utterly different from those of the past: our societies have ceased to be mainly rural; their economies are increasingly global. The old opposition between the cities and the countryside is vanishing. Studying the rural landscapes of the past remains a valuable target for research: it is intellectually important to fully understand their genesis and dynamics. With the new economic conditions of agricultural production, urban sprawl, rurbanization and the new tourist vocation of many rural areas, most of them have, however, ceased to be functional. New landscapes do appear, with larger fields, more buildings, playgrounds, sport facilities, etc.

Protecting the totality of past landscapes is impossible before so strong and rapid a spatial and economic change, but it is certainly worth preserving some of them (or some of their elements) as testimonies of former conditions. Since today many collective identities rely on the feeling people develop for the rural environment they live in, controlling the evolution of rural landscapes in order to maintain some of their more typical features – the presence of hedges or terraces, for instance – is also a valuable objective.

In an increasingly urbanized society, however, a growing part of the population has no direct experience of rural landscapes. Many have no feeling for them at all. As a result, it is urgent to teach urban dwellers the significance of rural landscapes and the need to keep them as harmonious as possible.

Two dynamics explain contemporary changes in the European rural landscapes: (i) modernization, which relied on the intensification of agriculture, the use of concentrated forms of energy, fertilizers and pesticides, higher mobility and the progressive urbanization of rural populations; (ii) sustainable growth, with its emphasis on renewable sources of power, lesser inputs of energy, fertilizers and pesticides, and a critical attitude towards mobility. Everyone is aware of what went wrong with modernization. Sustainable development is a more recent venture: to what extent do we have to control the proliferation of windmills or photovoltaic or photothermic equipments? How to balance amenities against the advantages coming from harnessing renewable energy? The problem has been tackled during this Conference. More reflection would be welcomed on this difficult topic.

Landscape analysis relied for long on a few assumptions. The explanation of landscape forms was mainly sought in local conditions. There was also an emphasis on their aesthetic quality. Thirty years ago, Peirce Lewis summarized the few axioms upon which the cultural interpretation of landscapes was based (Lewis 1979). These principles were adapted to the societies of the past. Our setting is quite different. Don Mitchell proposes six new axioms for understanding today's landscapes (Mitchell 2008). We no longer believe in the shaping of humanized landscapes by nature: 'landscape is produced; it is actively made' *(Axiom 1)*. In a world were transport and communication are easy and cheap, 'no landscape is local' *(Axiom 3)*. In order to understand landscape forms, 'history matters' *(Axiom 4)*. Other points are more debatable: Don Mitchell claims that 'landscape is [or was] functional' *(Axiom 2)*; it is true that all landscape features are encapsulated into running economic and ecological systems; in many studies, there is an overemphasis of their aesthetic qualities; but their functionality is never complete: many of their features have no direct functional significance and may persist over long periods. *Axiom 5* states: 'landscape is power'. The way property rights are distributed certainly has to be studied; in most landscapes, we encounter people who are deprived of any possibility to participate in their shaping. *Axiom 6* sounds more radical: 'landscape is the social form that social justice takes'. It is certainly an overstatement.

Don Mitchell proposes to replace the cultural approach to landscape studies by an analysis of their political economies. Even if his arguments are too controversial, we certainly need to introduce more curiosity in the economic forces, juridical principles and social constraints which operate in landscapes and constrain those who live them. As stated by Johnston and Westcoat:

> [...] landscapes do not just happen – they are produced, consumed, wasted, conserved, adapted and transformed. Landscapes are created to address problems of production, distribution, perceived security, lack of amenities, or maintenance of class or other distinctions. (2008: 211)

At a time when landscape experts are losing a part of their traditional authority – they have ceased to be considered the only people able to design landscapes – it is important to stress that we need other forms of expertise: the knowledge of the economic and social forces which contribute to shape visible forms and often create a severe imbalance of power and many inequalities among those who inhabit or visit them.

References

Berque, A. 1995. *Les Raisons du paysage. De la Chine antique aux environnements de synthèse*. Paris: Hazan.

Bunkse, E.V. 1999. God, Thine Earth is Burning: Nature Attitudes and the Latvian Drive for Independence, in *Nature and Identity in Cross-Cultural Perspective*, edited by A. Buttimer and L. Wallin. Dordrecht: Kluwer, 175-187.

Johnston, D.M. and J.L. Westcoat, Jr. 2008. Implications for Future Landscape Inquiry, Planning, and Design, in *Political Economies of Landscape Change. Places of Integrative Power*, edited by J.L. Westcoat Jr. and D.M. Johnston. Doordrecht: Springer, 195-214.

Lewis, Peirce. 1979. Axioms for reading the landscape: Some guides to the American scene, in *The Interpretation of Ordinary Landscape: Geographical Essays*, edited by D. Meinig. New York: Oxford University Press, 11-32.

Mitchell, D. 2008. New axioms for reading the landscape: Paying attention to political economy and social justice, in *Political Economies of Landscape Change. Places of Integrative Power*, edited by J.L. Westcoat Jr. and D.M. Johnston. Doordrecht: Springer, 29-50.

Mücher, C.A. and Wascher, D.M. 2007. European landscape characterization, in *Europe's Living Landscapes, essays exploring our identity in the countryside*, edited by B. Pedroli et al. Dordrecht: Kluwer.

Roger, A. 1997. *Court Traité du Paysage*. Paris: Gallimard.

Simoncini, R., de Groote, R. and Pinto-Correia, T. 2009. An integrated approach to assess options for multifunctional use of rural area. *Regional Environmental Change*, 3, Special Issue, September, 139-141.

Index